Lecture Notes in Computer Science 2052

Edited by G. Goos, J. Hartmanis and J. van Leeuwen

T0192037

Lecture Notes in Computer Science

Edited by G. Goos, J. Hartmanis, and J. van Leeuwen

3
Berlin
Heidelberg
New York
Barcelona
Hong Kong
London
Milan
Paris
Singapore
Tokyo

Vladimir I. Gorodetski Victor A. Skormin
Leonard J. Popyack (Eds.)

Information Assurance in Computer Networks

Methods, Models and Architectures for Network Security

InternationalWorkshop MMM-ACNS 2001
St. Petersburg, Russia, May 21-23, 2001
Proceedings

1 3

Current State of RBAC Models

To my knowledge the first use of the term RBAC is due to Ferraiolo and Kuhn [5] although there has been prior mention in the security literature of "roles" and "role-based security." Sandhu et al [14] subsequently published a seminal paper defining a family of models that has since come to be called RBAC96. A crucial insight of RBA96 was the realization that RBAC can range from very simple to very sophisticated so we need a family of models rather than a single model. A single model is too complex for some needs and simple for others. A graded family of models enables selection of the "correct" model for a particular situation. Publication of RBAC96 was followed by a flurry of research that has clearly established RBAC as the dominant access control model. Remarkably the basic concepts of RBAC96 have proved to be robust and no significant omissions have been identified. In many years of research following publication of RBAC96 we have had occasion to introduce only one new concept (role activation hierarchies [15]) which was not already present in RBAC96.

Let us now briefly review important achievements in recent RBAC research. The perspective given here is necessarily a personal one. As such the papers cited are those with greatest direct impact on our own understanding of RBAC models. There simply is not enough room to cite many other papers of considerable significance.

We feel that RBAC models have advanced in at least three respects in recent years, discussed below in turn.

Firstly, an important recent development is emergence of a consensus standard model which is supported by a major standards organization (the US National Institute of Standards and Technology or NIST). Following the publication of RBAC96 it became clear that many authors were pursuing very similar ideas but with differences in detail leading to confusion about the nature of RBAC. RBAC96 was unique in proposing the concept of a graded family of models. Once this family notion was accepted by the RBAC community consensus on a core set of RBAC concepts became feasible. To this end an initial attempt at a family of standard models was presented at the Berlin RBAC Workshop by Sandhu et al [18]. Workshop attendees reacted to this proposal with heated discussion [12]. The current proposal is to be published soon [6] and will then evolve into NIST publications. Deployment and use of RBAC in commercial products and systems will be facilitated by the NIST standard model.

Another important development is a deeper theoretical understanding of RBAC and particularly its relationship to MAC and DAC. There has been much confusion with some authors claiming that RBAC is a form of MAC while others arguing it is a form of DAC. Osborn et al [13] show that RBAC96 can be configured to do MAC or DAC as one chooses. So RBAC transcends the MAC-DAC distinction. Fundamentally it turns out that both MAC and DAC are just special cases of RBAC. For historic reasons MAC and DAC gained early dominance in the research community. MAC and DAC are easily unified within the framework of RBAC. This unification is more than coexistence. MAC systems also usually implement DAC but in these systems MAC and DAC simply coexist. The RBAC viewpoint is that MAC and DAC are just examples of policies to configure in a policy-neutral RBAC model.

The third significant development is a contextual understanding of the practical purpose of RBAC models . Sandhu [17] argues that the purpose of RBAC models is

Preface

This volume contains the papers selected for presentation at the International Workshop on Mathematical Methods, Models and Architectures for Network Security Systems (MMM-ACNS 2001) held in St. Petersburg, Russia, May 21–23, 2001. The workshop was organized by the St. Petersburg Institute for Informatics and Automation of the Russian Academy of Sciences (SPIIRAS) in cooperation with the Russian Foundation for Basic Research (RFBR), the US Air Force Research Laboratory (both the Information Directorate (AFRL/IF)) and the O ce of Scienti c Research (AFRL/OSR), and Binghamton University (USA).

MMM-ACNS 2001 provided an international forum for sharing original research results and application experiences among specialists in fundamental and applied problems of computer network security. An important distinction of the workshop was its focus on mathematical aspects of information and computer network security and the role of mathematical issues in contemporary and future development of models of secure computing.

A total of 36 papers coming from 12 di erent countries on signi cant aspects of both theory and applications of computer network and information security were submitted to MMM-ACNS 2001. Out of them 24 were selected for regular presentation. Five technical sessions were organized, namely: mathematical models for computer networks and applied systems security; methods and models for intrusion detection; mathematical basis and applied techniques of cryptography and steganography; applied techniques of cryptography; and models for access control, authentication, and authorization. Two panel discussions were devoted to the signi cant issues in the computer and information security eld. The rst sought to de ne the important open problems in computer security and to reach a conclusion as to mathematical methods and models can contribute, and the second focused upon security research and education in academia. The MMM-ACNS 2001 program was enriched by ve invited speakers: Dipankar Dasgupta, Alexander Grusho, Catherine Meadows, Ravi Sandhu, and Vijay Varadharajan.

An event like this can only succeed as a result of team e orts. We would like to acknowledge the contribution of the Program Committee members and thank the reviewers for their e orts. Our sincere gratitude goes to all of the authors who submitted papers.

We are grateful to our sponsors: the European O ce of Aerospace Research and Development (EOARD), the European O ce of Naval Research International Field O ce (ONRIFO), and the Russian Foundation of Basic Research (RFBR) for their generous support.

We wish to express our thanks to Alfred Hofmann of Springer-Verlag for his help and cooperation.

May 2001

Vladimir Gorodetski
Leonard Popyack
Victor Skormin

MMM-ACNS 2001 Workshop Committee

Program Committee

Program Co-chairmen:

Vladimir Gorodetski	St. Petersburg Inst. for Informatics and Automation, Russia
Leonard Popyack	Air Force Research Laboratory/IF, USA
Victor Skormin	Watson School of Eng., Binghamton Univ., USA

International Program Committee

Kurt Bauknecht	Univ. of Zurich, Dept. of Information Technology, Switzerland
Harold Carter	University of Cincinnati, USA
Peter Chen	Computer Science Dept., Luisiana State Univ., USA
Dipankar Dasgupta	Div. of Computer Science, Univ. of Memphis, USA
Jose G. Delgado-Frias	Electrical and Comp. Eng. Dept., Univ. of Virginia, USA
Lynette Drevin	Comp. Science and Inf. Systems, Potchefstroom Univ., South Africa
Jiri Fridrich	Watson School of Eng., Binghamton Univ., USA
Dimitris Gritzalis	Athens Univ. of Economics & Business, Greece
Alexander Grusho	Russian State Univ. for Humanity, Russia
Yury Karpov	St. Petersburg State Technical Univ., Russia
Igor Kotenko	St. Petersburg Inst. for Informatics and Automation, Russia
Martin Kutter	AlpVision, Les Paccots, Switzerland
Anatoly Maliuk	Moscow State Engineering Physical Inst., Russia
Catherine Meadows	Naval Research Laboratory, USA
Nikolay Moldovian	Spec. Center of Program Systems "SPECTR", Russia
Vladimir Orlov	Microtest Company, Moscow, Russia
Gyorgy Papp	V.R.A.M. Communication Ltd., Hungary
Hartmut Pohl	Fachhochschule Bonn-Rhein-Sieg, St. Augustin-Univ. of Applied Sciences, Germany
Ravi Sandhu	SingleSignOn.Net Inc. and George Mason Univ., USA
Igor Sokolov	Inst. for Informatics Problems, Moscow, Russia
Mikhail Sycheov	Bauman State Technical Univ., Russia
Leonid Ukhlinov	The State Customs Committee of Russia, Russia
Vijay Varadharajan	Div. of Inf. and Commun. Sciences Macquarie Univ., Australia
Minerva M. Yeung	Media and Internet Technology, Intel, USA
Louise Yngstrom	Dept. of Comp. and Systems Sciences, Univ. & Royal Inst. of Technology, Stockholm
Peter Zegzhda	St. Petersburg State Technical Univ., Russia

Reviewers

Kurt Bauknecht	Univ. of Zurich, Dept. of Information Technology, Switzerland
Kirill Bolshakov	St. Petersburg State Technical Univ., Russia
Dipankar Dasgupta	Div. of Comp. Science, Univ. of Memphis, USA
Gunther Drevin	Comp. Science and Inf. Systems, Potchefstroom Univ., South Africa
Lynette Drevin	Comp. Science and Inf. Systems, Potchefstroom Univ., South Africa
Dimitris Gritzalis	Athens Univ. of Economics & Business, Greece
Vladimir Gorodetski	St. Petersburg Inst. for Informatics and Automation, Russia
Yury Karpov	St. Petersburg State Technical Univ., Russia
Igor Kotenko	St. Petersburg Inst. for Informatics and Automation, Russia
Evgenii Krouk	St. Petersburg State Technical Univ., Russia
Catherine Meadows	Naval Research Laboratory, USA
Nikolay Moldovian	Spec. Center of Program Systems "SPECTR", Russia
Gyorgy Papp	V.R.A.M. Communication Ltd., Hungary
Vladimir Platonov	St. Petersburg State Technical Univ., Russia
Alexander Rostovtsev	St. Petersburg State Technical Univ., Russia
Ravi Sandhu	SingleSignOn.Net Inc. and George Mason University, USA
Michael Smirnow	GMD, FOKUS, Germany
Igor Sokolov	Inst. for Informatics Problems, Moscow, Russia
Vijay Varadharajan	Div. of Inf. and Commun. Sciences, Macquarie Univ., Australia
Dmitry Zegzhda	St. Petersburg State Technical Univ., Russia
Peter Zegzhda	St. Petersburg State Technical Univ., Russia

Table of Contents

An Intelligent Decision Support System for Intrusion Detection and Response

Dipankar Dasgupta and Fabio A. Gonzalez

Intelligent Security Systems Research Lab
Division of Computer Science
The University of Memphis
Memphis, TN 38152
{ddasgupt,fgonzalz}@memphis.edu

Abstract. The paper describes the design of a genetic classifier-based intrusion detection system, which can provide active detection and automated responses during intrusions. It is designed to be a sense and response system that can monitor various activities on the network (i.e. looks for changes such as malfunctions, faults, abnormalities, misuse, deviations, intrusions, etc.). In particular, it simultaneously monitors networked computer's activities at different levels (such as user level, system level, process level and packet level) and use a genetic classifier system in order to determine a specific action in case of any security violation. The objective is to find correlation among the deviated values (from normal) of monitored parameters to determine the type of intrusion and to generate an action accordingly. We performed some experiments to evolve set of decision rules based on the significance of monitored parameters in Unix environment, and tested for validation.

1 Introduction

The problem of detecting anomalies, intrusions, and other forms of computer abuses can be viewed as finding non-permitted deviations (or security violations) of the characteristic properties in the monitored (network) systems [12]. This assumption is based on the fact that intruders' activities must be different (in some ways) from the normal users' activities. However, in most situations, it is very difficult to realize or detect such differences before any damage occur during break-ins.

When a hacker attacks a system, the ideal response would be to stop his activity before he can cause any damage or access to any sensitive information. This would require recognition of the attack as it takes place. Different models of intrusion detection have been developed [6], [9], [10], [11], and many IDS software are available for use. Commercial IDS products such as NetRanger, RealSecure, and Omniguard Intruder alert work on attack signatures. These signatures needed to be updated by the vendors on a regular basis in order to protect from new types of attacks. However, no detection system can catch all types of intrusions and each model has its strengths and weaknesses in detecting different violations in networked computer systems. Recently, researchers started investigating artificial intelligence

V.I. Gorodetski et al. (Eds.): MMM-ACNS 2001, LNCS 2052, pp. 1-14, 2001.
© Springer-Verlag Berlin Heidelberg 2001

[3], genetic approaches [1], [6] and agent architectures [4], [5] for detecting coordinated and sophisticated attacks.

This paper describes the design and implementation of a classifier-based decision support component for an intrusion detection system (IDS). This classifier-based IDS monitors the activities of Unix machines at multiple levels (from packet to user-level) and determines the correlation among the observed parameters during intrusive activities. For example, at user level – searches for an unusual user behavior pattern; at system level – looks at resource usage such as CPU, memory, I/O use etc.; at process level – checks for invalid or unauthenticated processes and priority violations; at packet level – monitors number, volume, and size of packets along with source and type of connections. We developed a Java-based interface to visualize the features of the monitored Unix environment. We used some built-in tools (such as vmstat, iostat, mpstat, netstat, snoop, etc.), syslog files and shell commands for simultaneously monitoring relevant parameters at multiple levels. As the data collector sensors observe the deviations, the information is sent to the classifier system [7], [8] in order to determine appropriate actions.

2   Monitoring Data and Filtering

The behavior-based techniques of detecting intrusion or anomalies usually involve monitoring the network and system parameters continuously over a period of time and collecting information about the network's normal behavior [3], [6], [9]. Accordingly, some parameters of the system are identified as the important indicators of abnormalities. The detection is based on the hypothesis that security violations can be detected by monitoring a system's audit records for abnormal patterns of the system usage. Our prototype system collects historical data and determines the normal (activities) usage of the system resources based on various monitored parameters.

2.1   Multi-level Parameter Monitoring

Our prototype system currently monitors the parameters listed below, some of these parameters are categorical in nature, (e.g. type of user, type of connections) which are represented numerically for interpretation. However, the selection of these parameters is not final and may vary (based on their usefulness) in our future implementation. Various Unix commands are used and filtered the output to get the values of the selected parameters [10].

To monitor user-level activities, the following parameters are recorded as audit trail and analyzed by statistical methods to develop profiles of the normal behavior pattern of users:

  U1. Type of user and user privileges
  U2. Login/Logout period and location
  U3. Access of resources and directories
  U4. Type of software/programs use
  U5. Type of Commands used

Commands used to obtain packet level information:

- Netstat - shows network status i.e. displays the contents of various network-related data structures in various formats, depending on the option selected. Intrusive activities in most cases involve external connection into the target network by an outsider, though internal misuse is also crucial. So it is very important to monitor packets sent across the network both inbound and outbound along with packets those remain inside the subnet. Moreover, the number of external connections established and validity of each connection can be verified using these monitored parameters. The main processes for collecting data is a korn shell script that performs system checks and formats the data using awk, sed, cat, and grep and the appropriate parameters are filtered from the output of these commands for storing in a file.

2.2   Setting Thresholds

Historical data of relevant parameters are initially collected over a period of time during normal usage (with no intrusive activities) to obtain relatively accurate statistical measure of normal behavior patterns. Accordingly, different threshold values are set for different parameters.

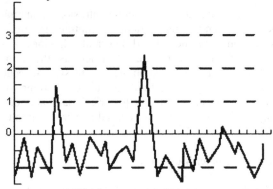

Fig. 1. Showing different threshold levels as a measure of degree of deviations

During monitoring, each parameter is checked for any deviation by comparing current parameter values (at different nodes) with the profile of the normal usage (and behavior pattern). However, the threshold settings of the monitored parameters need to be updated to accommodate legitimate changes in the network environment. Figure 1 shows an example of time series where the variations in the data pattern indicate the degree of deviation from the normal. It is to be noted that we used different threshold values for different parameters and allowed enough clearance to accommodate legitimate variations in system usage.

Setting of the thresholds for determining deviations, as a function of alert level is tricky. The detection system should not be alerting unnecessarily for trivial circumstances, but on the other hand, should not overlook real possibilities of serious

Commands used to obtain packet level information:

- Netstat - shows network status i.e. displays the contents of various network-related data structures in various formats, depending on the option selected. Intrusive activities in most cases involve external connection into the target network by an outsider, though internal misuse is also crucial. So it is very important to monitor packets sent across the network both inbound and outbound along with packets those remain inside the subnet. Moreover, the number of external connections established and validity of each connection can be verified using these monitored parameters. The main processes for collecting data is a korn shell script that performs system checks and formats the data using awk, sed, cat, and grep and the appropriate parameters are filtered from the output of these commands for storing in a file.

2.2   Setting Thresholds

Historical data of relevant parameters are initially collected over a period of time during normal usage (with no intrusive activities) to obtain relatively accurate statistical measure of normal behavior patterns. Accordingly, different threshold values are set for different parameters.

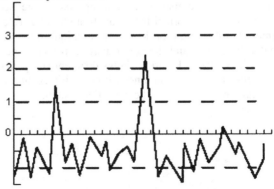

Fig. 1. Showing different threshold levels as a measure of degree of deviations

During monitoring, each parameter is checked for any deviation by comparing current parameter values (at different nodes) with the profile of the normal usage (and behavior pattern). However, the threshold settings of the monitored parameters need to be updated to accommodate legitimate changes in the network environment. Figure 1 shows an example of time series where the variations in the data pattern indicate the degree of deviation from the normal. It is to be noted that we used different threshold values for different parameters and allowed enough clearance to accommodate legitimate variations in system usage.

Setting of the thresholds for determining deviations, as a function of alert level is tricky. The detection system should not be alerting unnecessarily for trivial circumstances, but on the other hand, should not overlook real possibilities of serious attacks. Each parameter at different level is quantified and encoded by two-bit to represent a value between 0-3 as degree of deviation as shown in Table 1.

Table 1. Binary Encoding of Normalized Parameter values

0	00	Normal
1	01	Minimal
2	10	Significant
3	11	Dangerous

3   Designing Classifier-Based Decision Support Module

It is very difficult to develop intelligent decision support component for intrusion detection systems, as uncertainties and ambiguities. The best approach may be to design an evolvable system that can adapt to environment. A classifier system is an adaptive learning system that evolves a set action selection rules to cope with the environment. The condition-action rules are coded as fixed length strings (classifiers) and are evolved using a genetic search. These classifiers are evolved based on the security policy – this rule set forms a security model with which the current system environment needs to be compared. In our approach, the security policies are embedded while defining the normal behavior. In a classifier rule, the condition-part represent the amount of deviation (or degree of violation) in the monitored parameters from the normal values (0-3) and the action-part represent a specific response (0-7) according to the type of attack. Figure 2 shows different components of the prototype system. The data fusion module combines (discussed in section 3.2.2) the parameter values and put as an input template to the classifier system.

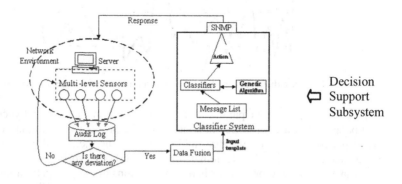

Fig. 2. Different modules of the classifier-based intrusion detection system

3.1   Creating a High-Level Knowledge Base

The degree of importance of each level (of monitored parameters) is hypothesized based on the domain knowledge. The purpose is to generate rules from a general knowledge base designed by experts. Though the accuracy of this knowledge base will result in more realistic actions, the heuristic rule set that we used can provide similar detection ability. Table 2, gives an example of such a knowledge base, a higher value of action indicates stronger response.

Table 2. Estimates of affects of intrusive activities at different levels and proposed actions

Hypothesis Number	User Level	System Level	Process Level	Packet Level	Action
1	0.2	0.0	0.0	0.1	1
2	0.4	0.0	0.0	0.4	3
3	0.0	0.1	0.2	0.8	6

Symbolically, each hypothesis is a 5-tuple (k_1, k_2, k_3, k_4, a), where k_i [0.0, 1.0] and a {0,...,7}. Suppose that the input template to the classifier contain the message (m_1, m_2, m_3, m_4), this messages match the hypothesis rule (k_1, k_2, k_3, k_4, a) if and only if:
$$k_1 <= m_1 \text{ and } k_2 <= m_2 \text{ and } k_3 <= m_3 \text{ and } k_4 <= m_4,$$
That is, the condition part of a hypothesis expresses minimum values of deviation at different level.

Accordingly, if we have two hypotheses: $K = (k_1, k_2, k_3, k_4, a_k)$ and $L = (l_1, l_2, l_3, l_4, a_l)$, and for i=1,2,3,4 $k_i <= l_i$. That means that hypothesis L is more specific than hypothesis K, therefore, any message that matches L will match K. In order to maintain the consistency of the knowledge base, the action of the hypothesis L (a_l) has to be stronger than the action of the hypothesis K (a_k): $a_l >= a_k$.

This characteristic gives to the knowledge base a hierarchical structure and increases the robustness of the system.

3.2   Classifier Systems in Rule Discovery

Classifier systems are dynamical systems that evolve stimulus-response rules or classifiers (using Genetic Algorithms), each of which can interact with the problem-solving environment. A classifier system receives inputs (as messages) from the monitored environment in the form of binary strings and determines an action (as output). The environment then provides some feedback in terms of a scalar reinforcement for the action and the classifier system updates that rule's strength accordingly. When a classifier system gets a message from the environment it places that message on a message board, it then searches the classifier list to find a rule that matches the message. If there exist a number of competing rules, the winner is selected based on the strength of the rules (tie is broken probabilistically). However, if no rule in the classifier list matches the current message, the classifier system evolves new classifiers using genetic algorithms. A classifier rule has the form:
<condition> : <action>

The condition part is a string from the alphabet {'0', '1', '#'}, where '#' is a wildcard that can match either '0' or '1'. The action part indicates the suggested response if the rule is activated, in our case, 7 possible actions are considered. Here is an example of a rule:

#1 0 0 1 1 0 1 #1 1 0 #0 0 #0 1 ###0 0 0 1 ##1 1 1 1 1 0 #1 1 0 #:5

Figure 3 illustrates the use of the Classifier-Based Decision Support Subsystem. When an input arrives to the subsystem (information about of the level of deviation of the different parameters), it goes to the message list; then, the Rule Selection Module determines the matching rule using the bucket brigade algorithm, and finally the action of the winner rule is sent to the environment.

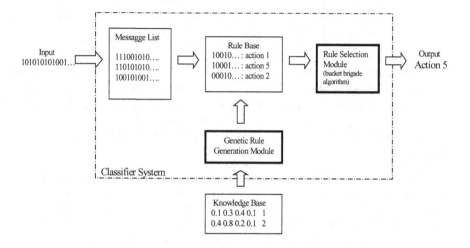

Fig. 3. Components of Classifier Based Decision Support Subsystem

3.2.1   Genetic Algorithm

Genetic Algorithms (GAs) represent a class of general-purpose adaptive search methods that mimic the metaphor of natural biological evolution. Genetic algorithms operate on a population of candidate solutions applying the principle of survival of the fittest to produce better and better approximations to a solution. In general, in order to successfully employ GA to solve a problem, one needs to identify the following four components:

1. Syntax of the chromosome – represent the problem space
2. Interpretation of the chromosome – decoding and checking the feasibility
3. Fitness Measure of the chromosome – determine the quality of solutions
4. Genetic Operator s – crossover and mutation operators are used to manipulate candidate solutions

Genetic algorithm works with a population of candidate solutions (chromosomes). The fitness of each member of the population (point in search space) is computed by an evaluation function that measures how well an individual performs with respect to the problem domain. The constraints can be incorporated in the evaluation function in the form of penalty terms. The better-performing members are rewarded and individuals showing poor performance are punished or discarded. So starting with an initial random population, the genetic algorithm exploits the information contained in the present population and explores new individuals by generating offspring in the next generation using genetic operations. By this selective breeding process it eventually reaches to a near-optimal solution with a high probability.

Several different genetic operators have been developed, but usually two (crossover and mutation) are extensively used to produce new chromosomes. The crossover operator exchanges portions of information between the pair of chromosomes. The mutation operator is a secondary operator, which randomly changes the values at one or more genes of a selected chromosome in order to search for unexplored space. The workings of simple GAs have been described in detail elsewhere [13], [15].

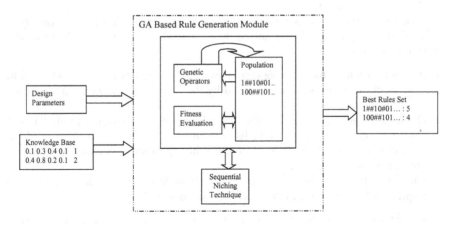

Fig. 4. The details of GA Rule Generation mechanism

While evolving the rule set in the classifier system, the genetic algorithm evaluates each rule to determine a fitness value based on the performance (as shown in figure 4). To generate a diverse set of rules, we use a sequential niching technique, which is discussed in a subsequent section.

3.2.2   Data Fusion (Encoding of Monitored Parameters)

Each classifier rule is evolved through the genetic search process. In our implementation, the genetic representation of a candidate solution is called a chromosome and is made up of the parameters at the various levels mentioned above and the encoding is shown in figure 5. The objective is to find correlation among the deviated values (from normal) of monitored parameters to determine the type of intrusion and to generate action accordingly.

Levels	User Level	System Level	Process Level	Packet Level
Parameters	U1 U2 U3 U4 U5	S1 S2 S3 S4 S5	P1 P2 P3 P4 P5	N1 N2 N3 N4 N5
An Example	00 11 10 11 11	10 01 00 11 10	00 01 10 11 11	10 11 10 01 00

Fig. 5. Chromosome layout (in binary) – encoding different monitored parameters at multiple levels

3.2.3   Fitness Evaluation

The purpose of the fitness function is to measure how good each chromosome (rule) is in solving the problem. In our approach, we consider the following elements while generating the rule set:

ú How well a rule reflects the domain knowledge.   The deviation of each level has to be as closer as possible as specified by the knowledge base (given in equation 2).

ú The level of generality of the rule . A rule can be very specific (with a few '#' characters) or very general (with a large number of '#' characters). A good rule has to be closer to the optimum level of generality specified in the design parameters (given in equation 3).

ú The diversity in the best rule se t. This item is taken into account using the Sequential Niching Algorithm, which will be discussed later.

In order to illustrate the role of the fitness function, let us consider a chromosome, which is represented as $C =<c_1,c_2,..., c_{40}>$, with c_i $\{'0','1','\#'\}$. However, the chromosome (C) can be structured in subparts as shown below,

P^1		P^2		P^3		P^4	
P^1_1 ..... P^1_5		P^2_1 ..... P^2_5		P^3_1 ..... P^3_5		P^4_1 ..... P^4_5	
$c_1 c_2$ ... $c_9 c_{10}$		$c_{11} c_{12}$ ... $c_{19} c_{20}$		$c_{21} c_{22}$ ... $c_{29} c_{30}$		$c_{31} c_{32}$ ... $c_{39} c_{40}$	

where,

3 : Portion of the chromosome corresponding to the level i (i=1 user; i=2 system; i=3 process; i=4 packet)

3 : Portion of the chromosome corresponding to the parameter j of the level i (i=1,..,4)

For designing the fitness function we also need the following definitions:

Min δ : Minimum value that δ can match, as shown in table 3.

Table 3. Definition of Min δ

δ	Min δ
00	0
01	1
10	2
11	3
0#	0
#0	0
1#	2
#1	1
##	0

Accordingly,

$$\mathrm{MinValue}(P^i) = \frac{1}{5}\sum_{j=1}^{5}\frac{\mathrm{Min}(P_j^i)}{3} \tag{1}$$

and NumWC δ : number of '#' characters of δ .

As discussed in section 3.1, an hypothesis is represented by a list of real values (k_1,k_2,k_3,k_4,a). Then, the KnowledgeDist function can be define as the distance from the minimum value of each level of the rule to the one specified by the knowledge base.

$$\mathrm{KnowledgeDist}(C) = \frac{1}{4}\sum_{i=1}^{4}|k_i - \mathrm{MinValue}(P^i)| \tag{2}$$

The Generality function measures the difference between the number of wildcards ('#') at each level in the rule compare to the optimal value (optNumWC), where GenCoef is a value between 0.0 and 1.0 and specifies the importance of the generality on the evaluation of the fitness.

$$\mathrm{Generality}(C) = \mathrm{GenCoef} * \frac{1}{4}\sum_{i=1}^{4}|\mathrm{OptNumWC} - \mathrm{NumWC}(P^i)| \tag{3}$$

The fitness is now calculated by subtracting the KnowledgeDis and the Generality values from 2, since the maximum fitness value is considered as 2.0.

$$\mathrm{Fitness}(C) = 2 - \mathrm{KnowledgeDist}(C) - \mathrm{Generality}(C) \tag{4}$$

3.2.4   The Sequential Niching Algorithm

In order to evolve a diverse set of rules we use a sequential niching algorithm, which repeatedly run GA module with a modified fitness function as described below.

Let B=<b_1,b_2,...,b_n> be the best individuals of the previous runs of the genetic algorithm, the idea of this niching algorithm is to deprecate the fitness value of an individual depending in the distance of that individual to the best individuals.

Let f be an individual, let be dist(f,b_j) be the distance between the individual f and the best individual b_j, the modified fitness of f (fitness'(f)) is defined as follows:

$$\text{fitness'}(f) = M_n(f) \tag{5}$$

where,

$$M_0 = \text{Fitness}(f)$$
$$M_i = M_{i-1} * G(f, b_i)$$

$$G(f, b_i) = \begin{cases} (\text{dist}(f, b_i) / R)^{\text{alpha}} & \text{if } \text{dist}(f, b_i) < R \\ 1 & \text{otherwise} \end{cases} \tag{6}$$

R and alpha are the parameters of the sequential niching algorithm.

4   Preliminary Experiments and Some Results

We performed several experiments with different data sets in order to test the performance of the classifier-based decision support system. These experiments are conducted in a simulated laboratory environment. In all experiments, we used the high level knowledge (as discussed in section 3.1) based on the following hypothesis:

1. If U>= 0.2 or S>= 0.2 or P>= 0.2 or N>= 0.2 then execute action 1
2. If N>= 0.5 or U>= 0.5 or P>= 0.5 then execute action 2
3. If (U>= 0.4 and S>= 0.4) or (P>= 0.4 and N>= 0.4) or (P>= 0.4 and N>= 0.4) then action 3
4. If U>= 0.8 or S>= 0.8 or P>= 0.8 or N>= 0.8 then execute action 4
5. If (U>= 0.4 and P>= 0.6) or (S>= 0.4 and N>= 0.6) then execute action 5
6. If (U>= 0.5 and S>= 0.6) or (P>= 0.5 and N>= 0.6) then execute action 6
7. If (U>= 0.6 and S>= 0.6 and P>= 0.6) or (N>= 0.6 and P>= 0.6) then execute action 7
8. If U>= 0.7 and S>= 0.7 and P>= 0.7 and N>= 0.7 then execute action 8

Some of the above hypotheses are experimented with to generate a set of classifier rules. In particular, the first four hypotheses are used which generated 1440 rules from 2^{40} possibilities, however, rules for other hypotheses can be evolved in a similar fashion. The goal of the experiment is to test the performance of this evolved rule set to detect any arbitrary input pattern. To accomplish this, we generated a random list of 40-bit messages (representing values of monitored parameters) as input to the classifier system. For each message, we compared the output of the classifier (an action) with the action suggested for the knowledge base (hypothesis).

Accordingly, the system is tested with simulated data (10.000 messages) having different types of deviation to examine the coverage and robustness in detection. In our preliminary experiments four types of actions were chosen. The results of these experiments are shown in the Table 4.

Table 4. Experimental results showing the performance of the classifier based decision supprot system. The shaded cells represent the correct answers of the system

Correct Action	Classifier System Output			
	1	2	3	4
1	78%	22%	0%	1%
2	2%	96%	0%	2%
3	7%	38%	53%	3%
4	0%	4%	17%	79%

In table 4, the first row (correct action 1) establishes that the classifier system identified action 1 on the 78% of the test messages, action 2 on the 22% of the messages, action 3 on the 0% of the messages and action 4 on the 1% of the messages. The higher values on the diagonal indicate that the classifier system could correctly determine the appropriate action in most cases.

Figure 6 also reflects that in most of the test cases, the classifier rules depicted the correct action.

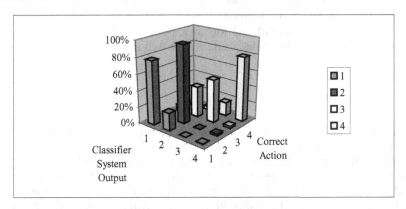

Fig. 6. The percentage of correct actions determined by the classifier rules

However, the higher level of incorrect responses was for the action 3, where the classifier system outputs 53% of the cases action 3, and 38% of the cases action 2. This fact indicates that a high number of rules are necessary to represent more specific rules.

We are currently experimenting to evolve a comprehensive set of rules based on the significance of monitored parameters at the various levels. Accordingly, we predefined eight different action types, but it is the classifier system, which will choose the appropriate action depending on the input message. The classifier decision is to be passed to response agents to undertake one or more of the following actions (A0 – A7):

A0. Take no action
A1. Informing the system administrator via e-mail or messaging system
A2. Change the priority of user processes
A3. Change access privileges of certain user
A4. Block a particular IP address or sender
A5. Disallow establishing a remote connection request
A6. Termination of existing network connection
A7. Restarting of a particular machine

The results reported (in table 4 and figure 6) are the part of our on-going research in developing intelligent decision support subsystems for intrusion detection. We expect some interesting classifier rules to evolve that are more meaningful, for example,

ú If connection is external and CPU usage > Threshold3 and free memory significantly low swap space is minimum then perform Action A4
ú If connection is external and average number of packets received >Threshold2 and duration of the connection >Threshold 2 then perform Action A6
ú If number of connections > Threshold1 and number of processes running > Threshold1 and amount of swap space in Minimum then perform Action A5
ú If connection is internal and CPU usage > Threshold1 and memory usage > Threshold2 and average number of packets sent/ received > Threshold1 respectively then perform Action A1
ú If CPU usage > Threshold1 and Memory usage is > Threshold2 and number of blocked processes > Threshold2 then perform Action A2

5   Concluding Remarks

Most existing intrusion detection systems either use packet-level information or user activities to make decisions on intrusive activities [6], [9], [10]. In this paper, we described an intrusion detection system that can simultaneously monitor network activities at different levels (such as packet level, process level system level and user level), it can detect both inside misuse and outside attacks. The main emphasis of this work is to examine the feasibility of using a classifier-based intelligent decision support subsystem for robust intrusion detection.

The proposed system has some unique features of simultaneous monitoring at multi-level to detect both known and unknown intrusions and generate specific response. The developed system will perform real-time monitoring, analyzing, and generating appropriate response to intrusive activities. This work is a part of a larger research project on developing an intelligent intrusion detection system [2]. In this paper, we emphasized on the design and implementation of classifier system as decision support component. We are currently experimenting in a simulated environment as a part of an early development. We anticipate that a more effective and practical rule base will emerge after the implementation and observation of the network activity over a period of time during the testing phase.

References

1. Crosbie, M., Spafford, G.: Applying Genetic Programming to Intrusion Detection. COAST Laboratory, Purdue University, (1997) (also published in the proceeding of the Genetic Programming Conference)
2. Dasgupta, D.: Immunity-Based Intrusion Detection System: A General Framework. In the proceedings of the National Information Systems Security Conference, (October, 1999)
3. Frank, J.: Artificial Intelligence and Intrusion Detection: Current and future directions. In Proceedings of the 17th National Computer Security Conference, (October, 1994)
4. Balasubramaniyan, J., Fernandez, J.O.G., Isacoff, D., Spafford, E., Zamboni, D.: An Architecture for Intrusion Detection using Autonomous Agents, COAST Technical report 98/5, Purdue University, (1998)
5. Crosbie, M., Spafford, E.: Defending a computer system using autonomous agents. In Proceedings of the 18th National Information Systems Security Conference, (October, 1995)
6. Me, L., GASSATA,: A Genetic Algorithm as an Alternative Tool for Security Audit Trail Analysis. in Proceedings of the First International Workshop on the Recent Advances in Intrusion Detection, Louvain-la-Neuve, Belgium, (September, 1998) 14–16
7. Zhang, Z., Franklin, S., Dasgupta D.: Metacognition in Software Agents using Classifier Systems. In the proceedings of the National Conference on Artificial Intelligence (AAAI), Madison, (July, 1998) 14-16
8. Boer, B.: Classifier Systems, A useful approach to machine learning? Masters thesis, Leiden University, (August 31, 1994)
9. Mukherjee, B., Heberline, L.T., Levit, K.: Network Intrusion Detection. IEEE Network (1994)
10. Axelsson, S., Lindqvist, U., Gustafson, U., Jonsson, E.: An Approach to UNIX security Logging, Technical Report IEEE Network (1996)
11. Lunt, T.F.: Real-Time Intrusion Detection. Technical Report Computer Science Journal (1990)
12. Debar, H., Dacier, M., Wepspi, A.: A Revised Taxonomy for Intrusion Detection Systems. Technical Report Computer Science/Mathematics (1999)
13. Goldberg, D.E.: Genetic Algorithms in Search, Optimization & Machine Learning. Addison–Wesley, Reading, Mass. (1989)
14. Back, T., Fogel, D.B., Michalewicz, Z.: Handbook of Evolutionary computation. Institute of Physics Publishing and Oxford university press (1997)
15. Dasgupta, D., Michalewicz, Z. (eds): Evolutionary Algorithms in Engineering and Applications. Springer-Verlag (1997)

Mathematical Models of the   Covert Channels

Alexander Grusho

Russian State University for Humanity, Kirovogradskaia 25, Moscow, Russia
aaotee@mail.infotel.ru

Abstract. The analysis of the covert channels is necessary for construction of the protected computer systems, stenography, including methods of construction of labels for the intellectual property protection and secure labels of the copies of information objects. In some conditions it is possible to prove the invisibility by the supervising subject of the covert channel. The review of cases is given, when in some absolute sense using methods of the theory of probability and statistics it is possible to prove the "invisibility" of the covert channel by the supervising subject.

1   Introduction

The analysis of the covert channels is necessary for construction of the protected computer systems, stenography, including methods of construction of labels for the intellectual property protection and secure labels of the copies of information objects. All covert channel models are defined by the basic set, the method of the covert embedding of the message in the basic set and the observer, that is the supervising subject who can see some function of the basic set with the built - in latent message. The quality of a covert channel is defined by the next three requirements:

1) The information of the covert message should be unclear (supervising subject can not find out anything about the contents of the covert message)

2) The presence of the covert message should be invisible (supervising subject does not "see" the existence of the covert message in the basic set)

3) The restoration of the information from the covert message on the reception end can be done, despite of the fact that the supervising subject can transform the basic set with any transformation from some class.

These conditions make think that the existence of the covert channels is a problem. For its solution it is necessary to enter restrictions on opportunities of the supervising subject to solve tasks 1) -3). For mathematical modeling of the covert channels we can consider, that a finite automaton represents any computer system. In some conditions it is possible to prove the invisibility by the supervising subject of the covert channel. In the report the review of cases will be given, when in some absolute sense using methods of the theory of probability and statistics it is possible to prove the "invisibility" of the covert channel by the supervising subject.

V.I. Gorodetski et al. (Eds.): MMM-ACNS 2001, LNCS 2052, pp. 15-20, 2001.

2   The Models for Invisibility

Probability methods allow constructing substantial models for the analysis of the covert channels in computer systems.

Computer system •1 can be presented by the finite automaton $(X, S, Y, d, 1)$, where X is a finite set of the entrance messages, S is a finite set of states of system, d: $X \hat{o} S$ ì S is a function of transitions from one state to another, $y = 1 (x, s)$, x   X, s   S, is a function of outputs, which the subject U can observe. U is the subject supervising a correctness of the tasks in the computing system •1.

Let's consider the problem of the covert channel construction to control an agent introduced by some adversary into our computer system. The adversary wants the agent to carry out illegal actions in our computer system. Let's simulate of such agent by the finite automaton.

The illegal automaton •2 $= (Z, S_1, d_1)$, receives on an input the same message • from •, as automaton •1. However on an input of the automaton •2 there is an interpreter to translate • into a control signal z of the automaton •2 that is the function g, g: X ì Z, displaying a common input of both automata in the input alphabet Z of the automaton •2.

Let's assume, that the distribution of probabilities $P_X(x)$ is defined on X. It generates through functions $1 (x, s)$ and $g(x)$ the distribution $Q_s(y)$ on Y and the distribution $P_Z(z)$ on Z. Channels from •to Y and Z thus are determined which we shall designate

$$X a \ Y, X a \ Z.$$

Let's assume

$$\{y, z\}_s = \{x: 1 (x, s) = y, g(x) = z\}.$$

Definition 1. The subject U does not see in a state s the covert channel $X a \ Z$ iff the distribution P_X satisfies to a condition: at all y   Y, z   Z

$$P_X(\{y,z\}_s) = Q_s(y)P_Z(z). \tag{1}$$

We are interested with a question of construction of the channel invisible for U in all states s   S.

In [1] some properties of such models are investigated. The necessary and sufficient conditions are received that the observer U does not see in any state s the channel $X a \ Z$.

Theorem 1. Let for any s   S, y   Y, z   Z

$$¢ \{y, z\}_s ¢ = 1. \tag{2}$$

The covert channel $X a \ Z$ is invisible for U in all states s   S iff the following conditions are carried out.

(i) There exists s_0 S, in which the channel $X a \ Z$ is invisible for U.

(ii) For any s û s_0 from S the next sets of values of probabilities coincide:

$$\{Q_s(y)\} = \{Q_{s_0}(y)\}. \tag{3}$$

The model of invisible interaction between the agents through the observable controllable channel was considered in [2]. We shall mention here simplified variant of results of this work.

There are two pairs of users or subjects in computer system, (U_1, U_2) and (U_1', U_2'), each of them uses the same channel of communication without noise. For simplicity we suppose, that the transfer of the information is only in one direction, from U_1 to U_2 or from U_1' to U_2'. In these cases we write U_1 a U_2 or U_1' a U_2'. Second pair tries to organize transmission of their information through the channel U_1' a U_2' in such a way that externally it does not differ from transmission through the channel U_1 a U_2. There is some supervising subject U, which observes the transmission, but does not know, what pair subjects, (U_1, U_2), or (U_1', U_2'), conducts it. It is supposed, that volume of the transmitted information is great. So U can only observe of an output of some function l, which intercepts all transfer, processes it and the result gives out to the subject U as some message y. It is supposed, that the subjects (U_1, U_2) and (U_1', U_2') know the circuit of the control and the function l and can use these given for organization covert channel.

Let • be a set of the messages, which can be transmitted from U_1 to U_2, •. (•) be a distribution of probabilities on •, through which we simulate the channel U_1 a U_2. The function l: • i Y defines the target message y from $Y = \{y_1, y_2..., y_n\}$ for the supervising subject U. Thus, observing the channel U_1 a U_2, the subject U can receive the message y from set Y, chosen according to probability measure

$$Q(y) = P_x\{x: l(x) = y\}.$$

For the description of the channel U_1' a U_2' we shall consider the set Z of the possible messages from U_1' to U_2'. The choice of the message z from Z is defined by some probability measure •(z). For the message z some message • from • gets out and is transferred on the channel. On the reception end on x is restored z. On the reception end according to some unknown for U arrangement it is known which of the channels U_1 a U_2 or U_1' a U_2' work.

Let's consider the following way of a choice of the message • on z. U_1' and U_2' have the function

$$g: Z \, ô \, Q \, ì \, X,$$

where Q is some finite set. The restoration of the message z is determined by a condition, that for any z $\,$ Z, q $\,$ Q,

$$g^{-1}(g(z,q)) = z.. \tag{4}$$

The choice of q is defined by a conditional distribution P_Q (q | z). Then the distribution P(z, q) on Z ôQ and the function g(z, q) define a probability measure $P_X^/$ (x) on •. Measures P_x and $P_X^/$ may be different. Probability measure $P_X^/$ (x) and function l(x) determine a probability measure $Q^/$ (y) on set Y. Only in case of distinction Q and $Q^/$ the observer U can distinguish channels U_1a U_2 and $U_1^/$a $U_2^/$.

Definition 2. The subject U does not see the channel $U_1^/$ a $U_2^/$ iff for any distribution $P_z(z)$ and for every y Y

$$Q^/ (y) = Q(y).$$

In this case we shall speak that the channel $U_1^/$ is invisible for U.

It is possible to prove simple conditions, at which there is a channel $U_1^/$a $U_2^/$, invisible for U.
Theorem 2. Let for any y from Y such, that Q(y) > 0, the next condition is satisfied:

$$\left|1^{-1}(y)\right| \ddagger |z|.$$

Then there exists the channel $U_1^/$a $U_2^/$, invisible for U.

Till now probability models of the determinate processes helped to construct "invisibility" in absolute sense. Thus this approach did not assume essentially new constructions of the covert channels. However it is possible to offer model of the statistical covert channel, which cannot be defined but in the terms of statistical structures.

Let's consider the following circuit of communication

$$U$$
$$\downarrow$$
$$U_1 \xrightarrow{\hspace{2cm}} U_2$$

Where U_1 is a source of a message, U_2 is an addressee of a message and U is a supervising subject.

Let's assume, that the usual session of communication from U_1 to U_2 consists in transmission of values $X_1, X_2,..., X_N$ of the independent equally distributed random variables, each of which has distribution $•_0$ = N(0, 1) - normal distribution with parameters 0 and 1. Let U knows this distribution. Let's assume, that U_1 and U_2 have agreed upon the covert channel to transmit one bit b. For this purpose they secretly beforehand have chosen i (i = 1,..., N) and have defined, that if b = 0, the distribution X_i has distribution $•_0$. If b = 1, the distribution X_i is equal P_1 = N(•, 1). Allowable mistakes a and b of the first and second sort for reception of bit b are given. We assume, that U does not know i. U gets a sequence of values of random variables $X_1, X_2,..., X_N$, where X_i is distributed depending on value of bit b.

Let's consider actions U_2 on reception. U_2 knows i and can accept the decision on the value of bit b, basing on the best criterion. At organization of communication U_1 and U_2 choose parameter a so that the best criterion of a level a gives a mistake of the second sort not superior b.

As U does not know i, on supervision X_1, X_2,..., X_N it should check up a hypothesis H_0 that it is a sample from distribution N (0, 1) against alternative H_1, that there is such i, that X_i is distributed N (a, 1) with known a, and all others are distributed normally with parameters 0 and 1.

Let's consider a problem [3] of existence of consistent criterion to check of a hypothesis H_0 against alternative H_1 by the subject U, when • ì ¥ and N ì ¥ . The first condition guarantees correct transfer of bit, and second - opportunity to hide transfer of bit b.

Theorem 3. When $2a^2$ - lnNfi -¥ there is no consistent criterion to check H_0 against H_1. When $2a^2$ - lnNfi + ¥ there exists a consistent criterion check H_0 against H_1.

3   The Models of Unidirectional Channels

The previous examples do not exhaust variety of models of the covert channels. On the other hand the problem to prove absence of the covert channels is urgent. Let's consider some probability models concerning this problem. In the elementary case probability definition of a channel is connected to a matrix of conditional probabilities •of the size môn, where m is a number of elements in set • of the entrance messages of the channel, and n is a number of elements in set Y - target messages of the channel. The matrix • together with distribution P_X on the initial messages define joint distribution •(•,y) on XôY. On the other hand if the joint distribution •(•,y) on XôY is given and is not product of measures $•_x$ôP$_Y$, then not trivial channel from • to Y is determined. Thus set of all potential channels, including covert channels, is immersed in set of pairs dependent random variables. Let's note, that the channel from • to Y does not exist, when these random variables are independent. In real computer system these conditions often are not carried out. However in some models such approach becomes constructive. For example, we shall consider two computer systems High and Low and we shall put a task of the description of the unidirectional channel from Low to High. The proof of a one-orientation consists of two parts. Let Z is a set of sequences of states of the channel from Low to High and $•_z$ is the probability distribution on these states. Let's consider any pair of random variables • from Low and Y from High. Then fulfillment for all x, y and z of the condition

$$P(x,y¢z) = P(x¢z)P(y¢z)$$

means the absence of the channel from Y to X provided that z it is known to Low and does not carry the information about Y. If this equality is carried out for all • from Low and Y from High, then the unique opportunity to transfer information from High to Low is any stenography scheme with the basic set z from Z. Thus, the problem to prove the absence of covert channels from High to Low is possible to reduce to the problem of absence of channels in models considered earlier.

References

1. Grusho, A.: Covet Channels and Information Security in the Computer Systems. J. Discreet Mathematics (in Russian), Vol. 10(1), Moscow (1998) 3–9
2. Grusho, A.: On the existence of Covert Channels. J. Discreet Mathematics (in Russian), Vol. 11(1), Moscow (1999) 24–28
3. Grusho, A., Timonina, E.: Asymptotically Invisible Statistical Covert Channel. Revue of Applied add Industrial Mathematics (in Russian), Vol. 6(1), Moscow, TVP (1999) 135–136

Open Issues in Formal Methods for Cryptographic Protocol Analysis

Catherine Meadows

Center for High Assurance Computer Systems, Naval Research Laboratory,
Washington, DC, 20375
meadows@itd.nrl.navy.mil

Abstract.   The history of the application of formal methods to cryptographic protocol analysis spans nearly twenty years, and recently has
been showing signs of new maturity and consolidation. A number of specialized tools have been developed, and others have e ectively demonstrated that existing general-purpose tools can also be applied to these
problems with good results. However, with this better understanding of
the eld comes new problems that strain against the limits of the existing
tools. In this talk we will outline some of these new problem areas, and
describe what new research needs to be done to to meet the challenges
posed.

V.I. Gorodetski et al. (Eds.): MMM-ACNS 2001, LNCS 2052, p. 21, 2001.
c Springer-Verlag Berlin Heidelberg 2001

Future Directions in Role-Based Access   Control Models

Ravi Sandhu

Chief Scientist, SingleSignOn.Net Inc.
11417 Sunset Hills Road, Reston, VA 20190, USA
rsandhu@singlesignon.net,   www.singlesignon.net
&
Professor of Information Technology and Engineering
George Mason University, Fairfax, VA 22030, USA
sandhu@gmu.edu,   www.list.gmu.edu

Abstract. In the past five years there has been tremendous activity in role-based access control (RBAC) models. Consensus has been achieved on a standard core RBAC model that is in process of publication by the US National Institute of Standards and Technology (NIST). An early insight was that RBAC cannot be encompassed by a single model since RBAC concepts range from very simple to very sophisticated. Hence a family of models is more appropriate than a single model. The NIST model reflects this approach. In fact RBAC is an open-ended concept which can be extended in many different directions as new applications and systems arise. The consensus embodied in the NIST model is a substantial achievement. All the same it just a starting point. There are important aspects of RBAC models, such as administration of RBAC, on which consensus remains to be reached. Recent RBAC models have studied newer concepts such as delegation and personalization, which are not captured in the NIST model. Applications of RBAC in workflow management systems have been investigated by several researchers. Research on RBAC systems that cross organizational boundaries has also been initiated. Thus RBAC models remain a fertile area for future research. In this paper we discuss some of the directions which we feel are likely to result in practically useful enhancements to the current state of art in RBAC models.

Introduction

Research on access control models was started in the 1960s and 1970s by the two thrusts of mandatory and discretionary access control. Mandatory access control (MAC) came from the military and national security arenas whereas discretionary access control (DAC) had its roots in academic and commercial research laboratories. These two thrusts were dominant through the 1970s and 1980s almost to exclusion of any other approach to access control models. In the 1990s we have seen a dramatic shift towards pragmatism.

The dominant access-control model of the 1990s is role-based access control (RBAC). In this paper we make the case that RBAC will continue to be dominant for the next decade.

V.I. Gorodetski et al. (Eds.): MMM-ACNS 2001, LNCS 2052, pp. 22-26, 2001.
© Springer-Verlag Berlin Heidelberg 2001

Current State of RBAC Models

To my knowledge the first use of the term RBAC is due to Ferraiolo and Kuhn [5] although there has been prior mention in the security literature of "roles" and "role-based security." Sandhu et al [14] subsequently published a seminal paper defining a family of models that has since come to be called RBAC96. A crucial insight of RBA96 was the realization that RBAC can range from very simple to very sophisticated so we need a family of models rather than a single model. A single model is too complex for some needs and simple for others. A graded family of models enables selection of the "correct" model for a particular situation. Publication of RBAC96 was followed by a flurry of research that has clearly established RBAC as the dominant access control model. Remarkably the basic concepts of RBAC96 have proved to be robust and no significant omissions have been identified. In many years of research following publication of RBAC96 we have had occasion to introduce only one new concept (role activation hierarchies [15]) which was not already present in RBAC96.

Let us now briefly review important achievements in recent RBAC research. The perspective given here is necessarily a personal one. As such the papers cited are those with greatest direct impact on our own understanding of RBAC models. There simply is not enough room to cite many other papers of considerable significance.

We feel that RBAC models have advanced in at least three respects in recent years, discussed below in turn.

Firstly, an important recent development is emergence of a consensus standard model which is supported by a major standards organization (the US National Institute of Standards and Technology or NIST). Following the publication of RBAC96 it became clear that many authors were pursuing very similar ideas but with differences in detail leading to confusion about the nature of RBAC. RBAC96 was unique in proposing the concept of a graded family of models. Once this family notion was accepted by the RBAC community consensus on a core set of RBAC concepts became feasible. To this end an initial attempt at a family of standard models was presented at the Berlin RBAC Workshop by Sandhu et al [18]. Workshop attendees reacted to this proposal with heated discussion [12]. The current proposal is to be published soon [6] and will then evolve into NIST publications. Deployment and use of RBAC in commercial products and systems will be facilitated by the NIST standard model.

Another important development is a deeper theoretical understanding of RBAC and particularly its relationship to MAC and DAC. There has been much confusion with some authors claiming that RBAC is a form of MAC while others arguing it is a form of DAC. Osborn et al [13] show that RBAC96 can be configured to do MAC or DAC as one chooses. So RBAC transcends the MAC-DAC distinction. Fundamentally it turns out that both MAC and DAC are just special cases of RBAC. For historic reasons MAC and DAC gained early dominance in the research community. MAC and DAC are easily unified within the framework of RBAC. This unification is more than coexistence. MAC systems also usually implement DAC but in these systems MAC and DAC simply coexist. The RBAC viewpoint is that MAC and DAC are just examples of policies to configure in a policy-neutral RBAC model.

The third significant development is a contextual understanding of the practical purpose of RBAC models . Sandhu [17] argues that the purpose of RBAC models is two-fold. On hand they help us articulate access-control objectives in a mathematical and rigorous framework. On the other hand they help us understand how to actually architect a system with attendant trust, liability and authority responsibilities and obligations. The clear separation of a model from objectives (or policy) and architecture (and even deeper mechanism) is captured in the four layer OM-AM (for objectives, models, architecture, mechanism) framework of [17]. RBAC models are designed to be objective (or policy) neutral but can be configured to achieve a wide range of policies (including the extreme cases of MAC and DAC discussed above). In this paper our focus is on models and OM-AM allows us to clearly understand the two-faced nature of RBAC models. On one side models help us understand and articulate policy. On the other side a given model can be implemented in many different architectures (and with many different mechanisms).

Future Directions for RBAC Models

Now we consider aspects of RBAC models that need further research. Some of these have already been explored. Some are even rather mature but consensus in the community has not yet been achieved. Others have only been hinted at in the literature or only preliminary exploratory work has been published. So we can divide our discussion roughly into two categories: areas in which strong progress has been made but consensus needs to be developed to reach maturity such as embodied in standards, and areas in which only preliminary work has been accomplished.

One of the main omissions in the NIST standard model [6] is the authorization of administration of RBAC. Access control is basically simple so long as the permissions do not change. However a static model of access control is not very realistic. Sandhu et al [15] have argued that administration of RBAC in large scale systems must itself be decentralized and can profitably be managed using administrative roles. The ARBAC97 model shows how this can be done using RBAC96 as the underlying model. It would be desirable to develop standards in this arena because administration is often the place where security breaks. Moreover, the ARBAC97 model addresses RBAC administration from one point of view and one administrative paradigm.

Alternate administrative paradigms for RBAC have been recently discussed in the literature. Hildmann and Barholdt [8] and Herzberg et al [7] consider some issues in assigning roles to users in systems that cross organizational boundaries. Barka and Sandhu [2] have proposed a framework for modeling delegation of roles from one user to another. Huang and Atluri [10] discuss the dynamics of RBAC in workflow systems. Damianou et al [4] and Hitchens and Varadharajan [9] have proposed languages for specifying RBAC policy. Thomas and Sandhu [19] have argues the need for active authorization models which are self-adminstering. These papers present specific perspectives and viewpoints on adminstration of RBAC. However, we are far short of an integrated model around which community consensus can be developed.

Most RBAC research to date has been based on a single organization's point of view. This is a natural consequence of the initial motivation for RBAC which is concerned with managing access rights in large-scale systems. In the future we are likely to see greater interest in applications of RBAC to Business-to-Business and

Business-to-Consumer electronic commerce. RBAC is a natural technology for separating responsibilities in cross-organization systems. User-role assignment can be handled by one organization while permission-role assignment is handled by another.

Another consequence of the organizational emphasis in past RBAC research is that assigning a role to a user is generally considered an administrative act of some other user (or administrator). In the digital economy we can conceive of roles that are acquired due to payment, such as membership in a club or society, or as a reward or bonus, such as frequent flyer status. We can also have roles that are traded between users for some kind of a fee. Developing a comprehensive RBAC administrative model to cover this scope is a challenging research task.

While role hierarchies are well understood RBAC constraints have only received attention in recent years. Classically separation of duty has been seen as the main motivation for constraints in RBAC models. More recently their importance beyond separation of duties has been recognized in the literature [11, 3]. Ahn and Sandhu [1] have proposed the RCL2000 language for specifying RBAC constraints and have argued that prohibition and obligation constraints are both required with separation constraints being an example of prohibition.

Conclusion

We hope to have convinced the reader that research on RBAC models has just begun and much interesting and challenging work remains to be done. The RBAC arena is intrinsically dominated by practical considerations and offers an opportunity for good theoretical research to be translated into practical impact on products and practice.

Acknowledgement. The work of Ravi Sandhu is partially supported by the National Science Foundation.

References

1. Gail Ahn and Ravi Sandhu: Role-Based Authorization Constraints Specification. ACM Trans. on Information and System Security, V. 3, No 4 (November 2000)
2. Ezedin Barka and Ravi Sandhu: Framework for Role-Based Delegation Models. Proc. 16th Annual Computer Security Applications Conference, New Orleans (Dec., 2000)
3. Bertino, E., Bonatti, P., and Ferrari, E.: TRBAC: A Temporal Role-Based Access Control Model. ACM Transactions on Info. and System Security, 4:3, (Aug. 2001) to appear
4. Damianou, N., Dulay, N., Lupu, E., and Sloman, M.: The Ponder Policy Specification Language. Int. Workshop on Policy, Jan. 2001, Springer LNCS 1995
5. Ferraiolo, D. and Kuhn, R.: Role-Based Access Control. In Proc. of the NIST-NSA National Computer Security Conference. (1992) 554–563
6. Ferraiolo, D.F., Sandhu, R., Gavrila, D., Kuhn, D.R. and Chandramouli, R.: A Proposed Standard for Role-Based Access Control. ACM Transactions on Information and System Security, V. 4, No 3, (August 2001) to appear

7. Herzberg, A., Mass, Y., Mihaeli, J., Naor, D. and Ravid, Y.: Access Control Meets Public Key Infrastructure, Or: Assigning Roles to Strangers. IEEE Symposium on Security and Privacy, Oakland (May 2000)
8. Hildmann, T. and Barholdt, J.: Managing trust between collaborating companies using outsourced role based access control. In Proc. of 4th ACM Workshop on Role-Based Access Control. 1999 (105–111)
9. Hitchens, M. and Varadharajan, V.: Tower: A Language for Role Based Access Control. Int. Workshop on Policy, Bristol, UK, January 2001, Springer LNCS 1995
10. Huang, W., and Atluri, V.: A secure web-based workflow management system. In Proc. of 4th ACM Workshop on Role-Based Access Control (1999)
11. Jaeger, T.: On the Increasing Importance of Constraints. Proc. 4th ACM Workshop on Role-Based Access Control, Fairfax, Virginia (Oct. 28–29, 1999) 33–42
12. Jaeger, T. and Tidswell, J.: Rebuttal to the NIST RBAC model proposal. Proc. 5th ACM Workshop on Role-Based Access Control, Berlin, Germany. (July 26–28, 2000) 65–66
13. Osborn, S., Sandhu, R. and Munawer, Q.: Configuring Role-Based Access Control to Enforce Mandatory and Discretionary Access Control Policies. ACM Trans. on Information and System Security, V. 3, No 2, (May 2000) 85–106
14. Sandhu, R., Coyne, E., Feinstein, H. and Youman, C.: Role-Based Access Control Models. IEEE Computer, V. 29, No 2. (Feb. 1996) 38–47
15. Sandhu, R.: Role Activation Hierarchies. Proc. 3rd ACM Workshop on Role-Based Access Control, Fairfax, Virginia. (October 22–23, 1998) 33–40
16. Sandhu, R., Bhamidipati, V. and Munawer, Q.: The ARBAC97 Model for Role-Based Administration of Roles. ACM Trans. on Info. and System Security 2:1, (Feb. 99) 105–135
17. Sandhu, R.: Engineering Authority and Trust in Cyberspace: The OM-AM and RBAC Way. Proc. 5th ACM Workshop on RBAC, Berlin. (July 26–28, 2000) 111–119
18. Sandhu, R., Ferraiolo, D. and Kuhn, R.: The NIST Model for Role-Based Access Control: Towards A Unified Standard. Proc. 5th ACM Workshop on RBAC, 47–63
19. Thomas, R. and Sandhu, R.: Task-based Authorization Controls (TBAC): Models for Active and Enterprise-Oriented Authorization Management. In Database Security XI: Status and Prospects, Chapman & Hall 1998. 262–275

Secure Networked Computing

Vijay Varadharajan

Microsoft Chair Professor in Computing
Division of Information and Communication Sciences
Macquarie University, Australia
vijay@cit.uws.edu.au
vijay@ics.mq.edu.au

Abstract.   Mobile computing and agent technologies are receiving a
great deal of interest as they have a lot to o er towards achieving the
vision of usable distributed systems in a heterogenous network environment such as the Internet. The ability to move computations across the
nodes of a wide area network helps to achieve deployment of services
and applications in a more flexible, dynamic and customisable way than
the traditional client-server distribution paradigm. However fundamental challenges remain in the area of security. In this talk, we will address
some of the key issues in secure networked computing. We will rst address the developments that have taken place in distributed computing
which as led to greater mobility and networked applications in both local
area and wide area networks. Then we will consider some of the characteristics of network based computing and in particular the development
of mobile agent based applications. We will then briefly address the notion of trust and describe how mobile agents violate some of the common
trust assumptions made in the design of secure systems. Then we will
outline a security model for mobile agents and the provision of security
services for agent based applications. We will conclude by identifying
some of the open security issues that need to be investigated.

V.I. Gorodetski et al. (Eds.): MMM-ACNS 2001, LNCS 2052, p. 27, 2001.
c Springer-Verlag Berlin Heidelberg 2001

Composability of Secrecy

Jan Jürjens[?]

Computing Laboratory, University of Oxford, GB

Abstract. Modularity has been seen to be very useful in system development. Unfortunately, many security properties proposed in the literature are not composable (in contrast to other system properties), which is required to reason about them in a modular way.

We present work supporting modular development of secure systems by showing a standard notion of secrecy to be composable wrt. the standard composition in the specification framework Focus (extended with cryptographic primitives). Additionally, the property is preserved under the standard renement. We consider more ne-grained conditions useful in modular verication of secrecy.

Key words. Network security, cryptographic protocols, secrecy, modularity, composability, renement, formal specication, computer aided software engineering.

1   Introduction

Mathematical models have been used extensively to ensure security of computer networks through reasoning about formal specications [18]. In spite of several success stories, major challenges remain. An important one of them is the problem of scaling up the used methods to large systems [19].

For this, modularity is an important tool: One starts with considering components of a system that are so small that formal reasoning is still feasible. One proves that these components satisfy the required security property. If the security property under consideration is composable then (by denition) the whole system satises the security property.

Unfortunately, this is often not the case: various formulations of security properties in formal models are noncomposable (e. g. McLean's Applied Flow Model and Gray's Probabilistic Noninterference (in its original model) have been shown to be noncomposable in [9]; exceptions are [25,16]).

Not only does this make verication of specications harder; even worse, it is also not clear what a non-composable security property guarantees in practice, where one would in fact like to connect systems to each other [20].

In this work we consider composability of security properties, more specically of secrecy. Continuing earlier work [11] we use an extension of the general-purpose, tool-supported CASE-framework Focus [5] with cryptographic operations including symmetric and asymmetric encryption and signing to consider

---

[?] jan@comlab.ox.ac.uk - http://www.jurjens.de/jan – This work was supported by the Studienstiftung des deutschen Volkes and the Computing Laboratory.

V.I. Gorodetski et al. (Eds.): MMM-ACNS 2001, LNCS 2052, pp. 28–38, 2001.

composability of a notion of secrecy following a standard approach. We exhibit conditions under which it is composable and which make modular reasoning widely and usefully applicable. The secrecy property is also preserved under re- nement , another highly useful strategy in formal system development which is often not given for security properties (re nement paradox [15]). This seems to be the rst secrecy property shown to be preserved under composition and re nement. We also give more ne-grained notions of secrecy useful for modular development.

In the next subsection we give background on security and composability and refer to related work. In Section 2, we introduce the cryptographic extension of the speci cation framework Focus. In Section 3 we de ne the secrecy properties considered here. In Section 4 we consider composability and re nement. After that, we conclude.

Some of the proofs have to be omitted for lack of space; they will be given in the long version of this paper to be published.

1.1 Security and Composability

The approach of modular development (or "divide-and-conquer") has a long tradition in the formal development of systems. Therefore there has been some e ort to show that properties of interest are composable. For example, Abadi and Lamport give a proof method for properties falling within the Alpern-Schneider framework of safety and liveness properties. Unfortunately, many security properties do not fall within the Alpern-Schneider framework [15].

Many secrecy properties follow one of the following two approaches (discussed in [1]). One is based on equivalences: Suppose a process speci cation P is parameterised over a variable x representing a piece of data whose secrecy should be preserved. The idea is that P preserves the secrecy of this data if for any two data values d_0, d_1 substituted for x, the resulting processes $P(d_0)$ and $P(d_1)$ are equivalent, i. e. indistinguishable to any adversary, (this appears e. g. in [2]). This kind of secrecy property ensures a rather high degree of security. However, composability of it (also called the hook-up property) is related to the question if equivalence is a congruence which again is often not the case [23].

The secrecy property considered in this paper follows the second approach: a process speci cation preserves the secrecy of a piece of data d if the process never sends out any information from which d could be derived in clear, even in interaction with an adversary (this is attributed to [7] and used e. g. in [6]; a similar notion is used in [24]). In general, it is slightly coarser than the rst kind in that it may not prevent implicit information flow, but both kinds of security properties seem to be roughly equivalent in practice [1]. Since modularity is a highly useful property, a notion of secrecy that reveals only most weaknesses but is modular is well worth considering, especially since more ne-grained security properties may be hard to ensure in practice, as pointed out in [17].

Related Work. This line of work was initiated in [11], where secrecy was shown to be preserved under various standard re nements in the framework Focus and where it was used to uncover a previously unpublished flaw in a variant of the handshake protocol of TLS [1] proposed in [4], to propose a correction and to prove it secure.

An overview on the use of formal methods in security protocols is given in [18]. The need for composability is pointed out in [19].

[25] gives a hook-up property for information flow secure nets. [16] discusses composability of notions of secure information flow using traces based on procedure calls. [15] gives general results for composability of possibilistic notions of secure information flow. [9] shows two notions of secure information flow, namely McLean's Applied Flow Model (AFM) and Gray's Probabilistic Noninterference (PNI) (in its original model) to be non-composable, and provides a slight modification of Gray's model where PNI is composable.

In [14], threat scenarios are used to formally develop secure systems using Focus. Composability (and refinement) is left for further work. Further applications of Focus to system security are in [13].

2   Specification Language

We give a short overview on the specification language used here; for a more detailed account cf. [11]. In this work, we view specifications as nondeterministic programs in the specification framework Focus [5] (providing mechanical assistance in form of the CASE tool Autofocus [8]). Executable specifications allow a rather straightforward modelling of cryptographic operators such as encryption.

Specifically, we consider concurrently executing processes interacting by transmitting sequences of data values over unidirectional FIFO communication channels. Communication is asynchronous in the sense that transmission of a value cannot be prevented by the receiver.

Processes are collections of programs that communicate synchronously (in rounds) through channels, with the constraint that for each of its output channels c the process contains exactly one program p_c that outputs on c. This program p_c may take input from any of P's input channels. Intuitively, the program is a description of a value to be output on the channel c in round $n + 1$, computed from values found on channels in round n. Local state can be maintained through the use of feedback channels, and used for iteration (for instance, for coding while loops).

To be able to reason inductively on syntax, we use a simple specification language from [11,3]. We assume disjoint sets D of data values, Secret of unguessable values, Keys of keys, Channels of channels and Var of variables. Write Enc $\stackrel{\text{def}}{=}$ Keys [Channels [Var for the set of encryptors that may be used for encryption or decryption. The values communicated over channels are formal expressions built from the error value ?, variables, values on input channels, and data values using concatenation. Precisely, the set Exp of expressions contains the empty expression " and the non-empty expressions generated by the grammar

---

[1] TLS is the successor of the Internet security protocol SSL.

$$
\begin{array}{lll}
E ::= & & \text{expression} \\
\quad d & & \text{data value } (d \in D) \\
\quad N & & \text{unguessable value } (N \in Secret) \\
\quad K & & \text{key } (K \in Keys) \\
\quad c & & \text{input on channel } c \ (c \in Channels) \\
\quad x & & \text{variable } (x \in Var) \\
\quad E_1 :: E_2 & & \text{concatenation} \\
\quad \{E\}_e & & \text{encryption } (e \in Enc) \\
\quad Dec_e(E) & & \text{decryption } (e \in Enc)
\end{array}
$$

An occurrence of a channel name c refers to the value found on c at the previous instant. The empty expression " denotes absence of output on a channel at a given point in time. We write $CExp$ for the set of closed expressions (those containing no subterms in $Var \cup Channels$). We write the decryption key corresponding to an encryption key K as K^{-1} (and call it a private key). In the case of asymmetric encryption, the encryption key K is public, and K^{-1} secret. For symmetric encryption, K and K^{-1} may coincide. We assume $Dec_{K^{-1}}(\{E\}_K) = E$ for all $E \in Exp$, $K, K^{-1} \in Keys$ (and we assume that no other equations except those following from these hold, unless stated otherwise).

(Non-deterministic) programs are defined by the grammar (where $E, E' \in Exp$ are expressions):

$$
\begin{array}{lll}
p ::= & & \text{programs} \\
\quad E & & \text{output expression} \\
\quad \text{either } p \text{ or } p' & & \text{nondeterministic branching} \\
\quad \text{if } E = E' \text{ then } p \text{ else } p' & & \text{conditional} \\
\quad \text{case } E \text{ of key do } p \text{ else } p' & & \text{determine if } E \text{ is a key} \\
\quad \text{case } E \text{ of } x :: y \text{ do } p \text{ else } p' & & \text{break up list into head and tail}
\end{array}
$$

Variables are introduced in case constructs, which determine their values. The first case construct tests whether E is a key; if so, p is executed, otherwise p'. The second case construct tests whether E is a list with head x and tail y; if so, p is evaluated, using the actual values of x, y; if not, p' is evaluated. In the second case construct, x and y are bound variables. A program is closed if it contains no unbound variables. while loops can be coded using feedback channels.

From each assignment of expressions to channel names $c \in Channels$ appearing in a program p (called its input channels), p computes an output expression.

For simplification we assume that in the following all programs are well-formed in the sense that each encryption $\{E\}_e$ and each decryption $Dec_e(E)$ appears as part of p in a case E' of key do p else p' construct (unless $e \in Keys$), to ensure that only keys are used to encrypt or decrypt. It is straightforward to enforce this using a type system.

A process is of the form $P = (I, O, L, (p_c)_{c \in \tilde{O} \cup L})$ where
- $I \subseteq Channels$ is called the set of its input channels and
- $O \subseteq Channels$ the set of its output channels ,

and where for each $c \in \tilde{O} \overset{\text{def}}{=} O \cup L$, p_c is a closed program with input channels in $\tilde{I} \overset{def}{=} I \sqcap L$ (where $L \subseteq$ Channels is called the set of local channels). From inputs on the channels in $\tilde{I}$ at a given point in time, p_c computes the output on the channel c.

We write I_P, O_P and L_P for the sets of input, output and local channels of P, $K_P \subseteq$ Keys for the set of the private keys and $S_P \subseteq$ Secret for the set of unguessable values (such as nonces) occurring in P. We assume that different processes have disjoint sets of local channels.

2.1 Stream-Processing Functions

In this subsection we recall the definitions of streams and stream-processing functions from [5].

We write Stream $_C \overset{def}{=} (CExp^{!})^C$ (where $C \subseteq$ Channels) for the set of C-indexed tuples of (finite or infinite) sequences of closed expressions. The elements of this set are called streams , specifically input streams (resp. output streams) if C denotes the set of non-local input (resp. output) channels of a process P. Each stream $s \in$ Stream $_C$ consists of components $s(c)$ (for each $c \in C$) that denote the sequence of expressions appearing at the channel c. The nth element in this sequence is the expression appearing at time $t = n$.

A function $f :$ Stream $_I \to \mathcal{P}$ (Stream $_O$) from streams to sets of streams is called a stream-processing function .

The composition of two stream-processing functions $f_i :$ Stream $_{I_i} \to \mathcal{P}$ (Stream $_{O_i}$) $(i = 1, 2)$ with $O_1 \setminus O_2 = \emptyset$; is defined as
$$f_1 \otimes f_2 : \text{Stream}_I \to \mathcal{P}(\text{Stream}_O)$$

(with $I = (I_1 \sqcup I_2) \setminus (O_1 \sqcup O_2)$, $O = (O_1 \sqcup O_2) \setminus (I_1 \sqcup I_2)$)

where $f_1 \otimes f_2(s) \overset{def}{=} \{ t_O : t_I = s_I \wedge t_{O_i} \in f_i(s_{I_i})(i = 1, 2) \}$ (where t ranges over Stream $_{I \sqcup O}$). For $t \in$ Stream $_C$ and $C^0 \subseteq C$, the restriction $t_{C^0} \in$ Stream $_{C^0}$ is defined by $t_{C^0}(c) = t(c)$ for each $c \in C^0$. Since the operator $\otimes$ is associative and commutative [5], we can define a generalised composition operator $\bigotimes_{i \in I} f_i$ for a set $\{ f_i : i \in I \}$ of stream-processing functions.

2.2 Associating a Stream-Processing Function to a Process

A process $P = (I, O, L, (p_c)_{c \in O})$ is modelled by a stream-processing function $\llbracket P \rrbracket :$ Stream $_I \to \mathcal{P}$ (Stream $_O$) from input streams to sets of output streams.

For honest processes P, $\llbracket P \rrbracket$ is by construction causal, which means that the $n + 1^{st}$ expression in any output sequence depends only on the first n input expressions. As pointed out in [21], adversaries can not be assumed to behave causally, therefore for an adversary A we need a slightly different interpretation $\llbracket A \rrbracket$ (called sometimes rushing adversaries in [21]).

$[E](M) = \{E(M)\}$ where $E \in Exp$

$[either\ p\ or\ p^0](M) = [p](M) \cup [p^0](M)$

$[if\ E = E^0\ then\ p\ else\ p^0](M) = [p](M)$ if $[E](M) = [E^0](M)$

$[if\ E = E^0\ then\ p\ else\ p^0](M) = [p^0](M)$ if $[E](M) \neq [E^0](M)$

$[case\ E\ of\ key\ do\ p\ else\ p^0](M) = [p](M)$ if $[E](M) \in Keys$

$[case\ E\ of\ key\ do\ p\ else\ p^0](M) = [p^0](M)$ if $[E](M) \notin Keys$

$[case\ E\ of\ x :: y\ do\ p\ else\ p^0](M) = [p[h/x, t/y]](M)$ if $[E](M) = h :: t$

where $h \in$ " and h is not of the form $h_1 :: h_2$ for $h_1, h_2 \in$ "

$[case\ E\ of\ x :: y\ do\ p\ else\ p^0](M) = [p^0](M)$ if $[E](M) =$ "

Fig. 1.   Definition of $[p](M)$

For any closed program p with input channels in $\tilde{I}$ and any $\tilde{I}$-indexed tuple of closed expressions $M \in CExp^{\tilde{I}}$ we define a set of expressions $[p](M) \in P(CExp)$ in Figure 1, so that $[p](M)$ is the expression that results from running p once, when the channels have the initial values given in M.

We write $E(M)$ for the result of substituting each occurrence of $c \in \tilde{I}$ in E by $M(c)$ and $p[E/x]$ for the outcome of replacing each free occurrence of x in process P with the term E, renaming variables to avoid capture.

Then any program p_c (for $c \in Channels$) defines a causal stream-processing function $[p_c] : Stream_{\tilde{I}} \to P(Stream_{\{c\}})$ as follows. Given $s \in Stream_{\tilde{I}}$, let $[p_c](s)$ consist of those $t \in Stream_{\{c\}}$ such that

- $t_0 \in [p_c]("; :::; ")$
- $t_{n+1} \in [p_c](s_n)$ for each $n \in N$.

Finally, a process $P = (I, O, L, (p_c)_{c \in \tilde{O}})$ is interpreted as the composition
$$[P] \overset{def}{=} \overset{N}{\underset{c \in \tilde{O}}{}} [p_c].$$

Similarly, any p_c (with $c \in Channels$) defines a non-causal stream-processing function $[p_c]_r : Stream_{\tilde{I}} \to P(Stream_{\{c\}})$ as follows. Given $s \in Stream_{\tilde{I}}$, let $[p_c]_r(s)$ consist of those $t \in Stream_{\{c\}}$ such that $t_n \in [p_c]_r(s_n)$ for each $n \in N$.

An adversary $A = (I, O, L, (p_c)_{c \in \tilde{O}})$ is interpreted as the composition
$$[A]_r \overset{def}{=} \overset{N}{\underset{c \in \tilde{O}}{}} [p_c]_r \overset{N}{\underset{l \in \tilde{O} \setminus O}{}} [p_l].$$
Thus the programs with outputs on the non-local channels are defined to be rushing (note that using the local channels an adversary can still show causal behaviour).

We define composition of the (non-rushing) processes $P = (I_P, O_P, L_P, (p_c)_{c \in O_P \cup L_P})$ and $Q = (I_Q, O_Q, L_Q, (p_c)_{c \in O_Q \cup L_Q})$ with $O_P \setminus O_Q = L_P \setminus L_Q = ;$ to be
$$P \otimes Q \overset{def}{=} (I_{P \otimes Q}, O_{P \otimes Q}, L_{P \otimes Q}, (p_c)_{c \in O_{P \otimes Q} \cup L_{P \otimes Q}})$$
where $I_{P \otimes Q} = (I_P \cup I_Q) \setminus (O_P \cup O_Q)$, $O_{P \otimes Q} = (O_P \cup O_Q) \setminus (I_P \cup I_Q)$ and $L_{P \otimes Q} = L_P \cup L_Q \cup ((I_P \cup I_Q) \setminus (O_P \cup O_Q))$.

3   Secrecy

We say that a stream-processing function $f : \text{Stream}_{;}^{!} P$ (Stream $_o$) may eventually output an expression $E \in CExp$ if there exists a stream $t \in f(\ )$ (where denotes the sole element in Stream $_;$), a channel $c \in O$ and an index $j \in N$ such that $(t(c))_j = E$. When we say that f may eventually output an expression we implicitly assume that f has no input channels; in particular, if e. g. $f = [\![P]\!] \quad [\![Q]\!]$, we assume that $I_P \quad O_Q$ and $I_Q \quad O_P$ (and $O_P \setminus O_Q = ;$ $L_P \setminus L_Q = ;$ for composition to be well-defined).

Definition 1.   We say that a process P preserves a secret $m \in \text{Secret} [\text{Keys}$ if there is no process A with $m \notin S_A [K_A$ and $K_P \setminus K_A = ;$ such that $[\![P]\!] \quad [\![A]\!]$ may eventually output m.

    The idea of this definition is that P preserves the secrecy of m if no adversary who does not already know the private keys of P can find out m in interaction with P. In our formulation m is either an unguessable value or a key; this is seen to be sufficient in practice, since the secrecy of a compound expression can usually be derived from that of a key or unguessable value [1]. For a comparison with other secrecy properties cf. Section 1.

Examples.

- $p \stackrel{\text{def}}{=} \{m\}_K :: K$ does not preserve the secrecy of m or K, but $p \stackrel{\text{def}}{=} \{m\}_K$ does.
- $p_1 \stackrel{\text{def}}{=}$ case c of key do $\{m\}_c$ else " (where $c \in$ Channels) does not preserve the secrecy of m, but $P \stackrel{\text{def}}{=} (\ \{c\}, \{e\}, (p_d, p_e))$ (where $p_e \stackrel{\text{def}}{=} \{l\}_K$) does.

One can also define a rely-guarantee version of secrecy as in [11].

4   Composability

Definition 2.   A process P respects secrecy of $m \in \text{Keys} [\text{Secret}$ if for all processes Q that preserve the secrecy of m and all processes A with $m \notin S_A [K_A$ and $K_P \setminus K_A = ;$ such that $[\![P]\!] \quad [\![Q]\!] \quad [\![A]\!]$ may eventually output m, there is a process A^0 with $m \notin S_{A^0}[K_{A^0}$ and $K_P \setminus K_{A^0} = ;$ such that $[\![A^0]\!] \quad [\![Q]\!] \quad [\![A]\!]$ may eventually output m.

Examples.

- The process P that tries to decrypt any input with K and outputs the result if successful, does not respect the secrecy of any m.
- The process P that tries to decrypt any input with K and outputs the result if successful and if the result does not contain m_0 unencrypted, respects the secrecy of m_0.

Theorem 1.   Suppose that the p rocesses P and P^0 preserve and respect secrecy of m. Then P P^0 preserves and respects secrecy of m.

It is necessary to assume that   P and P^0 respect secrecy of m: Consider a process P that outputs m encrypted under K and a process P^0 that tries to decrypt any input with   K and outputs the result if successful. Both   P and P^0 preserve secrecy of m, but P P^0 does not.

For compositionality considerations it is very useful to have a more ne-grained notion of preservation of secrecy.

De nition 3.   Suppose we are given a p rocess P , a value m 2 Keys [Secret and a set of private keys K   Keys . We say that

- P protects the secrecy of m with the keys in K if for any process A with m 2 S_A [K_A such that $[\![P]\!]$ $[\![A]\!]$ may eventually output m, we have K K_A, and
- P respects the protection of m by K if for all processes Q that protect the secrecy of m with K and all processes A with m 2 S_A [K_A and K 6 K_A such that $[\![P]\!]$ $[\![Q]\!]$ $[\![A]\!]$ may eventually output m, there is a p rocess A^0 with m 2 S_{A^0} [K_{A^0} and K \ K_{A^0} = ; such that $[\![A]\!]^0$ $[\![Q]\!]$ $[\![A^0]\!]$ may eventually output m.

The idea is that an adversary can nd out the secret only if she has access to the keys in K. One can give an even more ne-grained de nition saying that   P protects the secrecy m with the combinations of keys in K for a set K P (Keys) of key sets, where the idea is that an adversary can nd out the secret only if she has access to one of the sets of the keys in   K. Then P preserves the secrecy of m if and only if P protects the secrecy of m with the combination of keys in K consisting of all singleton sets {k} with k 2 K_P.

Examples.

- p $\overset{\text{def}}{=}$ {m}$_K$ :: K does not protect the secrecy of m with {K }, but p $\overset{\text{def}}{=}$ {m}$_K$ does.
- The process that tries to decrypt any input with   K and outputs the result if successful, does not respect the protection of any   m by {K }, but the process that tries to decrypt any input with   K and outputs the result if successful and if the result does not contain m_0 unencrypted, respects the protection of m_0 by {K }.

Theorem 2.   Suppose that the p rocesses P and P^0 protect the secrecy of m with K and that they respect the protection of   m by K. Then P P^0 protects the secrecy of m with K and respects the protection of   m by K.

To apply this result, we consider an intensional version of secrecy protection.

De nition 4.   A process P protects m intensionally with the combinations of keys in K if there exists a set of local channels H_P L_P such that

- an output p_c to a channels c may contain subexpressions of the form $Dec_e(E)$ only if
  - c 2 H_P, or
  - e is protected (by suitable conditionals) from being evaluated at run-time to a key in the union $\cup_{K 2 K}$ K of key sets in K, or
  - E is protected from containing m as a subexpression,
- an output p_c to a channel may contain a subexpression m or d (for a channel d 2 H_P) only if • c 2 H_P, or
  - the subexpression occurs as the key used in an encryption or decryption, or
  - it is encrypted under all keys in one of the sets K 2 K.

One can enforce these conditions with a type system; this has to be left out here.

One may also partition the set of Channels into a set of trusted and a set of untrusted channels (e. g. the communication channels within a system and those across a network). Then the above denition can be extended by requiring the set H_P to include the trusted channels.

Theorem 3. If a process P protects m intensionally with the combinations of keys in K then P protects m with the combinations of keys in K.

Renement. We dene the standard renement from [5] and show that it preserves secrecy. Further renement operators are considered in [11].

Denition 5. For processes P and P^0 with $I_P = I_{P^0}$ and $O_P = O_{P^0}$ we dene P ; P^0 if for each s 2 Stream $_{I_P}$, $[\![P]\!](s)$ $[\![P^0]\!](s)$.

Theorem 4. For each of the secrecy properties p dened above (possibly wrt. m or K) the following implication holds:
If P satises p and P ; P^0 then P^0 satises p.

5   Conclusion and Future Work

We presented work towards a framework for modular development of secure systems by showing a notion of secrecy (that follows a standard approach) to be composable and preserved under renement in the specication framework Focus extended by cryptographic primitives. We gave more ne-grained notions of secrecy that are useful for verication purposes.

Future work will address other security properties such as integrity and authenticity and the integration into current work towards using the Unied Modeling Language to develop secure systems [12,10].

Acknowledgements.   Many thanks go to Mart'n Abadi for interesting discussions. Many thanks also to Zhenyu Qian for providing the opportunity to give a talk on an early version of this work at Kestrel Institute (Palo Alto) and to the members of the institute for useful feedback. Part of this work was performed during a visit at Bell Labs Research at Silicon Valley / Lucent Technologies (Palo Alto) whose hospitality is gratefully acknowledged.

References

[1] Abadi, M.: Security protocols and their properties. In: Bauer, F. and Stein-
    brueggen, R., (eds.):   Foundations of Secure Computation   . IOS Press, 2000

[2] Abadi, M. and Gordon, A.D.: A calculus for cryptographic protocols: The spi
    calculus. Information and Computation   , 148(1):1–70 (January 1999)

[3] Abadi, M. and Jan J¨urjens: Formal eavesdropping and its computational inter-
    pretation, 2000. Submitted

[4] Apostolopoulos, V., Peris, V., and Saha, D.: Transport layer security: How much
    does it really cost? In: Conference on Computer Communications (IEEE Infocom)
    New York (March 1999)

[5] Broy, M. and Stølen, K.:   Speci cation and Development of Interactive Systems
    Springer (2000) (to be published)

[6] Cardelli, L., Ghelli, G., and Gordon, A.: Secrecy and group creation. In: CONCUR
    2000. (2000) 365–379

[7] Dolev, D. and Yao, A.: On the security of public key protocols.   IEEE Transactions
    on Information Theory   , 29(2):198–208 (1983)

[8] Huber, F., Molterer, S., Rausch, A., Sch¨atz, B., Sihling, M., and Slotosch, O.: Tool
    supported Speci cation and Simulation of Distributed Systems. In: International
    Symposium on Software Engineering for Parallel and Distributed Systems   . (1998)
    155–164

[9] Jan J¨urjens: Secure information fbw for concurrent processes. In: CONCUR
    2000 (11th International Conference on Concurrency Theory)   , Vol. 1877 of LNCS .
    Pennsylvania, Springer (2000) 395–409

[10] Jan J¨urjens: Object-oriented modelling of audit security for smart-card payment
     schemes. In: Paradinas, P., (ed.):   IFIP/SEC 2001 – 16th International Conference
     on Information Security   , Paris, (11–13 June 2001) Kluwer

[11] Jan J¨urjens: Secrecy-preserving re nement. In: Formal Methods Europe (Interna-
     tional Symposium)   , LNCS. Springer (2001)

[12] Jan J¨urjens: Towards development of secure systems using UML. In: Hußmann, H.,
     (ed.): Fundamental Approaches to Software Engineering (FASE/ETAPS, Interna-
     tional Conference)   , LNCS. Springer (2001)

[13] Jan J¨urjens and Guido Wimmel: Speci cation-based testing of rewalls. Submit-
     ted (2001)

[14] Lotz, V.: Threat scenarios as a means to formally develop secure systems.   Journal
     of Computer Security 5   (1997) 31–67

[15] McLean, J.: A general theory of composition for a class of ”possibilistic” proper-
     ties. IEEE Transactions on Software Engineering   , 22(1):53–67 (1996)

[16] Meadows, C.: Using traces based on procedure calls to reason about composability.
     In: IEEE Symposium on Security and Privacy   . (1992)

[17] Meadows, C.: Applying the dependability paradigm to computer security. In: New
     Security Paradigms Workshop   . (1995)

[18] Meadows, C.: Formal veri cation of cryptographic protocols: A survey. In: Asi-
     acrypt 96 . (1996)

[19] Meadows, C.: Open issues in formal methods for cryptographic protocol analysis.
     In: DISCEX . IEEE (2000)

[20] Millen, J.: Hookup security for synchronous machines. In: Computer Security
     Foundations Workshop III   . IEEE Computer Society (1990)

[21] P tzmann, B.: Higher cryptographic protocols, 1998. Lecture Notes, Universit¨at
     des Saarlandes

[22] Pohl, H. and Weck G., (eds.):   Internationale Sicherheitskriterien   . Oldenbourg Verlag (1993)

[23] Ryan, P. and Schneider, S.: Process algebra and non-interference. In: IEEE Computer Security Foundations Workshop   . (1999)

[24] Sewell, P. and Vitek, J.: Secure composition of untrusted code: Wrappers and causality types. In: CSFW . (2000)

[25] Varadharajan, V.: Hook-up property for information fow secure nets. In: CSFW . (1991)

Agent-Based Model of Computer Network Security System: A Case Study

Vladimir I. Gorodetski[1], O. Karsayev[1], A. Khabalov[1], I. Kotenko[1],
Leonard J. Popyack[2], and Victor Skormin[3]

[1] St. Petersburg Institute for Informatics and Automation, Russia
{gor,ok,khab, ivkote}@mail.iias.spb.su
[2] USAF Research Laboratory, Information Directorate, USA
Leonard.Popyack@rl.af.mil
[3] Binghamton University, USA
vskormin@binghamton.edu

Abstract. The paper considers a multi-agent model of a computer networks security system, which is composed of particular autonomous knowledge-based agents, distributed over the hosts of the computer network to be protected and cooperating to make integrated consistent decisions. The paper is focused on an architecture, implementation and simulation of a case study aiming at exploration distinctions and potential advantages of using such an architecture for the computer network protection. The paper describes the conceptual model and architecture of the particular specialized agents and the system on a whole as well as implementation technology. Simulation scenario, input traffic model and peculiarities of the distributed security system operation are described. The major attention is paid to the intrusion detection task and agents interactions during detection of an attack against the computer network. The advantages of the proposed model of a computer networks security system are discussed.

1 Introduction

During several last years the computer network and information security are the problems of a big concern within information technology research area. Increasing of computer networks scale and intensive emerging of new information technologies, significant growth of computer networks traffic and other factors enhance the number of possible targets for attacks. Malefactors become armed by variety of sophisticated tools to break in the security systems as well as to hide an occurrence of attacks. All above factors negatively influence upon the efficiency of the existing computer networks protection systems and enable research and development of new protection models and technologies.

The recent security models focus on an idea of cooperation of the distributed network security components situated both inside and outside of a particular network host. Along with conventionally used security tools like firewalls, access control, authentication and identification systems, intrusion detection systems (IDS) are becoming of great significance. IDS is a component of a security system collecting information from different entry points of a computer network being protected, and analyzing the collected information to detect features and facts both of a suspicious

V.I. Gorodetski et al. (Eds.): MMM-ACNS 2001, LNCS 2052, pp. 39–50, 2001.

user behavior (interpreted as attempts of intrusion) and intrusions aimed at compromising the network security. IDS performance is based on the input traffic monitoring. While monitoring the input traffic, IDS detects events, patterns consisting of the predefined sequences of events. The detected patterns form input facts of knowledge-based components of IDS responsible for the intrusion detection. These components make decisions on the basis of rules, whose truth-values depend on input facts (detected patterns). Using IDS allows to improve integrity and efficiency of the security system as a whole, on-line monitoring of users' activity, recognition of, notification about and, possibly, fixing corrupted system files and data as well as system configuration altering, etc. ([1], [3]).

The most known representatives of exploratory IDS, which contributed pioneer solutions to the information security field, are Haystack, MIDAS, IDES, NADIR DIDS, ASAX, USTAT, DPEM, IDIOT, NIDES, GrIDS, Janus, JiNao, EMERALD, NetSTAT, Bro, etc.

The main disadvantages of the currently existing commercial and majority of research IDS are as follows ([1], [4], [10], [15]):

(1) too high probability of false positives and false negatives,
(2) weak ability in detection of new previously unknown types of attacks;
(3) use of passive techniques for traffic analysis, as a rule;
(4) unfeasibility of the real-time intrusions detection, in particular, in high-speed networks;
(5) restricted range of detection techniques able to cope with a misuse detection rather than an anomaly detection
(6) isolated mode of operation of particular host-based IDS, which leads to weak abilities in the detection of distributed coordinated attacks against computer network as a whole
(7) easily possible deception, avoidance and disturbance of IDS by malefactors familiar with the main IDS realization principles;
(8) remarkable overhead of the real-time systems when IDS operates.

Recently a number of new models and innovations for network security system assurance were proposed. Common ideas are:

(1) use of knowledge-based frameworks, for example, rule-based systems ([14]), neural networks ([5]), genetic algorithms ([7]), human-like immunology systems ([8], [17]), etc., and
(2) use of cooperation of the particular distributed components of the network security system that should allow to detect unknown distributed attacks against computer network on a whole. The most promising framework for such an idea implementation is multi-agent system model.

Multi-agent model in IDS design attracted a great attention during several last years ([2], [4], [6], [9], [10], [11], [12], [13], [16], [18], etc.). It was ascertained that such a model allows enhancing IDS performance and reliability as compared with conventional approaches. However, the majority of known research using the multi-agent model of IDS and computer network security system on a whole only considers the simplified version of agents' architecture and their cooperative behavior and do not employ the capabilities of multi-agent technology. In particular, the proposed models and architectures use agents at the pre-processing phase of the protection task. Here the agents are not knowledge-based, managed by a high-level software manager, restricted by solving intrusion detection task and do not interact with the access control and other security system components. In fact, they do not use a security component cooperation that can provide for a most significant contribution a quality of computer network protection.

The paper is focused on the architecture and implementation of a case study aiming at the exploration of potential advantages of multi-agent architecture for the computer network protection. The paper describes the conceptual model and architecture of particular specialized agents and the system on a whole, the implementation technology as well as the simulation procedure. The major attention is paid to the intrusion detection task and agents interactions during detection of an attack against the computer network.

The rest of the paper is structured as follows. Section 2 gives an outline of the high-level architecture of the developed multi-agent security system, which integrates distributed host-based security systems. The architecture of the particular host-based security system implemented the case study is considered in Section 3. Section 4 describes the architecture of a particular security agent. The external environment of the case study is formed by distributed attacks model that is described shortly in Section 5. An example of the performance of the developed case study as a reaction on a particular distributed attack is outlined in Section 6. Section 7 presents the conclusion, which summarizes the paper results and plans for future research.

2   High-Level Architecture of the Case Study

The agent-based architecture of the computer network security system was proposed in [10]. The developed case study is an implementation of a particular simplified case of this architecture.

Below in the case study we consider the computer network consisting of four hosts (Fig. 1). This network can comprise several segments of a LAN and the input traffic can be both inside and outside LAN traffic. Each host is protected by an integrated agent-based security system comprising several specialized agent-demons (AD) monitoring the input traffic and extracting predefined significant events and patterns from the sequence of IP-packets, access control agent, authentication and identification agent and knowledge-based intrusion detection agents. These agents interact during their operation via messages exchange thus supporting cooperative multi-level input data processing.

Each message of an agent is represented in KQML language. The message provides instructions for the agent-receiver how to process the message and how to react to it, and specifies the content of the message. The message content is represented in terms of XML language and specifies an object with assigned attribute values. The object is understood as a significant event or a pattern with assigned attributes, facts about detected attempts to break in the accesses control system or authentication and identification system, and about other suspicious activities or attempts of attacks determined by an agent-demon.

Each agent processes the determined situation according to a predefined scenario and sends a message to an intrusion detection agent and/or other agent-demon. The messages proceeding from an intrusion detection agent can be twofold. The first type o f messages is addressed to an agent inside the host instructing it to break a connec-

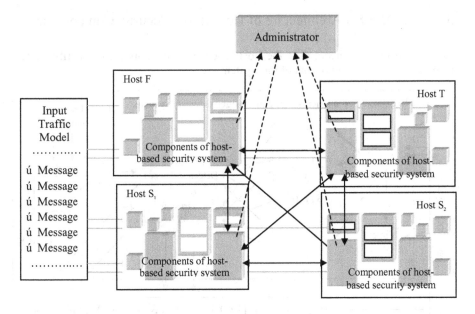

Fig. 1. General architecture of the case study

tion or to focus attention on a current connection if a suspicious activity has been determined. The second type of messages is addressed to the intrusion detection agent of other host. In this case the message content ("object") is the information about detected suspicious user's behavior or about the intrusion. An example of KQML message sent by "Input Traffic Model" component to the agent AD-E is

Performative: tell
    Sender: Input Traffic Model
    Receiver: AD-E (host S1)
    Content: <content>
    Ontology: ISS
    Language: XML

The content of this message in XML language is presented in the upper-left corner window in Fig. 6.

There is no particular upper-level managing software that coordinates host-based agent operation as well as intrusion detection agents of different hosts of the computer network. The multi-agent security system operates in the distributed way.

The "Input Traffic model" software (see Fig. 1) simulates the input traffic. The output of this program is a random sequence of IP-packets that is a mixture of normal and abnormal streams of events. The software code is written on the basis of Visual C++, JAVA 2 and XML software development kits.

3 More Detailed Architecture of Host-Based Security Components

Let us consider architecture of the host-based part of multi-agent security system (Fig. 2). The basic components of this system are:

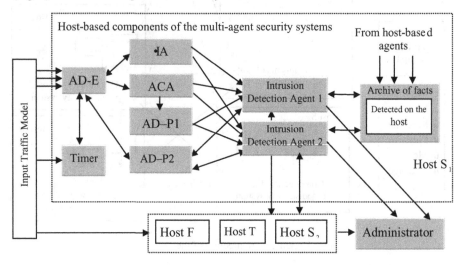

Fig. 2. Architecture and interaction of host-based part of the multi-agent security system

Agent-demon AD-E is responsible for the input traffic preprocessing. It monitors traffic aiming at extraction of sequences of so-called "events" that are semantically meaningful and associated with the current connections. These events are sorted, stored in AD-E database and each of such events is sent for the posterior processing to one or several agent-demons (see Fig. 2).

Agent-demons for identification and authentication (AIA) and access control (ACA) perform their conventional functionalities, record the results into their data bases and send messages to the Intrusion detection agent if a suspicious behavior or attempt of an attack has been detected.

Agent-demons AD-P1 and AD-P2 are responsible for extraction of the predefined ("interesting", "meaningful") patterns of "events" and making decisions (both preliminary and final) about attempts of attacks or suspicious behavior of users connected with the host. In the implemented case study the agent AD-P1 is responsible for extracting patterns that corresponds to the suspicious behavior or attempts of attacks like port scanning (on application level), finger search and buffer overflow. The agent AD-P2 is intended to extract patterns corresponding to the denial of service attack, syn-flood attack and port scanning (on transport level).

Intrusion detection agents (IDA) are responsible for high-level input data processing, and make rule-based decisions on the basis of input facts contained in received messages. They can receive messages about detected suspicious behavior and attempts of the attack from agent-demons of the same host as well as messages from security agents of the other hosts. The distinction between these IDAs is that they deal with different situations. IDA1 processes situations arising due to combined spoofing attacks whereas IDA2 performs high level processing of facts aiming at detection of distributed multi-phase attacks. Both IDA1 and IDA2 have similar architecture and the only distinction is the content of their knowledge bases (KB) comprising the rules, used to make decision based on input facts.

Notice that each agent of the host have a local database which is accessible for every host-based agent. In fact, altogether they form distributed database of the host-based security system.

4   Agent Architecture

All host-based security agents are implemented on the basis of a standard architecture depicted in Fig. 3. The distinctions are only in knowledge bases contents. The agent demons' KB contain only simple rules for extracting events or patterns whereas intrusion detection agents have more comprehensive KBs. Let us outline the generic architecture of agents.

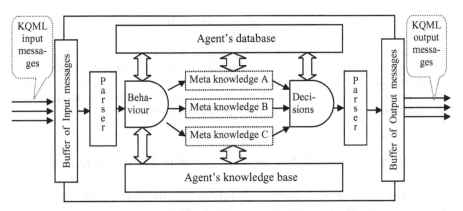

Fig. 3. Security agent architecture

Buffer of input messages is used for input messages storing, ordering and synchronization of their processing. Let us remind that all messages are represented in KQML format, whose content is specified in terms of XML language.

Parser performs syntactical analysis and interpretation of messages. The similar program performs generation and sending of output messages. To interpret the input and to generate the output message Parser has to access to the domain ontology of the agent, which is a part of the Agent's database.

The interpreted message content is the input of the component Behavior, which has to determine the way of input message processing. It generates a scenario on the basis of tree structured Meta-knowledge base, which if necessary derives a decision via accessing Local knowledge bases associated with the nodes of the meta-knowledge base structure. If necessary the decision-making procedure uses data about pre-history of the agent behavior. This data are contained in the Agent's database. The sequence of steps of the scenario to be executed is determined by the nodes that lie along the realized way from the root to the leaf of the meta-knowledge base structure. An example of the meta-knowledge base structure and its links to local knowledge base is given in Fig. 4.

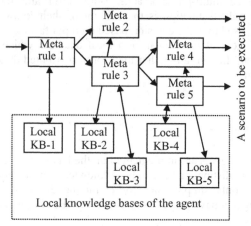

Fig. 4. An example of the agent's knowledge base architecture

A responsibility of the Decision component is to determine a reaction of the agent in the result of the scenario execution. There are two possible reactions: (1) generation of an output message(s) and (2) recording the result of the scenario execution in the agent's database. The latter as a rule corresponds to an intermediate decision made by the agent that is supposed to be used later.

5 Input Traffic Model: Scenarios of Combined Distributed Attack

The model of the input traffic corresponds to a number of scenarios used by the malefactor. A scenario can consist of a subset of attacks that are

1. Scanning of host ports aimed at a determination of the active ports and available services (ftp, telnet, etc.).
2. Attempts of connection with the host on the service like ftp, telnet, etc. using different source IP-addresses and guessing password.
3. Combined spoofing attack. This attack against the computer network comprises several steps. First, the malefactor establishes a tcp-connection with a host S and gets an initial number of the connection. Next, he realizes the denial of service attack against a trusted host T and during the host T hang-up he establishes a tcp-connection with the host S on behalf of the trusted host T. The host S sends the confirmation receipt to the host T. The concluding operation of the malefactor is sending the reply receipt to the host S from its host X instead of the host T, thus assigning the extended access rights of the trusted host T to the host X.
4. Attempts to get unauthorized access to files of a host.
5. Buffer overflow attack against a host aiming at boosting of the access rights.
6. Attempts to connect with the host services like telnet, ftp, etc.

1. The input traffic model can realize a reasonable sequence of these attacks using different entry points (hosts). These attacks form the upper-level specification of a distributed multi-phase attack. Each attack (each phase of a distributed attack) in turn is simulated by a sequence of IP-packets of the input traffic, which is the mixture of normal and abnormal users' activity. These sequences form the low-level model of every phase of the attack. Below the operation of the case study is demonstrated by a particular scenario of the distributed attack

6   Case-Study Simulation: An Example of a Distributed Attack and Multi-agent System Performance

The particular attack against the computer network is simulated by the sequence of input IP-packets. Some of them correspond to normal connections, other correspond to the attack simulation. The total length of input IP-packet sequence is of thousands among which several hundred correspond to the abnormal user activity. The upper-level randomization-based mechanism allows to simulate different order of phases of attacks.

Simulation was organized on the basis of the following strategy of an attacker. It is supposed that the attack source is the malefactor's host X and the attack is realized against the computer network depicted in Fig. 1. For each host-based protection system it is supposed that the host T is the trusted host, whose users have an extended access to the resources of the host S_1. In the developed case study the security system of each host comprises 7 agents as shown in Fig. 2. Every host is protected by the multi-agent subsystem of the same architecture comprising similar agents and similar structure of their interaction. The only distinction between different host-based security subsystems is the host-specific tuning of agents' attributes. The total number of agents of the network security system is twenty eight.

Within simulated attack a malefactor target is to get unauthorized access to the resources of the host S_2. The malefactor is going to act "indirectly", i.e. to use the access chain S_1–T–S_2, considering the hosts S_1 and T as the interim targets. It should be noticed that although the particular phases of attack are determined by the components of the multi-agent network security system, the connection with malefactor is not breaking in even if the attack has already been suspected or detected. The reason is to involve in cooperation as many components of the security system as possible. It should be noticed that in real-life situation this assumption corresponds to the case when it is necessary to become more informed about malefactor and his purposes or when some of the particular phases of attack are not detected.

In more detail, the simulated combined distributed attack comprises seven phases as follows:

1. Scanning of the ports of the host S_1.
2. Attempts of connection with the host S_1 to access its ftp or telnet services using various source IP-addresses and guessing password. It is supposed that this attack is unsuccessful.
3. Realization of the combined spoofing attack against host S_1 through host T. It is supposed that this attack is successful.

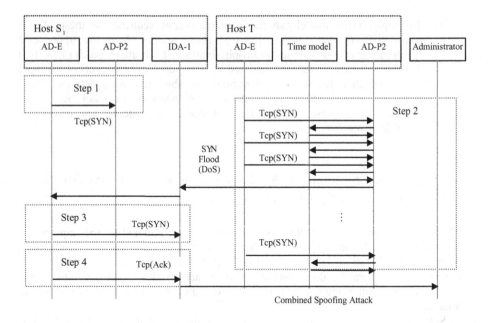

Fig. 5. Agents' interaction during processing of the Combined Spoofing Attack

4. Attempts to get the non-authorized access to the files of the host S_1 (unsuccessful).
5. Buffer overflow attack to boost access rights to the host S_1 resulting in getting access to the password file of the host S_1 (successful).
6. Reading the passwords and getting access to the files of the host S_2 (successful).
7. Connection with the ftp or telnet service of the host S_2 (successful).

Let us consider the operation and interactions of the security system agents during the described attack development. The input traffic is preprocessed by the agent-demon AD-E, which re-addresses the resulting messages to the specialized agent-demons AD-P1 and AD-P2. In the first phase of the attack the agent-demon AD-P2 of the host S_1 is playing a leading role. The second phase involves agents AIA of the same host. The third phase is monitored by the agents AD-P2 and IDA1 of the host S_1 and AD-P2 of the host T. During the fourth phase the agents AIA and ACA of the host S_1 are operating. The agent AD-P1 of the host S_1 detects the next phase of attack. During the sixth phase the agents ACA and IDA2 of the host S_1 are operating. During the final phase of attack the agents AIA and ACA of the host S_2 are working. An example of the interaction scenario of the security agent community is depicted in Fig. 5. It corresponds to the third phase of the attack development when the malefactor is realizing the Combined Spoofing Attack.

One more example of agents' interaction is given in Fig. 6. This figure is printout of the user visual interface window representing agent interaction during the port scanning attack processing. Left-hand pair of windows (top and bottom) reflects a pair of records of time-log files of agents AD-E and AD-P2 of the host S_1 which interact at this step of operation via exchange of messages ## 41–43. This interface depicts the

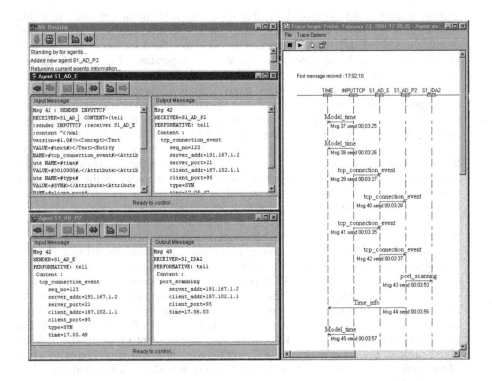

Fig. 6. Printout of the visual interface representing agents' interaction

content of input and output messages in short or in full notations. Notice that full
notation (input message #41 of agent AD-E) corresponds to a message content
represented in XML language. The right-hand area of Fig. 6 visualizes the process of
the agents' messages exchange in a time-ordered mode. Such visualization is a
functionality of a software component called Tracer.

7   Conclusion

As a rule, development of the distributed attack can affect several hosts compromising
a number of them at the same time. Malefactors can cooperate within the attack in
such a way that the particular malefactor attacks a single host, thus, masking the
combined distributed nature of the attack. A centralized security system should be too
"heavy" to process the entire batch of large-scale data about a prehistory and current
states of many concurrent connections between many hosts.

The case study simulation displayed a number of advantages of the multi-agent
architecture for protection of the computer network against distributed attacks. The
most significant one is a capability of comparatively "light" components of the multi-
agent security system to cooperate. At present the only way to detect efficiently the
distributed attack against the computer network is to establish a cooperation of

security software entities (agents) distributed over the hosts of the network and within each host. A sample of an advantage of agents cooperation can be seen within the example described in the previous section. In particular, the detection of the combined spoofing attack has become possible only due to cooperation of agents AD-P2 and IDA1 of the host S_1 and AD-P2 of the host T, i.e. due to the cooperation of software entities situated on different hosts. One more example of the necessity and advantage of the cooperation of security agents is the operation of the knowledge-based agent IDA2. It is intended to collect information about suspicious behavior of users from many entry points of the native host and exchange this information with similar agents of the other hosts to make integral decision about status of the connections. In the developed case study the IDA2 knowledge base is a comparatively poor to play a significant role in the intrusion detection. In further developments of the case study a significant accent will be made on expanding of its knowledge base and on the increasing of its role in the distributed attack detection.

The case study was implemented as an application of the so-called Multi-agent System Development Kit (MASDK), developed by authors of this paper. MASDK is a software tool to support the technology of multi-agent systems formal specification and implementation. It consists of a number of generic procedures, generic database structures and editors which make it possible to specify components of the multi-agent system in the formal specification language and to further generate the multi-agent system architecture, its communication components, to clone knowledge-based agents with the needed functionality, etc. Visual C++, JAVA2 and XML software were used to develop this tool kit. Notice that development of the model and the object-oriented project of the case study were the most time and effort consuming part of the work whereas its software implementation within MASDK required much less time and efforts.

Further research will concern the development of richer Input Traffic Model and knowledge bases of particular agents and mechanisms for decision making. Development of software for off-line agent learning to improve security system performance and to make possible its adaptability to new kinds of attacks is planned as well.

Acknowledgment. The authors are grateful to the Air Force Research Laboratory at Rome NY for funding this research.

References

1.   Allen, J., Christie, A., Fithen, W., McHugh, J., Pickel, J., Stoner E.: State of the Practice of Intrusion Detection Technologies. In: Technical Report CMU/SEI-99-TR-028. Carnegie Mellon Software Engineering Institute (2000)
2.   Asaka, M., Okazawa, S., Taguchi, A., Goto S.: A Method of Tracing Intruders by Use of Mobile Agents. In: Proceedings of INET'99 (1999)
3.   Bace R.: Intrusion Detection. Indianapolis, Macmillan Computer Publishing (1999)

4.   Balasubramaniyan, J.S., Garcia-Fernandez, J.O., Isacoff, D., Spafford, E., Zamboni D.: An Architecture for Intrusion Detection Using Autonomous Agents. Coast TR 98-05. West Lafayette, COAST Laboratory, Purdue University (1998)
5.   Bonifácio Jr., Cansian, A., Moreira E., and de Carvalho A.: An Adaptive Intrusion Detection System Using Neural Networks. Proceedings of the IFIP World Computer Congress–Security in Information Systems, IFIP-SEC'98, Chapman & Hall, Vienna, Austria (1998)
6.   Conner, M., Patel, C., Little M.: Genetic Algorithm/Artificial Life Evolution of Security Vulnerability Agents. In: Proceedings of 3rd Annual Symposium on Advanced Telecommunications & Information Distribution Research Program (ATIRP). Army Research Laboratory, Federal Laboratory (1999)
7.   Crosbie, M., Spafford G.: Applying Genetic Programming to Intrusion Detection. In: Proceedings of the AAAI Fall Symposium on Genetic Programming Cambridge, Menlo Park, CA, AAAI Press (1995)
8.   Dasgupta, D.: Immunity-Based Intrusion Detection System: A General Framework. In: Proceedings of the 22nd National Information Systems Security Conference, USA (1999)
9.   Helmer, G., Wong, J., Honavar, V., Miller, L.: Intelligent Agents for Intrusion Detection. In: Proceedings of the 1998 IEEE Information Technology Conference, Environment for the Future, Syracuse, NY, IEEE (1998)
10. Gorodetski, V., Kotenko, I., Skormin, V.: Integrated Multi-Agent Approach to Network Security Assurance: Models of Agents' Community. In: Information Security for Global Information Infrastructures, IFIP TC11 Sixteenth Annual Working Conference on Information Security, Qing, S., Eloff J.H.P (eds), Beijing, China (2000)
11. Jacobs, S., Dumas, D., Booth, W., Little, M.: Security Architecture for Intelligent Agent Based Vulnerability Analysis. In: Proceedings of 3rd Annual Symposium on Advanced Telecommunications & Information Distribution Research Program (ATIRP). Army Research Laboratory, Federal Laboratory (1999)
12. Jansen, W., Mell, P., Karygiannis, T., Marks D.: Mobile Agents in Intrusion Detection and Response. In: Proceedings of the 12th Annual Canadian Information Technology Security Symposium Ottawa, Canada (2000)
13. Karjoth, G., Lange, D., Oshima, M.: A Security Model for Aglets. In: IEEE Internet Computing (1997)
14. Lee, W., Stolfo, S.J., Mok, K.: A Data mining Framework for Building Intrusion Detection Model. In: Proceedings of the IEEE Symposium on Security and Privacy (1999)
15. Ptacek, T.H., Newsham, T.N.: Insertion, Evasion, and Denial of Service: Eluding Network Intrusion Detection. Secure Networks, Inc. (1998)
16. Queiroz, J., Carmo, L., Pirmez, L.: MICAEL: An Autonomous Mobile Agent System to Protect New Generation Networked Applications. In: Proceedings of Second International Workshop on the Recent Advances in Intrusion Detection, West Lafayette, USA. (1999)
17. Somayaji, A., Hofmeyr, S., Forrest, S.: Principles of a Computer Immune System. In: Proceedings of the 1997 New Security Paradigms Workshop (1998)
18. White, G., Fisch, E., Pooch, U.: Cooperating Security Managers: A Peer-Based Intrusion Detection System. In: IEEE Network, Vol. 10(1) (1996)

Security Considerations and Models for Service Creation in Pre mium IP Networ ks[1]

Michael Smirnov

GMD FOKUS, Global Networking,Kaiserin-Augusta-Allee, Berlin 10589, Germany
smirnow@fokus.gmd.de

Abstract. Service creation for premium IP networks facilitates customized and configured network services dynamically at run time by establishing recursive group communication (event notification) channels between involved service components residing in various network entities. Security, as a feature of danger discovery appears in our service creation design as additional benefit. Same principles of group communication guide us to identify danger patterns in distributed system behaviour and to propagate these patterns to relevant entities for event correlation and reaction.

1   Introduction

Since 1995 when the NSF Internet backbone has been disbundled we see ever growing concerns about Internet security. This is recognised by the IETF as one of the key issues of the Internet [1]. Security concerns are maybe even more familiar to businesses, which link these concerns to increasing complexity and programmability of the Internet [2]. The visison of traditional telcommunication services providers at future Internet – the Next Generation Networks paradigm [3] – acknowledges the departure from the heaven of switched networks with the dominant principle "trust by wire" towards the brave new world of connectionless IP networks. With billions of mobile users, e-business transactions and the need for QoS assurance rapid service deployability becomes a dominant requirement.

Service Creation (SC), as known from Intelligent Networks is the process of interrupting the basic call chain in a switch and consulting with additional intelligence, remote to the point of interrupt. Additional intelligence returns to the switch a routable entry thus allowing the call chain to be completed. There is no straightforward mapping of IN like service creation to IP networks: the basic call chain does not exist in connectionless networks, additional intelligence in IP networks has a per protocol nature and is distributed between multiple devices. These devices – middle boxes [7] – are: firewalls, network address and protocol translators, realm specific IP gateways, QoS enforcement devices, policy enforcement and policy decision points, tunnel terminators, proxy servers, bandwidth brokers, signature

[1] Research outlined in this paper is partially funded by the IST project CADENUS IST-1999-11017 "Creation and Deployment of End-User Services in Premium IP Networks" http:// www.cadenus.org/

V.I. Gorodetski et al. (Eds.): MMM-ACNS 2001, LNCS 2052, pp. 51–62, 2001.

management, authentication, authorisation and accounting servers, multimedia buffer managers, application-aware caches, load balancers, third-party secure associations provisioning, SMTP relays, enhanced routers, etc.

In some sense, service creation in Internet does already exist, however in a form of numerous client server protocols controlling different aspects and phases of an "IP call" establishment. Middle boxes provide QoS guarantees, security, accountability of Internet services, they support business relations of ISPs and ASPs with their customers. The Internet engineering community recognizes that there is a need for basic framework for middle box communication. Besides the obvious advantage of a standard for various control interactions during datagram flow processing, such a framework has a potential to restore the end-to-end service model [8] broken by families of middle boxes which populate the Internet in uncoordinated manner.

We concentrate on interactions between service creation layer (a novel task for network control plane) and premium IP layer, the one which is populated by middle boxes. The two major requirements for this – simplicity and scalability – practically mean that network complexity is to be manageable [8] for both service creation and danger discovery. Traditional approach – to develop a new client server protocol each time there is a need to establish a communication between entities – makes the Internet already look like a patchwork, and when applied to middle box communication would increase enormously the complexity of nodal processing. Our paradigm is entity-group protocol, which has relatively complex communication pattern (recursive group communication) but scales well because the nodal complexity remains constant. Service creation and danger discovery in our approach are based on event notification system which merges originally disjoint distributed states maintained on a per-protocol and on per-service basis in various network nodes by means of group communication. We examined in [4] how safety conditions can be achieved in this decentralised and distributed state management system. Conformance of a service under creation to service level agreements is analysed in [5], while a taxonomy of possible architectures is given in [6].

The paper is structured as follows. First, we introduce our service creation model with all important definitions. Then we discuss how danger patterns can be identified locally, aggregated on a per service basis and propagated throughout the system. We introduce group event notification protocol and discuss its performance. Discussion and related work conclude the paper.

2   Service Creation Model

2.1   Definitions

An entity is a network element or a piece of software capable of sending and receiving messages, for example client and server programs running at the same or different machines. Type of an entity is analogous to a type of a protocol the entity supports. A network is analogous to an autonomous system or administrative domain in the Internet.

Group of entities is a collection of entities. Protocol group is a collection of network entities supporting the same protocol. Service group is a collection of entities involved in provisioning of an End-to-End service [8]. When more than one protocol is involved in service provisioning, we call those entities service components. Service creation is a process of a needed co-ordination set-up between entities from different protocol groups. Protocol group type is defined by the protocol, while service group type is defined by the service. While protocol groups are homogeneous, and it is clear how to define a protocol group type, service groups are heterogeneous, and it is not obvious how to define a service group type. We define a service group type as a combination of protocol group types involved in service creation.

Service creation is a process of a needed co-ordination set-up between entities from different protocol groups. Co-ordination means that heterogeneous protocol entities have now new relations between them: communicate with and dependency relation. Communicate with relation is achieved by subscribing relevant mediators to relevant event notification groups. Dependency relation is established by ordering of accepted events and forming new conditions out of them. To support these newly established relations is the main function of mediators and group event notification protocol running between them.

Mediator is a SC specific entity. Mediators are associated with middle boxes; for back compatibility with legacy middle boxes they can intercept all messages of an associated middle box to form relevant event notifications to other mediators; it also stimulates middle boxes by issuing relevant messages on its local interface triggered by SC conditions and policies. Mediators substitute multiple client parts which otherwise would be needed for peer to peer communication between heterogeneous protocol entities of various middle boxes.

Auditor is a SC specific mediator, which assists in identification of danger models based on observed events and in propagation of relevant events for event correlation logic, it is not associated with any of the protocol entities; an auditor is subscribed to all security groups; at the start of the network operation there is only one security group (discovery group) to which all auditors are subscribed.

Danger model is an interpretation of observed violation of performance of a service and service components, that is violations of per service communication and dependency relations established between protocol modules during a service creation.

Security groups (SG) are groups of mediators only. Middle boxes are assumed to implement their own security mechanisms. SG are meant to identify danger models and prevent insecure communication between mediators during the service creation.

2.2   Notation

We use the following notation for service creation definition:curly braces denote a group of elements to appear in a comma-separated list between { and }; we use Latin alphabet for physical entities and Greek alphabet for types of entities; both Latin and Greek capitals denote sets, while regular letters with potentially some indices denote instances of entities or types.

We use the following notation for group event notification protocol specification:

ú   G – a group of protocol entities with members g, g_0, g_s, ...   G, where g_0 is recognized as current notifier, which triggers group event notification; g_s is current server member – it controls processing of service datagrams; semantically group G is defined as a set of all protocol entities of all types which are required for provisioning of the requested service;

ú   G\ g_i – sub-group of G produced by excluding member g_i from G;

ú   * – any network node; this symbol is used in protocol description when we are not interested in, or cannot know in advance the source address of an IP datagram, which arrival at current network node is nevertheless an important event:

ú | – logical OR, & – logical AND;

ú $\{a\}^n$ – protocol a (message exchange and corresponding state machine transitions) executed n times; n is positive integer;

ú $(\{a\}\|\{b\})$ – concurrent execution of protocols a and b;

ú $\{a\},\{b\}$ – sequential execution of protocols a and b;

ú s_x:xì m ì   y: s_y – an atomic message exchange: a protocol entity x, obtaining or keeping state data s_x sends a message m to a protocol entity y, which then keeps a state data s_y; note, that while s_y and s_x serve protocol entities of different types they may differ only in formats of service data representations;

ú s_x(S1+S2) – denotes that state data kept by member x consists of items S1 and S2;

ú   ′ – denotes empty state, i.e. no state data pertaining to a service;

ú $m_{proto,1}(...)$, – message m_1 of the protocol proto with the identified content;

ú v_S – size of state data S;

ú D_{SSDB} – service payload pertaining to service state data block -SSDB,

ú $m(D_{SSDB})$ – a datagram carrying [a part of] D_{SSDB};

ú   N+1 – size of group G.

2.3   Service Creation Example

Let us consider an example in which a network has the following sets of entities:

ú   a protocol group B={ b_1, b_2, b_3, b_4} of 4 Differentiated Services capable border routers with protocol group type b;

ú   a protocol group F={ f_1, f_2, f_3 } of 3 firewalls with protocol group type f;

ú   a protocol group P={ p_1, p_2} of 2 SIP[2] proxies with protocol group type p,

ú   a protocol group A={ a_1, a_2, a_3} of 3 Auditors with protocol group type a .

For simplicity we assume that there is no further decomposition of these protocol groups during service creation process[3].

Let us assume that the network creates dynamically the following services:

ú   Service 1 (S_1) is a best effort quality IP telephony with firewall traversal, thus involving entities from F and P groups;

ú   Service 2 (S_2) is high quality IP telephony with firewall traversal, thus involving entities from F, P, and B groups.

Both services assume SIP signaling, thus entities from P group are involved in both S_1 and S_2. During the SC phase network protocol groups are configured to provide coordinated behaviour for a particular service instance, which, when ready,

---

[2] SIP – Session Initiation Protocol, Internet standard for IP telephony signaling.

[3] It will be reasonable to partition a large network into a number of service zones and address corresponding protocol subgroups when a service creation request arrives from a host belonging to a service zone. These considerations however are not within the scope of this paper.

becomes a service group. Let us assume that during some period of observation a three service instances have been created:

1. One instances of S_1 involving f_1 and p_2, denoted $S_{1,1}=\{f_1, p_2\}$ of type $(f+p)$,
2. Two instances of S_2 involving p_2, f_3 and b_3 denoted $S_{2,1}=\{p_2, f_3, b_3\}$ and p_1, f_1, and b_1 denoted $S_{2,2}=\{p_1, f_1, b_1\}$, both of type $(p+f+b)$.

The above example has three service groups of two different types. We define a service group type by combining types of protocol groups participated in creation of service instances. Mentioning auditor type would be redundant – each service group has its own auditor, e.g.

$$a_2 \quad S_{1,1}, a_2 \quad S_{2,1}, a_3 \quad S_{3,1}.$$

To account for available resources we consider protocol groups to be complects rather than sets of entities; each entity is assigned a capacity, i.e. the number of times this protocol entity can participate in creation of different service instances. If, e.g. a capacity of p_1 is set to m, that means that a proxy p_1 can support up to m different instances of services requiring SIP proxy. Without any loss of generality we can assume that each protocol group entity has a capacity of m=1.

2.4   Mediators

A mediator, associated with a protocol entity c_i, controls the processing engine of c_i based on a Service State Data Block (SSDB) defined below. Processing engine of a protocol entity c_i is modeled as $<R_E, P_E>$, where R_E is a set of rules defining protocol messages processing, P_E is a set of policies (filters) modifying the R_E[4]. We define a mediator to be in association with a protocol entity; this binding provides that the mediator type is the same as for the associated entity.

A mediator's internal control state is decomposed into conditions, policies and messages. Conditions are represented as a set $\{ïT_{j,p}\}$, where each member is p^{th} precondition of jth event. Policies are represented as a set of rules $\{P_j^M\}$, which gives for each jth event additional mediator policy rule providing fine tuning of SC. Messages are a set of mediator event triggering messages $\{ M_j^M\}$. Address to which a message is sent is always a group addresses. Thus a mediator is

$$M_i = \{ïT_{j,p}, P_j^M, M_j^M\}.$$

Per service communication and dependency relations between mediators assisting involved protocol entities are established by subscribing mediators to relevant groups, or group event notification channels.

---

[4] Protocol policies are introduced here to assure back compatibility of service creation with a popular policy enabled networking by which a limited configurability of a legacy network can be achieved.

2.5   Events

An event is something we are interested in, and, at the same time, something countable, maybe via sets of probabilities. In general, event is an atomic metric of a discrete system. Usually, an event handle (policies and procedures) and event object (parent action and environment definition) are associated with an event. We also define pseudo-events when it makes things easier to understand, control or manage. In particular systems design events are characterized by event class, by event properties and interface, by event registration and alerts, by event modifiers and filters. The most common model of event propagation is Event Notification Channel (ENC), which connects event listeners (subscribers) with event creators (notifiers). A particular class of event propagation is a hierarchy of ENCs, in which specific bubbling event can be defined, the one, which is propagated upstream the hierarchy. Let us also distinguish mutation events – those changing the behaviour of event handlers and possibly changing event objects.

An event E is defined here as

$$E = \{A, B, T°, a, t\}, \tag{1}$$

where A and B – denote the action A which did happen in the network element B; $T°$– denotes a set of post conditions produced by A at B; a – denotes aging condition which is to be used by mediators to define the validity period of the event E, t – a timestamp of A. A more elaborated definition of an event and of safety conditions for event based service creation systems can be found in an earlier paper [4].

Event is valid until it represents a valid information. For example, aging condition can be in a form of maximum allowed reaction latency (L):

$$a := \{ L < t_m - t \}, \tag{2}$$

where t_m is a reaction completion time estimated by a particular mediator.

As soon as the information value is below certain threshold the event data may be discarded either completely or in part. Each such discard event is reported to a service group, the group Auditor matches it against its own $\{iT_a\}$ to determine whether all auditors group is to be notified on this. That is some preliminary correlation of warning events is done already in a totally distributed fashion. Conditions $iT_{j,p}$ are associated with aging timers, which values are set based on aging conditions from a received event notification message and mediator policies. At any moment in time, every condition may be in one of the following states: set and not aged, set and already aged, unset. While semantics of many events make no differentiation between the latter two, they are however distinguished in the mediator security model as potentially important information.

Event notification interface of a mediator is divided into input and output parts. The input part has important function of authentication checks: if a sender of a received message can not be authenticated (e.g. based on statically configured table of trusted senders, or via dynamic remote authentication from a trusted server) then corresponding data item is marked as untrusted. Event policy can be used to resolve trustworthiness conflicts, happening if one or several conditions were communicated by untrusted parties. If, while the event is in unset state and one of its conditions is set as untrusted, and a trusted message arrives setting the same condition, then event processor rewrites the condition status as trusted. If the opposite event notification occurs then untrusted message is ignored.

The function of output interface is to insert a timestamp into the message template and to forward it to a particular socket.

Meditor realizes isolation of a protocol entity processing from event notification service. For the design of this interface we have adopted the padded cell approach described in [9]. With padded cells interactions between untrusted (mediator from the viewpoint of a protocol entity) and trusted code (protocol entity's own code) occur only through well defined aliases.

3   Danger Models: Identification and Propagation

Locally identified danger is to be tailored by aggregation to the service creation system for proper reaction (at upper layer providing needed event correlation services) and then propagated throughout the SC system via dynamic creation of security groups. SG are controlled by join and leave messages sent by mediators to group addresses reserved for SG[5].

3.1   Dynamic Assignment of Attributes

A typical way to express attributes of an object is to associate them with the object, which complicates entity-relationship diagrams, because normally relations are aware only of some of objects' attributes. Similarly, a client-server relation between two network nodes always needs additional specification: to which protocol (attribute) this relation is valid. Another way of expressing attributes is to define groups of objects with similar attributes. If attributes have non-binary values, then it is possible to define sub-groups within the attribute group in which all members will have the same (or somewhat similar for fuzzy subgroups) value of the attribute. Dynamic creation of SG is equivalent to dynamic assignment of danger attributes to member mediators. Auditors assist in both danger aggregation in each SG and propagation to another SG over the discovery group.

The algorithm of danger knowledge propagation is as follows. When a mediator discovers a new danger pattern it distributes this pattern to its service group, the service group auditor checks its cache for matching patterns and then forwards it to the discovery group. An auditor which finds a similar danger pattern in its cache replies to a discovery group with a message which carries the previous danger pattern, and a group address of a new ad hoc security group. The mediator which has discovered the danger pattern must join this new security group.

Danger patterns discovered by mediators are nothing more than abnormalities in conditions states, these discoveries are pre-programmed by security rules. Security rules are setting allowed transitions in the condition state machine, which can be instantiated per attribute, i.e. per mediator group. Abnormalities are consequences of

---

[5] If native IP multicast is going to be used as layer 3 transport of group communication messages, then special care is to be taken for multicast addresses management; one solution based on private multicast address space is proposed in [15]

wrong (may be insecure) mapping of conditions states to event state. When this abnormality is discovered through the newly created SG then security rules for a particular attribute group will be changed in all mediators. In addition to this, any violation of a group identity (group size, group type, group behavior) is reported by an auditor to the discovery group and subsequently to event correlation service.

3.2   Defining Danger Patterns

We propose to use maximum rate performance hypothesis for the definition of danger patterns for involved protocols within the SC: if enough resources are provided, and if there are no damages to both the environment and to a protocol itself, then the protocol should operate at its maximum rate. We consider any violation from a maximum rate performance as potential danger pattern, which should be distributed for correlation with other group members. We see three possible approaches for the maximum rate definition.

First, to define conditions of optimal performance of a protocol is possible via simulation, which is not very reliable due to multiple assumptions, which are inevitable in a simulation study. Second, we propose an analytical technique, which gives very robust results. The protocol machine can be modeled after extended Petri nets with timing conditions for all operations. It is possible then to calculate frequencies of firing of all transitions belonging to protocol invariant expressions, that is only those conditions and transaction will be taken into account, which in collective behavior define protocol performance. The basic methodology for such analysis can be found in [10], a particular example of application of this methodology to the Internet group management protocol is given in [11].

Finally, we propose an over-simplistic analysis of the semantics of involved protocols to identify the following set of dangerous behavior patterns: when normal behavior patterns are not happening; and abnormal behavior patterns are happening. Protocol messages and their timing conditions can easily be associated with the two patterns above. A mediator observing message flow from/to an associated protocol entity can be then instructed to map this protocol specific danger patterns to messages sent to discovery group or to established security groups. Unfortunately, this semantic analysis is not easy for a single mediator – it may be unaware of service creation logic. We propose to use in this case simple state register associated with each of the pre-conditions triggering event notification. The state register is to be analyzed every time a pre-condition state changes. If, after such a change the event does not happen because a majority of pre-conditions are "waiting" for few still unset ones, then this can be considered as a danger pattern. This approach can lead to over classification of behaviour patterns, and can be recommended at initial phases of danger models understanding.

4   Generic Group Event Notification Protocol

A generic Group Event Notification (gGEN) protocol is a skeleton of any (event specific, or service specific or group specific Group Event Notification protocol (GEN). In GEN one member of a group (notifier) distributes event notification messages to all group members (subscribers). GEN is an example of an entity-group protocol. Like in a client-server in which one entity (client) is always responsible for triggering the communication with its peer (server), in an entity-group protocol there is always one entity, defined dynamically, which, being itself a group member, notifies other group members on certain event. We introduce a convenient syntax for protocol expressions in which all important state information is clearly identified while being at protocol parties or "on the wire".

Any protocol state machine is defined based on message content. For control protocols message content in turn can be subdivided into end-to-end state and transient state data. After examining a number of candidate protocols and on-going efforts we suggest the following generic items for the end-to-end part:

1. Service identification (Id) – a unique identifier of an end-to-end service; in case of a composite service the Id should pertain to a particular service level agreement between an end user and service provider;
2. Authentication data (Au) – either end-to-end or, related to the end-to-end but domain specific authentication data (passwords, sequence numbers, timestamps);
3. Service Requirements Specification (Sp) – can be any, including QoS, reliability, encryption, etc., controlling the intermediate datagram processing;
4. Timer data (T_{resv}) – a soft state refresh parameter, allowing receiving party to age all the above data items after the period of time unambiguously defined by T_{resv}.

We call the above data items Service State Data Block (SSDB). There is one SSDB per service invocation instance, and, omitting invocation index, we write:

$$SSDB = <Id, Au, Sp, T_{resv} >$$ (3)

4.1   Protocol Expressions

We write gGEN as the following protocol expression:

$$\{s_0(SSDB):g_1 \grave{1} \quad m_{GEN,1}(SSDB)\grave{1} \quad G:s_g (SSDB)\}^g,$$ (4)

where n $=]T_{sesson}/ T_{resv}$ [, i.e. n· T_{resv} à T_{sesson}, and T_{sesson} is duration of service session; s_g denotes that SSDB is tailored to fit the type of each group member's protocol entity. Expression (4) means, that an initiator sends a message with service specific data to all members of a group, which data is then stored by each member as a service specific state information. If session duration is larger than a timer value then a SSDB is refreshed to all group members by repeating the triggering message.

Clearly, there is certain overhead associated with distribution of SSDB to all members of the group, while the data will be used by a single member[6] of the group, we shall call this member a server. If session starts at time t_{start} and ends at time t_{end} then (n·T_{resv} - (t_{end} - t_{start}))·v is a state data overhead in the server, called hereafter a reservation overhead. At the same time, all N group members[7] keeping the SSDB will experience total overhead of n· T_{resv} ·N·v, which we call state overhead.

It is probably not possible to reduce the reservation overhead without decreasing refresh interval, but thus increasing the network control load. On contrary, state

---

[6] In case of unicast services.
[7] Recall that in total a group has N+1 members, but we do not count initiator's data as an overhead.

overhead can be significantly reduced by adding, ironically, even more state data. The extended gGEN model we write as:

$$\{\{ \; s_0 \,(SSDB){:}g_0 \,\grave{\imath} \quad m_{GEN,,1}\,(SSDB)\grave{\imath} \quad G{:}s_g\,(SSDB)\}^{n1} \,\|$$

$$\{\{ {*}\grave{\imath} \quad m_{serv}(D_{SSDB})\grave{\imath} \quad g_s{:}s(Fwd)\}, \tag{5}$$

$$\{ \; s_s\,(Fwd){:}g_s \,\grave{\imath} \quad m_{GEN,,2}\,(g_s\,Fwd)\,\grave{\imath} \quad g_0{:}s_0\,(Fwd)\}\}\},$$

$$\{ \; s_0\,(Fwd{+}SSDB){:}g_0 \,\grave{\imath} \quad m_{GEN,,3}\,(SSDB)\grave{\imath} \quad g_s{:}s_s\,(SSDB)\}^{n2}.$$

Expression (5) means that as soon as relevant for requested SSDB server is selected among group members, it notifies a trigger member on this; all subsequent refresh messages are sent in a unicast mode to the server member only. As the protocol expression (5) shows it may take n1 refreshes of SSDB before the actual data (message $m_{serv}(D_{SSDB})$) will reach a server member, thus $n = n1{+}n2$. The state overhead in this case is $T_{resv} \cdot N \cdot v_{SSDB} + v_{Fwd}$ under an assumption that service data is detected by the server member within the first refresh period, i.e. that $n1{=}1$. We claim, that if $n1 > 1$ then this overhead should be attributed as the reservation one.

We have reduced in (5) the state overhead n times in comparison to (4), by introducing one additional type of state data and by introducing two additional atomic message exchanges. The extended gGEN above is a mixture of multicast ($m_{GEN,,1}$) and unicast ($m_{GEN,,2}$, $m_{GEN,,3}$) messages. We write now another extension of gGEN, which is multicast only, as:

$$\{\{s_0(SSDB){:}g_0\grave{\imath} \quad m_{GEN,,1}(SSDB)\grave{\imath} \quad [G{:}s_g(SSDB{+}Status_g) \mid (g_s{:}s_s(SSDB{+}Status_s) \tag{6}$$
$$\&\; G\backslash g_s {:}(\,'\;)]\}^n \,\| \,\{{*}\;\grave{\imath} \quad m_{serv}(D_{SSDB})\grave{\imath} \quad g_s{:}s(Status_s)\}\}.$$

Expression (6) means that as soon as every group member receives a SSDB it establishes locally additional state $Status_g$; it denotes whether g is the server member for the service. If "yes", i.e. service data has reached a server member and $g = g_s$ then $Status_s$ ü Fwd_s as in the mixed gGEN extension, and all group members will silently discard any subsequent $m_{GEN,,1}$ messages if SSDB timer did not expire yet[8].

The state overhead of the multicast only gGEN is $T_{resv} \cdot N \cdot v_{SSDB} + (N{+}1) \cdot v_{Fwd}$ which is slightly more than for the mixed gGEN (5). However, taking into account that $\cdot v_{Fwd}$ is 1 bit only, we suggest to neglect the overhead of additional N bits (1 bit per each of N protocol entities in the group G) in favor of simplicity of multicast only gGEN.

4.2   Discussion

We argue that multicast only extension of gGEN, expression (6), meets both requirements of efficiency (minimal overhead) and simplicity (minimal complexity). A protocol overhead can be classified as communication and state overhead. The former can be further roughly estimated as a number of protocol message types and, in the case of entity group protocols, as a group size and group diversity, while the latter can be estimated as reservation state overhead and service state data overhead.

---

[8] If the SSDB timer T_{resv} has already expired, then the refresh message will obviously be considered as a new SSDB request, and this unwanted situation is to be avoided for all three flavours of gGEN by appropriate estimation of a refresh period.

We claim that complexity of gGEN mainly depends on a number of message types and on service state overhead.

As an illustration of multicast only extension of gGEN let us introduce a protocol metric, dubbed protocol volume (V_{proto}), as a product of protocol message types and service state overhead and consider this metric as a function y of x, $x = T_{resv} / T_{session}$. Then for the three discussed flavors of gGEN this function will be:

Original (4): $V_{proto} \sim const$;
Mixed (5): $V_{proto} \sim 3x$;
Multicast only (6): $V_{proto} \sim x$.

Clearly, for all cases where $x < 1$ protocol expression (6) is preferable in terms of both simplicity and efficiency, while x∂1 does not look realistic (session time is shorter than soft state refresh period).

5   Related and Future Work

We make no extensive comparison to related work intentionally: Internet experiment was started as fully trusted community of hosts and message processing intermediate nodes. On the other hand, switched networks are based on the "trust by wire" principle which is not applicable to IP networks. When Internet started experiencing lack of trust between autonomous systems the security work concentrated on adding security features to original protocols. These additions were constrained by already deployed technology. In our work we try to show how we can build a system which has a security features inherited from the system design, without later add-ons. When our service creation system of mediators interfaces legacy systems – community of protocol entities – we suggest to use the alias principle, also known as padded cells [9].

Recently a lot of attention in computer science has been drawn to principles learnt from immune systems, and this work was also inspired to some extent by biological principles interpreted for self-organising networks [12]. Traditional self/non-self discrimination paradigm in the immunology is no longer the dominant one. The notion of danger models was already used in this research as a working term when we learned that the danger models paradigm is also in use in immunology. Proposed by P. Matzinger, this paradigm states that: self/non-self does not matter, mutant behaviour matters, there is no danger if there is no damage. "Imagine a community in which the police accept anyone they met during elementary school and kill any new migrant. That's the Self/Nonself Model. In the Danger Model, tourists and immigrants are accepted, until they start breaking windows." (from [13]). The danger model operation in immunology is explained surprisingly similar to our vision: "The ... cell, awakened by the ... alarm signals then starts the chain of events that sets an immune response in motion." [14].

We investigated the use of recursive group communication for managing the distributed state for synchronisation of separately developed client-server protocols, and for identification and propagation of danger patterns within groups of mediators. This new paradigm of a network control plane is still work in progress: we plan to prototype our system in generic way and also to map it to various premium IP technologies, like MPLS, DiffServ, which will allow us to understand how to build event correlation service needed to react on yet unknown danger models and to achieve network evolvability and self organisation.

References

1.   Baker, F.: Internet Directions. Keynote presentation at QofIS'2000, available at URL http://www.fokus.gmd.de/events/qofis2000/html/programme.html
2.   Regli, William C. National Institute of Standards and Technology, IEEE Internet Computing 1997 January-February 1997 (Vol. 1, No. 1)
3.   Modaressi, A.R., Mohan, S.: Control and Management in Next-Generation Networks: Challenges and Opportunities. IEEE Communications Magazine (October, 2000) 94–102
4.   Smirnov, M.: QoS ad hoc interworking: dynamic adaptation of differentiated services boundaries. Lecture Notes in Computer Science, Springer, vol. 1938. 186–195
5.   Romano, S.P., Esposito, M., Ventre, G., Cortese, G.: Service Level Agreements for Premium IP Networks. Work in progress, Internet draft, draft-cadenus-sla-00.txt, (November, 2000) available at URL http://www.cadenus.org/papers/
6.   Smirnov, M.: Service Creation in SLA Networks. Work in progress, Internet draft, draft-cadenus-slan-screation-00.txt, November, 2000, available at URL http://www.cadenus.org/papers/
7.   Kuthan, J., Rosenberg, J.: Middlebox Communication: Framework and Requirements. IETF work in progress, Internet Draft draft-kuthan-midcom-framework-00.txt, November 2000, available with relevant resources at URL http://www.fokus.gmd.de/research/cc/glone/employees/jiri.kuthan/private/fcp/
8.   Clark, D.D., Blumenthal, M.S.: "Rethinking the design of the Internet: The end to end arguments vs. the brave new world" - workshop "The Policy implications of End to end", Stanford University, December, 2000, available at URL http://www.law.stanford.edu/e2e/papers.html
9.   Ousterhout, J.K., Levy, J.Y., Welch, B.B.: The Safe Tcl Security Model. Sun Microsystems, Technical Report TR-97-60 (March, 1977) 19
10.   Sifakis, J.: Use of Petri Nets for Performance Evaluation. Acta Cybernetica, Vol. 4, No 2. (1978) 185–202
11.   Smirnov, M.: IGMP Performance Model. GMD FOKUS, 1996, Technical Report TRF-96-03-01, 22 pp., available from URL http://www.fokus.gmd.de/glone/publ/trf-96-03-01.pdf
12.   Gold, R.: "SONATA: Self-Organizing and Adaptive Active Networking", Stockholm Active Networks Day, 28th August 2000, Available at URL http://www.fokus.gmd.de/glone/employees/richard.gold/private/publications.html
13.   Matzinger, P.: A quotation from the interview given to C. Dreifus, N.Y. Times, available at URL http://www.nytimes.com/library/national/science/061698sci-immune-theory.html
14.   Dunkin, M.A.: A Maverick Researcher Bucks the Establishment. Available at URL http://www.arthritis.org/ReadArthritisToday/1999_archives/1999_03_04polly.asp
15.   Smirnov, M., Sanneck, H., Witaszek, D.: Programmable Group Communication Services over IP Multicast. Proceedings of the IEEE Conference on High Performance Switching and Routing, June 2000, Heidelberg, IEEE, 281–290

Secure Systems Design Technology

Peter D. Zegzhda and Dmitry P. Zegzhda

Information Security Centre of Saint-Petersburg Technical University,
195273, Saint Petersburg, K-273, P/B 290
ZEG@ssl.stu.neva.ru

Abstract. Authors introduce a novel approach to the problem of secure system design, based on consistent and correct implementation of information flows and flow controls, fitted well with security policy, and on practical experience of security violation analysis. On this basis, concept of secure system design and development are proposed, forming a basis for secure information technologies development. Proposed technology and concepts can be applied from application tools and information systems design to operating system.

1 Introduction

Having impetuously flowed over Russia, a flush of computerization generated a need to use information technologies in those areas, where security is the essential feature required of information processing automation. Since information technologies were late in coming into use in Russia, we are to master all the stages of computerization simultaneously. Information security is such an area where one can't do without domestic implementations, the more so we have highly skilled specialists and advances in the area of theoretical information protection, especially in cryptography. On the other hand, it is impossible (and is not necessary) to isolate this area from total information technologies progress and discard mechanisms and technologies, being in use all over the world. But computerization makes us to maintain equilibrium between the necessity to support crucial system security with domestic special-purpose services and natural aspiration (and often necessity) to use the last world achievements in information technologies.

In Russia, special purpose systems designers are to face two conflicting problems:
- to meet high domestic security requirements; and
- to keep compatibility with widely used imported unprotected application services.

The main contradiction is that compatibility with popular imported mechanisms implies their obligatory use as a part of protected system, although it is absolutely obvious that these applications are unsuitable for security-related systems. The designers of these applications hadn't considered such problem and inherently intended their products for mass market and for small business. So, it is useless to integrate security services into existing application information processing systems.

Thus, the protected systems creation on the basis of popular components is more than an actual problem in such circumstances. Hence, to build an adequately protected information processing system, it is necessary to combine security mechanisms with

V.I. Gorodetski et al. (Eds.): MMM-ACNS 2001, LNCS 2052, pp. 63–71, 2001.

popular applications within a single system. Such problems need new approaches as protected system design solutions, new security techniques.

In Information Security Centre of St. Petersburg Technical University (ISC SPbTU) the investigations and developmental work have been carried out in this area within already six years and has resulted in the proposed technology of secure information processing system design implemented in the development of secure operating system.

2   Concept of Secure System Design

The following principles are used as the basis for the proposed concept (see fig. 1):

1. Integrality principle. Since security is an integral system characteristic, security mechanisms must form an integral complex and be integrated into information processing system at the basic services level.

2. Invariantness principle. Security models and techniques, used in the protected system design, must be solely information process oriented and independent of system and application mechanism architecture.

3. Unification principle. Security mechanisms must be flexible for use in different security models and in different information processing systems, no matter what the destination and inherence of information are processed. Security mechanisms must be intended for minimum administrations and be transparent for an ordinary user.

4. Adequacy principle. In the design of secure system architecture, it is necessary to take into account experience of existing systems operating, in order to clear sources of known security violations. Only in this case security mechanisms will be really adequate to real-world threats.

5. Correctness principle. Principles, security models and mechanisms, used in secure system architecture design, must be implemented properly, i.e. mechanisms implementing them should provide exactly inner possibilities of information flow control. Thus, on the one hand, all theoretical models and techniques used must be reflected in practice, and on the other hand, security mechanisms are to be in agreement with declared and proved properties of these models.

These principles define the design of secure system architecture as the development and implementation techniques.

3   Technology of Secure Information Processing System Design

The foregoing five principles of secure system design produce five components of technology of their design, each determining sequence of operations and conditions, under which the desired secure system features can be obtained. Let us consider the fundamental tenets of supposed technologies and the benefits of their application to secure system design.

3.1   Invariantness of Security Models

New interpretation of "secure system" concept [1], being developed in ISC SPbTU, formulates its main task as to induce computer system to implement adequately information flows and rules of their control, which have been in existence long before automated information processing services were used. It means that secure system must, firstly, implement only those information flows, which have been in existence before it was used, and not to create the new ones, and, secondly, allow for information flows control in accordance with given security policy rules.

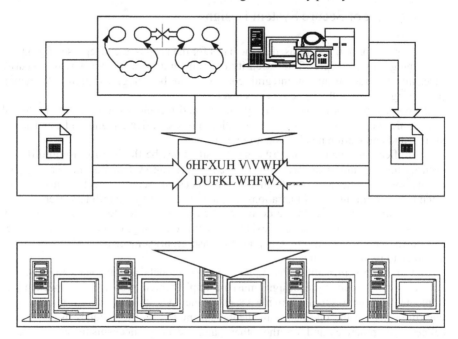

Fig. 1. Application of secure system design concept scheme

Complex security model of protected information system was developed in ISC SPbTU with this approach, reflecting information flow passing and control and based on the general notion of security policies and on generic system component interaction model. The main object being modelled is not execution of one or other operation (file access, message exchange, etc.), but underlying information processes. Fig. 2 represents the main stages of information system security model design with regard to technology proposed.

Such an approach allows to perform subject and object interaction walkthrough as information flows are passed and to detect the location points of access control services. Moreover, without going into peculiarities of computing system architecture one can use unified protection techniques, such as access control services independent of security policy, identification and authentication, independent of application service functionality peculiarities, etc.

Subject and object interaction model, which reflects information flow passing, can be applied to a wide range of systems, from operating systems to distributed computing systems and complex application-oriented applied software.

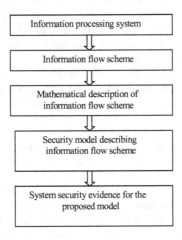

Fig. 2. Information flows modeling

3.2   Analysis and Elimination of Prerequisites to the Formation of Vulnerabilities

For protected system to be truly secure, it must be able to resist various security threats, which are purposeful and operate in expanses of current information technologies. To the authors mind, the success of security violations in information processing system is due to the features of the system itself, because in each successful violation certain peculiarities of information processing system design or functionality or security services drawbacks are necessarily used [1].

Thus, to solve the problem and to design a "really" secure system, we are to overcome the existing systems flaws responsible for successful attacks. To do this we must establish the reasons for weaknesses (vulnerabilities), which cause such violations, and to clear them away. For this purpose the conditions for security threats successful implementation were analyzed and systematizes, and prerequisites to their formation were established in ISC SPbTU [2]. Investigations demonstrated that the vast of security violations is due not to poor programming (as it is commonly supposed), but to basics principles, such as security model and protection architecture. Note that access control performance and checking of its implementation are of primary concern.

It follows from the analysis that security system developers must focus their efforts on implementation of such access differentiation systems, which could successfully withstand attacks and would not have detected flaws.

Given security violation prerequisites and the causes of vulnerabilities, we can evaluate actual system ability to resist security threats. After eliminating detected architectural flaws and thereby clearing away the causes of vulnerabilities during secure system designing we can get secure systems, able to resist security threats.

3.3   Implementation Correctness

Advanced security models and effective security techniques, used in secure system developing, do not yet assure true security. A complete agreement of practical system implementation with theoretical grounding is necessary before formally proved features of security model and parameters of methods and algorithms in use can provide the required level of protection. But the case in point is not only program implementation errors, but correct enforcement of security model and protection techniques within information processing system as well. The security violation analysis demonstrated that distortion of models and principles built into secure system architecture is one of the main prerequisites to formation of security flaws. Therefore, to eliminate prerequisites of security violations we must provide correct implementation.

This problem appeared at the meeting-point between security theory with its abstract models and their practical implementation in application systems. The matter is that security mechanism designers are to adapt security models and mechanisms within the limits of specific system architecture. Numerous deviations from theoretical principles of mandatory access control model during its implementation, especially in administration and remote communication mechanisms, offer an example of such adaptation. In practice it results in exceptions to the overall security concept in the form of privileged programs, which are not controlled by access differentiation mechanisms and, therefore, are very sensitive to program errors, failures and input monitoring; therefore, as soon as intruder captures these programs, he gains the overall system control.

Our technique of information processes orientation of security models and policies provides partially solution of these problems, giving the opportunity to separate security service implementation problems of specific system from the features of security services themselves. But information flows scheme and the rules of flow control can be incomplete or hardly to formalize, whereas any system includes objects (for example, service files, resources, etc.) ignored by security policy. Furthermore, users have no opportunity to handle information directly and, in consequence, use software, which can produce undesirable and uncontrolled information flows (Trojan horse attack, for instance) without the user's consent.

To solve the problems mentioned, we suggest to apply the following principles used at different stages of secure system design(fig 3):

1. Security models, used for a real system design, are to be complete and non-conflicting; that is for all combinations of operations called and communicating entity types the model rules are to produce practical applicable result and not to give collisions. This prevents occurrence of exceptions to the total assess control rules.

2. Security mechanisms are to be developed with the tools, setting up a correspondence between protection system project and software use. Formal high-level specifications can act as such tools. Using such mechanisms, we can avoid design errors, regulate coding process and apply formal verification methods.

3. Since it is impossible to avoid programming errors, the amount of program code is to be minimized during secure system design and security mechanism implementation, as program code errors are crucial for security. Hence, it is necessary to minimize the amount of code functions of TCB (Trusted Computing Base).

4. Since any secure system contains program code, errors in which discredit all the protection system (access control or encryption subsystems, for instance), trusted programming and verification technologies are needed for such components.

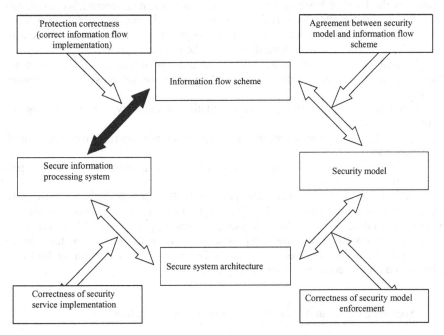

Fig. 3. Correctness of secure system design

The implementation of these methods, which are at the meeting-point between information security theory, information processing automation and programming and verification technologies makes possible correct implementation of security models and protection techniques.

3.4   Generic Security Tools

Although different approaches are used for secure systems and security services design, all of them are based on the same set of functions, which is the cornerstone of security, namely information flows control and system resource access control functions. So, the same controls can be used to implement a wide range of secure systems and underlying security models.

When using standard generic security mechanisms, we can get efficient secure system design, obviate unneeded redundancy and gain in security services safety by program code reuse. But whereas identification, authentication and encryption tools are quite generic and widely used in various systems, the development of access control tools, versatile to various security models and policies, is not a trivial task and requires its own investigation methods. Access control services are traditionally related to operating system architecture. We consider this situation unjustified, because quite cumbersome component, which needs careful verification, becomes importable.

Proposed technology of secure system design makes a provision for access controls location at the level of basic system primitives, implementing information exchange between its components. To solve this problem generic access control tools were developed during investigations, suitable to implement a wide range of security models. These tools operate in accordance with the following principles.

1. Unified security model representation as a set of subject and object security attributes and access differentiation rules, defined at the set of security attributes values.

2. Generalized subject and object representation as uniquely identified vectors of security attributes values.

3. Special language, which formulates access differentiation rules with the help of logic predicates.

4. Separate implementation of access control and managing subject and object security attributes values.

On the basis of these principles generic tools were designed, allowing to implement different security models, which in turn can serve as the base for various systems, depending on requirements posed and security policy used. Moreover, such an approach allows to update finished product systems, as dictated by the tasks being accomplished, and if need be, to enforce different policies for different kinds of subjects and objects communications and types.

3.5   Security Systems and Widespread Applications Integration

As we have already mentioned, today's protected information processing systems development and exploitation have such a peculiarity that all the attempts to promote products, incompatible with popular mechanisms, are inherently doomed to failure. At first glance situation looks like hopeless: popular mechanisms are unfit for private information processing, they cannot be updated without producer support, which is hardly to get, and the updating needed is so cardinal that has no even technical justification, not to mention economic one. The most radical solution, from the security standpoint, is to build secure system from the ground (operating system) up to the top (applications), but this has no economic justification. So the only way is to compose secure systems of available components, remedying their inherent flaws by special system architecture and specially designed additional components.

Such a problem often faces those crucial systems developers, who try to compose trusted systems of untrusted components. It is assumed, that system capacity for work under any conditions is provided through repeated replication and safety factor.

But to compose protected system of untrusted components one must use quite different approaches, because security service redundancy does not always strengthen security (for example, reencryption does not always add cryptographic strength). It is obvious that we can not solve the problem, simply adding security services to ordinary systems, for example, adding network traffic encryption to Windows based systems, because the system security is always defined as the security of its weakest link, and in the case being considered, it is operating system, reach in vulnerabilities and not intended for such problems solution. So, from the practical standpoint, this aspect of technology of secure system design is the most important, because it guides the way to resolution of existing contradiction.

The solution is not to remedy the inherent flaws of application services, but to use secure system architecture, which provides popular applications integration in such a way that their flaws could not manifest themselves as vulnerabilities and compromise the overall system.

As a basis for such architecture we suggest to use hierarchical structure peculiar to all today's software systems, where high-level mechanisms rely upon services, provided by lower-level components. When designing secure system, we can use hierarchical architecture, where system entities, representing users, communicate under the control of lower-level security services. We call it "control level". Scheme of fig. 4 illustrates this architecture.

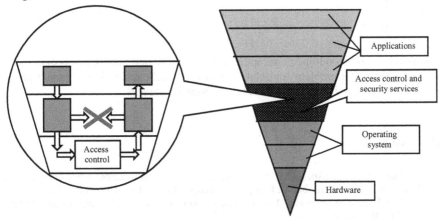

Fig. 4. Communications control in secure system

Since users can be represented by various applications, operating at several hierarchic levels, several control levels can be taken, but the condition of lower-level services control should be met. In this case, the security of any level communications relies upon lower-level interactions security [3]. The only condition to be met in this architecture is impossibility for information communications to bypass secure mechanisms (bypass low-level service).

Thus, security mechanisms and wide-spread applications integration amounts to development of such intermediate program layer that existing applications could operate on its base. This layer is primarily intended to protect all kinds of information communications by information access control, communicating parties identification and authentication. Moreover, this software is to perform other security functions, such as security management, audit, etc. Note, that the level of performing is of no concern: it could be operating system processes, drivers, network services, internetworking, etc. What mattered is only satisfying the requirements outlined.

In order to apply the approach proposed and to design secure operating system, where popular applications could operate, we are to take the following steps.

1. Specify information resources to be protected. It can be file, document, database and network connection, internetwork traffic, etc.

2. Specify a level, where control over these information resources access is to be performed to suit the requirements outlined.

3. Design secure system architecture with security services at the level specified.

4. Implement security services at this level with regard to technologies proposed.

5. Integrate security services and application services into a single system to suit architecture designed.

The most reasonable decision for such a "control level" implementation is domestic secure special-purpose operating system, which could become a platform for solution of all the problems regarding information security, still keeping compatibility with wide-spread products and enabling to design secure information processing systems, based of this operating system.

4   Resume

The development of outlined concept of secure information processing system design allows to solve the problem formulated and to find radical solution of secure private information processing problem under today's circumstances.

References

1. Zegzhda, D.P., Ivashko, A.M.: On creation of secure information processing systems. "Information security problems. Computer systems". 1 (1999) 99–105 (in Russian)
2. Zegzhda, D.P.: The analysis of Internet security accidents causes. "Vesnik Svyazi". 4 (1999) 95–99 (in Russian)
3. Zegzhda D.P., Ivashko A.M.: Information systems security basics. "Goryachaya linia – Telekom". Moscow (2000) 449 (in Russian)

A Privacy-Enhancing e-Business Model Based on Infomediaries

Dimitris Gritzalis[1], Konstantinos Moulinos[1,2], and Konstantinos Kostis[1,3]

[1]Dept. of Informatics Athens University of Economics & Business, 76 Patission Ave.,
Athens, GR-10434 Greece
{dgrit,kdm,kikigus}@aueb.gr
[2]Hellenic Data Protection Authority, Omirou 8 St., PC 10564, Greece
[3] Hellenic Army General Staff, Research and Information Systems Division/CCIS
kostis@ccis.army.gr

Abstract . Rapid evolution of Internet may largely depend on gaining and maintaining the trust of users. This possibility may especially rule enterprises, whose financial viability depends on electronic commerce. Neither customers will have the time, the ability or the endurance to work out the best deals with vendors, nor will vendors have time to bargain with every customer. In order for customers to strike the best bargain with vendors, they need a privacy supporter, an information intermediary or infomediary. Infomediaries will become the custodians, agents, and brokers of customer personal information exchanged via Internet, while at the same time protecting their privacy. There is a scale between security and privacy that currently leans towards security; security adopts strong user authentication mechanisms in order to control access to personal data, while privacy requires loose authentication in order to provide user anonymity. In this paper we introduce a new infomediaries-based, privacy-enhancing business model, which is capable of providing anonymity, privacy and security, to customers and vendors of e-commerce. Using this model, customers of e-commerce can buy goods or services, without revealing their real identity or preferences to vendors, and vendors can sell or advertise goods or services without violating the privacy of their customers.

1   Introduction

The increasing popularity of Internet has generated significant interest in the development of electronic retail commerce. Internet has the potential to evolve into an interconnected marketplace, facilitating the exchange of a wide variety of products and services. Internet effects on the common commercial activities include, inter alia [1], by shifting power from sellers to buyers, by reducing the cost of switching suppliers and freely distributing a huge amount of price and product information. It also, reduces the transactions costs, the speed, range and accessibility of information,, the cost of distributing and capturing personal information, in order to create new commercial possibilities.

In order to identify customer preferences, and customize products and services, marketers are looking for new ways of capturing, processing and exchanging customer data. They collect data every time customers visit their web sites; they also

V.I. Gorodetski et al. (Eds.): MMM-ACNS 2001, LNCS 2052, pp. 72–83, 2001.

use numerous techniques to analyze that data and create mature user profiles. A user profile includes personal data, which may identify in a unique way a customer consuming behavior. Such a collection and processing of personal data may lead to private and family life violation, thus discouraging the public from using new technologies. According to a 1998 Harris poll, the lack of privacy and security in communications is the main reason of a consumer being off the Internet, and this is true for the great majority of potential users. Consumers are worrying about how their personal data will be used and how this data can be protected against unauthorized access [2].

A defense against online privacy infringement consists of business models and approaches that do not reveal the identity of the communicating parties. Such approaches are called anonymous. Internet operation may be based on anonymity. Should individuals wish to maintain the same level of privacy they enjoy in the real world, they should be given the choice for anonymity over the Internet.

Infomediaries (I/M) are business entities supporting the development of anonymous business models. Their basic role is to accumulate information about web users, and deal products and services on behalf of them. On the one hand, I/M can protect privacy by hindering marketers from collecting customer personal data, while they offer services, which maximize the value of a customer profiles. This paper presents such a business model, capable of supporting secure and anonymous electronic transactions. The use of I/M enable customers to increase their bargain capability without revealing personal data and, at the same time, enable vendors to promote products and services without violating customers' privacy. The paper is organized as follows: In section 2, a brief description of the privacy, anonymity and I/M notion is given, while in section 3 the major threats against digital trust and anonymity are presented. In section, 4 the existing privacy-enhancing technologies (PET) are presented. In section 5, the proposed model is presented. Section 6 refers to how most common threats are dealt with, by the proposed model. Finally, section 7 refers to the concluding remarks of the paper.

2   Definitions

Within the context of this paper privacy is the right of individuals, groups, or institutions to control, edit, manage, and delete information about them, and decide when, how, and to what extent that information may be communicated to others. On the other hand, anonymity is the ability of an individual to prevent others from linking his identity to his actions. Anonymity is examined as a service offered and ensured by communication networks. Confidentiality, as a service, is a means to offer privacy (usually by deploying encryption). The basic difference between confidentiality and privacy is related with the subject of information. Although information is confidential when the owner can control it, information is private when its subject can control it. Due to the fact that anonymous information has no subject, anonymity can ensure privacy. Confidentiality is the prerequisite for anonymity provision. An Infomediary (I/M)is a business entity whose (sole or main) source of revenue derives from collecting consumer information, and developing detailed profiles of individual customers for use by selected third-party vendors [3]. I/M basic operation is based on matching customers consuming preferences with vendor products and services offerings. In order to do so, customers send their preferences to I/M and the latter develop a customer personal profile. On the other hand, vendors send their offers to I/M, without establishing a direct communication channel with customers. The matching between customers preferences and vendors offers is compiled at I/M premises. Supposing that I/M are operating in a trusted environment, security and anonymity of customers' personal data are ensured.

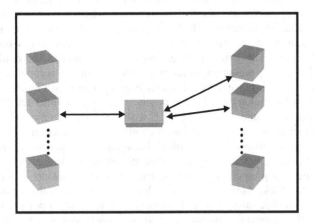

Fig. 1. Operation of infomediary

In a networked economy, customers' ability to collect information about their behavior and preferences implies that they can also choose to withhold this information from vendors. At, the same time, the accessibility of such information has raised concerns about privacy. These characteristics of the new economic activities may lead to a status, where companies should have to negotiate with customers, should they wish to gain access to customer personal data. This process demonstrates the need of I/M that can handle negotiations and payments, and add value to customer personal information, while at the same time ensuring privacy [3,4]. The communication channels (Fig. 1) are vulnerable to various security and anonymity threats between the communicating parties. This is especially true when I/M are to operate in an insecure environment, such as the Internet.

3   Threats Against Security and Anonymity

A network of interconnected I/M is expected to evolve in a distributed global environment, such as Internet. Should an I/M gain customer trust, it has to be capable of defending against threats to digital trust and user anonymity. In the sequel, two categories of threats will be examined. Threats against digital trust in transit include [5,6]: Monitoring of communication lines, Shared key guessing, Shared key stealing, Unauthorized modification of information in transit, Forged Network Addresses, Masquerade, Unauthorized access, Repudiation of origin, Private key stealing and Private key compromise. Threats against anonymous communication channels include [7]: Message coding, Timing, Message volume, Flooding, Intersection and Collusion.

4   Privacy and Anonymity Supporting Technologies and Models

"Anonymous re-mailers" allow e-mail messages to be sent without revealing the identity of the sender. Some operate through Web pages where an e-mail is created and sent without any information identifying the sender. Other re-mailers are designed to receive an e-mail message from one party, re-address it and send it to a second party. In the process, header information that would identify the sender is removed [8]. They suffer by given disadvantages when it comes to implement a global e-commerce infrastructure. The basic drawbacks of remailers are: a) users must rely on the security of the operation site to resist intruders who would steal the identity table, b) attackers who could eavesdrop on Internet traffic could match up incoming and outgoing messages to learn the identity of the acronyms/pseudonyms and c) some remailers impose substantial delays and performance limitation.

Rewebber is a technology used for anonymous surfing the Internet. Firstly, a user visits the rewebber site and strokes the URL wishes to visit. Then rewebber manages, using various techniques, to substitute with another URL or encrypt the real URL. As a result the real communication channel between the requester address and the requested URL, is not revealed [9]. The basic disadvantage of this technology is that it often provides no protection against traffic analysis by means of timing attack, message volume attack, intersection attack, flooding attack, or collusion attack.

TAZ servers provide marketers with an easy way to point to potential consumers, as well as to offer consumers an easy way to access vendors. A TAZ server consists of a public database mapping virtual hostnames ending in .taz to re-webber locators. The database is public, so there is no incentive to threaten them with legal, social, or political pressure, because any information that the operator can access is also publicly available. A major disadvantage of TAZ servers is that a locator that contains a simple chain of rewebbers looks complicated; there is also a naming problem, still not solved [9].

Onion Routing [10] is a flexible communications protocol, resistant to eavesdropping and traffic analysis designed to provide anonymous, bi-directional and near real-time connections. The disadvantages of Onion routing are that it does not provide protection between end-users against timing attacks, message volume attacks (but only between onion routers), intersection attacks and flooding attacks.

Crowds [11] is a system for protecting user anonymity while browsing the web. It uses a strategy similar to Mixmaster and Onion Routing. Crowds operation is based on grouping users into a "crowd". As a result, a user's request to a web server is first passed to a random member of the crowd. That member can either submit the request directly to the end-sever, or forward it to another randomly chosen member. In the latter case, the next member independently chooses to forward or submit the request. The disadvantages of Crowds are that it does not provide protection between end-users against timing attacks and message volume attacks, intersection attacks, or flooding attacks .

JANUS is a cryptographic engine that assists clients in establishing and maintaining secure and pseudonymous relationships with multiple servers. JANUS is hiding the identity of recipients but not the identity of the senders [12].

Web-Mixes[7] is an anonymity system for the Internet. It uses a modified Mix [13] concept with an adaptive chop-and-slice algorithm, a ticket-based authentication system that makes flooding attacks impossible or very expensive, and a feedback system giving the user information on his current level of protection. It also sends dummy messages whenever an active client has nothing to send. Web-mixes protects against coding attacks using public key cryptography, against timing and message volume attack using dummy traffic and chop-and-slice algorithm, but it provides no protection against intersection attack. It protects against flooding attacks by using "tickets"; it protects against collusion defending k-1 of k Web-Mixes.

The aforementioned technologies have the following disadvantages:

1. They provide encryption mechanisms as a means to provide anonymity, but fail to protect authenticity and integrity of the exchanged messages.
2. A subject enjoys privacy when he can control, on his own, the dissemination of his private information. These technologies exploit confidentiality mechanisms only to provide anonymity, but fail to control access to users personal data.
3. These technologies are more privacy-enhancing tools than models oriented to support global and anonymous e-commerce purchases. Thus, they do not integrate technologies, which help users explore and make use of the electronic marketplace.

5   A Privacy-Enhancing e-Business Model

Fig. 1 presents the basic operations of a single I/M. Additional functionalities should be described, should such a network of interconnected I/M emerge, operating in an untrusted environment. These functionalities include:

1. A secure acquaintance mechanism (e.g. I/M directory service) between different I/M. Thus, all new I/M members are capable of introducing themselves.
2. Users should trust I/M as they are agents of their personal information. Thus, a proper security infrastructure (i.e. I/M PKI) should be provided to implement this trust.
3. Integration of the basic I/M operations (i.e. customer profile creation, vendor offers collection, and matching) may violate users anonymity in case of collusion attack. In order to implement the "need to know" principle, these operations should be performed by different, independent, and special-purpose entities.

The suggested I/M model (Fig. 2) model takes into account the additional functionalities. It includes four distinct entities: the Customer, the Vendor, the Customer-oriented I/M and the Super-I/M (Table 1). For each entity there is a number of requirements and functionalities. The requirements refer to each entity expectations from the I/M PKI. The functionalities refer to the implemented functions, as well as to the roles played by each entity within the I/M PKI. Initially, Cx, I/Mcx, and Vx have to register to I/MS in order to join the PKI. A typical registration procedure may be followed using the web. At the end of this phase, I/MS issues digital certificates identifying each party within the PKI.

The protocol used for the implementation of the suggested model is the following:
    Case 1: Customer request:

1. Customer (C_x) encrypts product description (PD) with the I/M_{cx} public key ($P_{I/Mcx}$). Let M be the result of this procedure. Then, it sends a request to I/M_{sx}, containing personal information (P) and M, encrypted with I/M_s public key ($P_{I/Ms}$)

$$C \rightarrow I/M_s: \{P,M\}P_{I/MS}$$

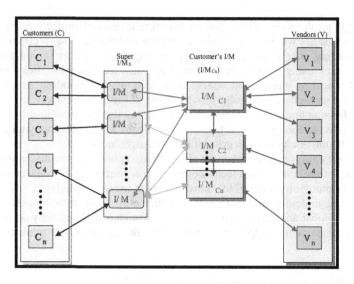

Fig. 1. The suggested model

Table 1. Model's entities, requirements, and functionalities

Entity	Requirements (expectations)	Functionality
Customer (C_x)	Buy goods from vendors, Retain privacy and anonymity	
Vendor (V_x)	Dispose products to customers, Increase their sales, Reduce the advertising cost, Gain more revenues, Aware of customers preferences	Delivery of products to I/M_s
Customer oriented Infomediary (I/M_{cx})	Act on customers' benefit, Own large databases, Increased marketing skills	Collection of product offerings, Building of profiles, Matching of profiles with vendors' products, Gathering of customers requests from I/M_s Reference to other I/M_{cx} when an I/M_x request does not match an entry in the local database.
Super Infomediary (I/M_s)	Trusted by vendors and customers	Supervision of model procedures, Setting up of the I/M PKI, Protection of anonymity, privacy, and authenticity, Collection of personal information, Collection of customer preferences

2. I/M_s decrypts the received message using the private key corresponding to $P_{I/MS}$, anonymizes the information contained within C_x request, substitutes P with a

random identifier (ID), and sends ID and M to I/M_{cx} encrypted with the I/M_{cx} public key ($P_{I/MCx}$).

$$I/M_s -> I/M_{cx}: \{ID, M\} \, P_{I/MCX}$$

3. I/M_{cx} decrypts the received message using the secret key corresponding to $P_{I/MCX}$. Then, it decrypts M using the secret key corresponding to $P_{I/MCX}$ and retrieves PD.

Case 2: Vendor offer:

1. Vendor (V_x) sends to I/M_{Cx} his personal random identification number (VID) accompanied with his product offer (PO), encrypted with I/M_{cx} public key ($P_{I/MCx}$).

$$V_x -> I/M_{Cx}: \{VID, PO\} \, P_{I/MCX}$$

2. I/M_{Cx} decrypts the received message using secret key corresponding to $P_{I/MCX}$, and appends VID and PO to its local database.

Case 3: Delivery:

1. C_x matches PD with PO (within the local database) and creates the response (R) corresponding to PD and accompanied with a response ID (RID). In case of an initial unsuccessful matching between PD and PO, a referral can be forwarded to another I/M_{Cx}. This procedure may be continued until initial request is satisfied, or a final matching failure is referred.

When a successful matching is performed, then C_x sends the following messages:

1.1 To V_x: It sends PD and RID encrypted with V_x public key (P_{Vx}).

$$I/M_{Cx} -> V_x: \{RID, PD\} P_{Vx}$$

1.2 To I/M_S: It sends ID and RID encrypted with I/M_S public key ($P_{I/MSx}$)

$$I/M_{Cx} \rightarrow I/M_S: \{ID, RID\} \, P_{I/MSx}$$

2. V_x decrypts the received message (message 1.1) using the private key corresponding to P_{Vx}, and delivers goods to I/M_S. I/M_S decrypts the received message (message 1.2) using the private key corresponding to $P_{I/MSx}$ and matches RID to P, through ID.

3. When goods corresponding to RID delivered to I/M_S, it constructs a message containing ID, RID and send it to C_x encrypted with C_x public key (P_{Cx}).

$$I/M_S -> C_x: \{ID, RID\} \, P_{Cx}$$

4. C_x decrypts the message using the private key corresponding to P_{Cx}, retrieves RID corresponding to ID and is waiting for goods to be delivered.

The services, which are offered by I/M PKI are depicted in Table 2.

The new major element, which has been introduced by the suggested model, is the evolution of an I/M network operating in a trusted environment, as a means to protect customer privacy. The cornerstone of this model is the notion of an I/M_S supervising the operation of an I/M PKI. This element is introduced to increase privacy, anonymity, and security of customers.

An I/M_S can provide additional protection against collusion attacks. This kind of attacks can be exploited in case of direct communication between an I/M_{cx}, which owns customers personal information, and V_x. An unscrupulous I/M_{cx} could collate customer personal information with V_xs' database and thus violate customers' privacy. This can be avoided by stripping off customers' personal information by I/M_S. This way, neither customers, nor vendors, directly communicate to each other but through a trusted I/M_S clearing every transaction. Furthermore, product description request is hidden even from I/Ms by encrypting it with I/M_{Cx} public key. This way, no party within I/M PKI can correlate identifiable personal information with customers' consuming behavior.

Table 2. Services offered by I/M PKI

Service	I/M$_c$	I/M$_s$
Registration		+
Key management		+
Cryptographic services		+
Digital Signatures		+
Non-repudiation (time-stamping)		+
Certificate management		+
Directory services	+	+
Camouflaging communications	+	+
User anonymity		+
Database management	+	+
Delivery		+

In order to avoid collusion attacks, personal information and matching operations are separated. According to the I/M model (Fig. 1), the acquaintance of customers personal information and its matching with a product offered by vendors is performed by the same entity. This may lead to collusion attacks if an I/M sells personal profiles to vendors. To avoid this, these operations have been assigned to different entities:

ú Personal data collection is performed by a trusted organization (e.g. a Data Protection Authority). This organization anonymizies the personal data by substituting the person's identity with a randomly generated number. Each customer request is assigned a different id number. I/Ms know nothing about customer's consuming preferences because these preferences are encrypted with I/M$_{cx}$'s public key.

ú I/MC$_x$ performs the matching between anonymous customer preferences and vendor offerings. I/MC$_x$ may be any company.

The use of a single I/MS, which supervises the I/M PKI, may lead to network traffic congestion problems. For this reason, a network of I/MS may be established. I/Mcx and Vx registrations are now distributed between local I/MS. Furthermore, each I/MCx occasionally communicates with its peers, in case matching referrals not appearing in its local database. To decrease network traffic payload, only changes to customers preferences and to vendors offerings database are transmitted. During the registration phase, I/MS and Vx send the instances of their local databases to I/MCx. From then on, only database instance changes are communicated with I/MCx. Only asserted parties may be involved in the I/M PKI. Thus, every asserted entity should be equipped with a digital certificate. Directory services may be used in order for a I/MCx to communicate with its peers and look up for a specific Cx request not satisfied by information included in its local database. Security of the model communication channels is achieved using digital certificates issued by I/MS during registration.

Two levels of end-to-end user anonymity may be distinguished:

(a) Customers personal profile anonymity, which refers to mechanisms adopted to protect personal data from revealing to unauthorized parties and it is achieved using the following mechanisms:

1. C_x communicates only with I/M_s, who strips off the customer's personal information from his requests to I/M_{Cx} by substituting P with ID (step 2 of communication protocol, customer request). This way no personal information is circulated within I/M PKI.
2. The intermediation of I/M_s during delivery of goods makes impossible the direct communication between C_x and V_x.

(b) Communication level anonymity, which refers to the mechanisms and technologies which may be adopted to camouflage communications among different parties. For communication level anonymity, Web-Mixes and Mix model mechanisms may be adopted. In detail, a chain of servers (Mixes-Chaum) can be used between C_x and I/M_{cx}. Each Mix in the chain strips off the identifying marks on incoming requests, and then sends the message to the next Mix, based on routing instruction which encrypted with its public key. C_x encrypts communication using the public keys of each Mix on the route. The Mixes store the requests (messages) they receive and - at designated intervals - randomly forward a request to its destination. If no message is waiting to be sent, then the Mix randomly generates a message to be sent.

Table 3 refers to the technologies and standards available for offering the services required by the suggested model.

6   How Common Threats Are Dealt with

Digital certificates and SSL protocol exploit public and symmetric key cryptography. These technologies are deployed to defend against most threats.

1. Monitoring of communication lines. Avoided by using Public Key Cryptography.
2. Shared key guessing. Avoided by using strong symmetric encryption.
3. Shared key stealing. Avoided by using public key encryption (protect transmission of shared keys in plaintext across a data network).
4. Unauthorized modification of information in transit. Avoided by using public key encryption during all communication steps between two parties.
5. Forged Network Addresses and Masquerade: We can distinguish two cases of forging:
ú   An unscrupulous user pretends to be a self-signed I/M_s. This is avoided by issuing certificates by a trusted I/Ms, only, and allow use within I/M PKI.
ú   An unscrupulous user pretends to be a trusted party. Avoided by using certificate-hashing mechanisms.
6. Unauthorized access: Avoided through the use of a sound access control policy.
7. Repudiation of origin: Avoided by time-stamping all messages between I/M_s and other entities.

Table 3. Services and candidate technologies and standards

Service	Candidate technologies
Registration	WWW + SSL, SSH + custom application, Secure e-mail (S/MIME, PGP), Postal mail
Key management	ISO 8732, ISO 11770, PKCS
Cryptographic services	RSA Cryptokit, Microsoft CryptoAPI, Open Group GCS/API
Digital Signatures	ISO 9594, ISO 9796, ISO 14888, ISO 9798, GSS-API, Microsoft CryptoAPI, GCS API, PKCS
Non-repudiation (Time-stamping)	U.S Patent 5,136,647, Annex to ISO 13888-3, PKIX, ISO 13888
Certificate management	X.509, SPKI
Directory services	X.500/LDAP, Z.39.50
Camouflaging communications	Web-Mixes, Mix, Onion Routing
User anonymity	Request strip off
Database management	Medium-class or high-end DBMS supporting SQL
Delivery	S/MIME, PGP, Postal mail

8. Private key stealing and Private key compromise: Avoided by using strong encryption and by storing cryptographic tokens in removable media.
9. Message coding. Avoided by using end-to-end Onion Routing.
10. Timing. Avoided by using Mixes. A Mix waits until a defined number n of messages has arrived from n users. After that time, all messages are put out together, but in a different order. Also, dummy messages are sent from the starting point (i.e. V_x, C_x, I/M_{Cx}, I/M_s) into the network to make traffic analysis harder. Large messages (and streaming data) are chopped into short pieces of a specific constant length ("slice"). Each slice is transmitted through an anonymous Mix channel. In addition, active users without an active communication request send dummy messages. Thus, nobody knows about the starting time and duration of a communication because all active users start and end their communications at the same time.
11. Message volume. Avoided by using Mixes. All incoming and outgoing messages of a Mix have the same length. To prevent replay attacks, a Mix will process each message only once.
12. Flooding. Avoided by using Mixes. Each user has to show that he is allowed to use the system at the respective time slice by providing a ticket valid for the certain slice, only. To protect the identity of the user the ticket is a blinded signature issued by the anonymous communication system.
13. Intersection. Use of dummy traffic makes intersection attacks harder but does not prevent them.
14. Collusion. Only I/M_s (trusted participant) issues digital certificates and public keys. In addition, the parties, which perform personal data collection and matching services, are different. Therefore, there is no way of direct communication neither between I/M_{Cx} and V_x, nor between C_x and V_x. As a second

level of prevention, customer's product description requests are encrypted with I/MCx's public key. This means that although I/Ms knows customer's personal information it cannot be correlated it with his consuming preferences. On the other hand, although I/MCx possess anonymous customer's consuming profiles, it cannot correlate them with any identifiable personal information. This way the "need to know" principle is enforced within I/M PKI.

Conclusions

Exploiting Internet services usually means leaving personal digital traces. Anonymity and un-observability on the Internet is tough to ensure. On the other hand, there is a substantial need for anonymous communication, as a fundamental building block of privacy in the information society. Although existing technologies offer, more or less, serious technical advances with regard to communication camouflaging, they present substantial disadvantages when it comes to be adopted as global anonymous e-commerce carriers. On the other hand, infomediaries is a promising technology on the financial area, which has not been exploited on anonymous communication field. In this paper we have described a business model, based on the infomediaries concept, which is suitable for anonymity-aware and privacy-concerned users when making electronic purchases. The model allows users not to reveal their personal data or preferences to vendors, and at the same time allows vendors sell or advertise goods or services, without violating the privacy of their customers.

The basic characteristics of the suggested model are.

ú The introduction of I/M as a privacy-enhancing agent in the e-commerce and e-government area. Although I/M have been mainly exploited as a means to establish new virtual enterprises and maximize user personal information value, they also offer new possibilities regarding privacy and anonymity.

ú The stand-alone I/M model is neither adequate, nor security and anonymity robust for large-scale e-economies. Thus, the evolution of an I/M network, with entities co-operating in a secure and anonymous way, is introduced, offering anonymous electronic transactions.

ú The traditional I/M model, which integrates customer preferences acquisition and vendor offerings, is vulnerable to collusion attacks. The separation of these operations is estimated to largely contribute in making anonymous global electronic purchases.

ú Suggested model's candidate technologies are well established not only in the application level but also in the international standards area. This makes model's implementation feasible and compatible with existing privacy enhancing technologies. As a result, no further technical and operational turbulence will hesitate the integration of our model within e-commerce infrastructures.

Trust is a key element in e-commerce. If I/M are to play a role within the digital economy, they should limit the searching cost of goods and services. They should also help determining the best price for goods or services, support protecting the customer from unwanted intrusions by vendors, while at the same time support alerting a customer in case of new product offerings that meet his needs and preferences. Finally, they should ensure customers privacy and anonymity.

References

1. Sarkar, Butler, Steinfield: Intermediaries and Cybermediaries: A Continuing Role for Mediating Players in the electronic marketplace. Journal of Computer-Mediated Communication, Vol. 1. 3 (December 1995)
2. Pfleeger C., Cooper, D.: Security and Privacy: Promising Advances. IEEE Software Magazine, Vol. 14. 5 (September/October 1997) 27–32
3. Hagel, Rayport: The new infomediaries. The Mc Kinsey Quarterly, 4 (November 1997)
4. Hagel, J., Singer, M.: Net Worth: Shaping Markets When Customers Make the Rules. HBS Press (1999)
5. De Vivo M., De Vivo G., Isern G.: Internet Security Attacks at the Basic Levels. Operating Systems Review ACM press, Vol. 32. 2 (April 1998) 4–15
6. Crijns, M., et. al.: Issues facing the secure link of Chambers of Commerce. European Commission, COSACC Project, Deliverable No. 3 (December 1998)
7. Berthold, O., Federrath, H., Köhntopp, M.: Project "Anonymity and Unobservability in the Internet"
8. Organization for Economic Co-operation and Development. Inventory Of Instruments and mechanisms contributing to the implementation and enforcement of the OECD privacy guidelines on global networks, DSTI/ICCP/REG(98)12/FINAL (19 May 1999)
9. Goldberg, Wagner: "TAZ Servers and the Rewebber Network: Enabling Anonymous Publishing on the WWW". First Monday Peer Reviewed Journal on The Internet, Vol. 3. 4 (April 1998)
10. Cranor, L.: Internet Privacy. Communications of the ACM, Vol. 42. 2 (February 1999) 29–66
11. Reiter, M., Rubin, A.: Crowds: Anonymity for Web Transactions – AT&T Labs Research, www.research.att.com/projects/crowds
12. Bleichenbacher, D., Gabber, E., Gibbons, P., Matias, Y., Mayer, A.: On secure and Pseudonymous Client-Relationships with Multiple Servers (May 1998)
13. Chaum, D.: Untraceable electronic mail, return addresses and digital pseudonyms. Com. of the ACM, 24(2) (February 1981)

Applying Practical Formal Methods to the Specication and Analysis of Security Properties

Constance Heitmeyer

Naval Research Laboratory (Code 5546)
Washington, DC 20375 USA
heitmeyer@itd.nrl.navy.mil
http://www.chacs.itd.nrl.navy.mil/SCR

Abstract.   The SCR (Software Cost Reduction) toolset contains tools
for specifying, debugging, and verifying system and software requirements. The utility of the SCR tools in detecting specication errors,
many involving safety properties, has been demonstrated recently in
projects involving practical systems, such as the International Space Station, a flight guidance system, and a U.S. weapons system. This paper
briefly describes our experience in applying the tools in the development
of two secure systems: a communications device and a biometrics standard for user authentication.

1   Introduction

In 1978, the requirements document for the flight program of the A-7 aircraft
[13,14] introduced a special tabular notation for writing specications. Part of
the SCR (Software Cost Reduction) requirements method, this notation was designed to document the requirements of real-time, embedded systems concisely
and unambiguously. During the 1980s and 1990s, SCR tables were used by several
organizations in industry and government, e.g., Grumman [19], Bell Laboratories [15], Ontario Hydro [21], the Naval Research Laboratory [7], and Lockheed
[5], to document the requirements of many practical systems, including a subma-
rine communications system [7], the shutdown system for the Darlington nuclear
power plant [21], and the flight program for Lockheed's C-130J aircraft [5].

While human eort is critical to creating requirements specications and human inspection can detect many specication errors, eective and widespread
development of precise, unambiguous specications in industry requires powerful, robust tool support. Not only can software tools nd specication errors
that inspections miss, they can do so much more cheaply. To explore what form
tools supporting the formal specication of requirements should take, we have
developed a suite of software tools for constructing and analyzing requirements
specications in the SCR tabular notation [8]. The tools include a   specication editor for creating the specication [9], a   simulator for validating that the
specication sates the customer's intent [8], a   dependency graph browser for
understanding the relationship between dierent parts of the specication [10],

V.I. Gorodetski et al. (Eds.): MMM-ACNS 2001, LNCS 2052, pp. 84–89, 2001.

and a consistency checker [11] to analyze the speci cation for properties such as
syntax and type correctness, determinism, case coverage, and lack of circularity.
The toolset also contains the   model checker Spin [16], a veri er   TAME [1], a
property checker based on decision procedures called Salsa [2], and an   invariant
generator [17], all of which may be useful in analyzing speci cations for critical
application properties, such as safety and security properties.

The utility of the SCR tools has also been demonstrated in several projects
involving real-world systems. In one project, NASA researchers used the SCR
consistency checker to detect several missing assumptions and instances of ambiguity in the requirements speci cation of the International Space Station [4]. In
a second project, engineers at Rockwell Aviation used the SCR tools to detect 28
errors, many of them serious, in the requirements speci cation of a flight guidance
system [20]. In a third project, our group at NRL used the SCR tools to expose
several errors, including a safety violation, in a moderately large contractor-
produced speci cation of a U.S. weapons system [12]. Recently, we have begun
using the SCR method and tools to analyze speci cations for security properties.
This paper briefly describes our experiences in applying the SCR tools to two
secure systems: a communications device called CD and a biometrics standard.

2 Applying the SCR Tools to Secure Systems

2.1 Applying SCR to a Communications Device

COMSEC (Communications Security) devices, devices which manage encrypted
communications, are vital to the correct operation of U.S. military systems. CD
is a COMSEC device that is designed to provide cryptographic processing for a
U.S. Navy radio receiver. In addition to generating keystreams compatible with
another cryptographic device and supporting multiple channels and multiple
cryptographic algorithms, CD downloads associated algorithms and keys into
working storage, assigns them to designated communication channels, maintains
the association between an algorithm and its keys, and clears algorithms and keys
from memory. CD, based on a technology for implementing COMSEC devices
in software as well as hardware, presents a new challenge in the development
of COMSEC devices. While a solid base of experience exists for implementing
trustworthy COMSEC devices in hardware, implementing COMSEC devices in
software is rare.

To provide a high degree of assurance in the correctness of CD's speci cation,
we applied the SCR tools [18]. Our results suggest that applying SCR in the
development of COMSEC devices of moderate size and complexity is practical,
e ective, and low-cost. the development In approximately one person-month, we
were able to represent a signi cant subset of a prose requirements document for
CD in the SCR notation and to establish that the SCR speci cation satis es a
set of security properties. The SCR speci cation of CD is moderately complex,
consisting of 39 variables (17 input variables, three auxiliary variables, and 19
output variables). Figure 1 provides a natural language formulation and a formal
representation of each of the seven security properties that we veri ed with the

SCR tools. Because the SCR requirements speci cation of CD has been validated using simulation and veri ed to satisfy seven critical security properties, the SCR requirements speci cation of CD can help guide both the development of the source code for CD and the development of test cases for evaluating the conformance of the source code with CD's requirements.

No.	Description	Property
1	If CD is tampered with, then key 1 in keybank 1 is zeroized	@T(mTamper) cKeyBank1Key1¢= 0
2	When the zeroize switch is activated, key 1 in keybank 1 is zeroized	@T(mZeroizeSwitch = on) cKeyBank1Key1¢= 0
3	No key can be stored in location 1 of keybank 1 before an algorithm has been loaded into the fi rst location of algorithm storage segment 1	cKeyBank1Key1 $\neq$ 0 cAlgStoreSegment1 $\neq$ 0
4	If backup power has an undervoltage when primary power is unavailable, the CD enters either Alarm mode or Off mode	@T(mBackupPower = undervoltage) WHEN mPrimaryPower = unavailable smOperation¢= sAlarm OR smOperation¢= sOff
5	If backup power is overvoltage then the CD is in Initialization, Standby, Alarm, or Off mode	mBackupPower = overvoltage smOperation = sInitialization OR smOperation = sStandby OR smOperation = sAlarm OR smOperation = sOff
6	If primary power has an overvoltage then either the CD is in Initialization, Standby, Alarm, or Off mode, or the CD enters Initialization mode	@T(mPrimaryPower) = overvoltage smOperation = sStandby OR smOperation = sAlarm OR smOperation = sOff OR smOperation¢= sInitialization
7	If primary power has an undervoltage then either the CD is in Initialization, Standby, Alarm, or Off mode, or the CD enters Initialization mode	@T(mPrimaryPower) = undervoltage smOperation = sStandby OR smOperation = sAlarm OR smOperation = sOff OR smOperation¢= sInitialization

Fig. 1.   Sample properties of CD

2.2   Applying SCR to a Biometrics Standard

Positive identi cation and authentication of personnel is a critical issue for many systems. For example, U.S. government personnel must often interact with the commercial sector in situations where reliable personnel identi cation is critical for limiting access to sensitive information systems. While biometrics technology addresses the critical issue of personnel identi cation and authentication, prior to deployment, a biometrics product must be certi ed to satisfy published assurance standards. However, the labor-intensive process of evaluating and validating a biometrics product is very expensive and time consuming. One way to reduce these problems is to use automated methods to support product evaluations. Applying such methods should not only lead to a less costly and shorter process for evaluating biometrics and other security products but should also produce a more e ective process.

To assess the utility of the SCR method and tools for evaluating a biometrics product for correctness, we applied SCR to the BioAPI speci cation [3], a standard which de nes the interface between an authentication device that uses biometrics data and an application program. The goal of the biometrics API (application program interface) is to enable rapid development of biometrics applications, the flexible deployment of many biometrics devices across platforms and operating systems, and an improved ability to exploit price and performance advances in biometrics. From the BioAPI standard, we produced an SCR tabular speci cation, which captures the behavior of six major operations in the standard. The SCR speci cation consists of 20 variables: 10 input variables, one mode class variable, and nine output variables. In about two weeks, we were able to create the speci cation, to demonstrate with the consistency checker that the speci cation contained no missing cases and no ambiguity, and to verify a critical security property. The goal of this property is to demonstrate that "the system shall successfully authenticate a user before mediating actions initiated by that user."

3   Observations

The SCR method and tools contributed to the speci cation and analysis of these two systems in a number of ways. We describe these ways below:

- Requirements Capture.   Developing a formal requirements speci cation from the prose requirements document for CD was di cult, largely because the prose document was organized very di erently than an SCR speci cation. Moreover, even though the prose document was high quality, a number of questions about the required behavior of CD arose. Two SCR tools were useful in correcting and extending our initial SCR speci cation of CD's required behavior. First, we used an automatic invariant generator to construct state invariants from the draft speci cation. Analyzing these invariants identi ed a number of missed requirements and some incorrectly captured requirements. After correcting these problems, we used our simulator and a GUI builder to construct a simulation of CD. Because the CD program manager was very busy, he did not have the time to review our speci cation. Instead, we showed him several scenarios using our CD simulator. By viewing the simulation, he was able to quickly identify a number of errors in our CD speci cation which we subsequently corrected.
{ Formal Veri cation.   To verify the seven security properties listed in Figure 1, we ran TAME, a user-friendly interface to the theorem prover PVS. TAME was able to prove four of the seven properties directly. To prove the remaining properties, TAME needed several supporting invariant lemmas. Fortunately, each of the required lemmas belonged to the set of state invariants that we were able to construct with our invariant generation algorithm [17].
- Detecting Incorrect Properties.   We were unable to prove that the CD speci cation satis es an eighth security property. Although we tried apply-

ing the model checker Spin to the CD speci cation, Spin repeatedly ran out
of memory due to the large state space of the CD speci cation and thus was
unable to verify or refute any of the eight security properties. The false property was detected by running TAME and studying the problem transitions
returned by TAME. By experimenting with the CD simulator, we were able
to construct a counterexample that ended in one of the problem transitions
and hence demonstrated that the eighth property was false.

- Correct Formulations of Security.   Formulating a correct formal statement of a given security property can be di cult. In our work on the biometrics standard, the correct formulation of the security property (see above)
required more time than verifying the property.

- Code Validation.   The most important open problem is how to validate the
source code that implements a secure system. While specifying the required
behavior of a secure system and formally proving that the speci cation satis es critical security properties can often be accomplished in a reasonable
time, one still needs to demonstrate that the source code operates securely.
One approach to code validation is speci cation-based testing. That is, one
can derive a set of test cases from the speci cation and automatically use
these test cases to determine whether the source code satis es the speci -
cation. Some initial progress in developing an automatic test case generator
from a requirements speci cation is reported in [6].

Acknowledgments.   Jim Kirby developed both the CD speci cation and the
speci cation of the BioAPI standard. Moreover, Jim, Myla Archer, and Ralph
Je ords used TAME and the invariant generator to verify that the CD speci -
cation and the BioAPI speci cation satisfy selected security properties. Ramesh
Bharadwaj also veri ed the properties in Figure 1 using Salsa and constructed
a counterexample for the eighth property of CD using the SCR simulator. I am
grateful to Myla Archer and Jim Kirby for their comments on an earlier draft
of this paper.

References

1. Archer, M., Heitmeyer, C., and Riccobene, E.: Using TAME to prove invariants
   of automata models: Case studies. In   Proc. 2000 ACM SIGSOFT Workshop on
   Formal Methods in Software Practice (FMSP'00)   (August 2000)
2. Bharadwaj, R. and Sims, S.: Salsa: Combining constraint solvers with BDDs for
   automatic invariant checking. In   Proc. Tools and Algorithms for the Construction
   and Analysis of Systems (TACAS '2000)   , Berlin (March 2000)
3. BioAPI Consortium. The BioAPI Speci cation. Version 1.00 (March 30, 2000)
4. Steve Easterbrook and John Callahan. Formal methods for veri cation and validation of partial speci cations: A case study.   Journal of Systems and Software
   (1997)
5. Faulk, S.R., Finneran, L., Kirby, Jr., Shah, S., and Sutton, J.: Experience applying the CoRE method to the Lockheed C-130J. In: Proc. 9th Annual Conf. on
   Computer Assurance (COMPASS '94)   . Gaithersburg, MD (June 1994)

6. Gargantini, A. and Heitmeyer, C.: Automatic generation of tests from require-
ments speci cations. In: Proc. ACM 7th Eur. Software Eng. Conf. and 7th ACM
SIGSOFT Symp. on the Foundations of Software Eng. (ESEC/FSE99)   , Toulouse,
FR (September 1999)

7. Heitmeyer, C.L. and McLean J.:   Abstract requirements speci cations: A new
approach and its application.   IEEE Trans. Softw. Eng.   , SE-9(5) (September 1983)
580–589

8. Heitmeyer, C., Kirby, Jr.J., Labaw, B., and Bharadwaj, R.: SCR*: A toolset for
specifying and analyzing software requirements. In   Proc. Computer-Aided Veri -
cation, 10th Annual Conf. (CAV'98)   , Vancouver, Canada (1998)

9. C. Heitmeyer, A. Bull, C. Gasarch, and B. Labaw. SCR*: A toolset for specifying
and analyzing requirements. In   Proc. 10th Annual Conf. on Computer Assurance
(COMPASS '95) , Gaithersburg, MD (June 1995) 109–122

10. Constance Heitmeyer, James Kirby, Jr., and Bruce Labaw. Tools for formal spec-
i cation, veri cation, and validation of requirements. In: Proc. 12th Annual Conf.
on Computer Assurance (COMPASS '97)   . Gaithersburg, MD (June 1997)

11. C. L. Heitmeyer, R. D. Je ords, and B. G. Labaw. Automated consistency checking
of requirements speci cations.   ACM Transactions on Software Engineering and
Methodology , 5(3) (April–June 1996) 231–261

12. Heitmeyer, C., Kirby, J., Labaw, B., Archer, M., and Bharadwaj, R.: Using abstraction and model checking to detect safety violations in requirements speci cations.
IEEE Trans. on Softw. Eng.   , 24(11) (November 1998)

13. Heninger, K., Parnas, D.L., Shore, J.E., and Kallander, J.W.: Software requirements for the A-7E aircraft. Technical Report 3876, Naval Research Lab., Wash.,
DC (1978)

14. Heninger, K.L.: Specifying software requirements for complex systems: New techniques and their application.   IEEE Trans. Softw. Eng.   , SE-6(1) (January 1980)
2–13

15. Hester, S.D., Parnas, D.L., and Utter, D.F.: Using documentation as a software
design medium.   Bell System Tech. J. , 60(8) (October 1981) 1941–1977

16. Holzmann, G.J.: The model checker SPIN.   IEEE Transactions on Software Engineering , 23(5) (May 1997) 279–295

17. Ralph Je ords and Constance Heitmeyer. Automatic generation of state invariants from requirements speci cations. In: Proc. Sixth ACM SIGSOFT Symp. on
Foundations of Software Engineering   . (November 1998)

18. Kirby, Jr.J., Archer, M., and Heitmeyer, C.: SCR: A practical approach to building
a high assurance COMSEC system. In   Procee dings of the 15th Annual Computer
Security Applications Conference (ACSAC '99)   . IEEE Computer Society Press
(December 1999)

19. Meyers, S. and White, S.: Software requirements methodology and tool study for
A6-E technology transfer. Technical report, Grumman Aerospace Corp., Bethpage,
NY (July 1983)

20. Miller, S.: Specifying the mode logic of a flght guidance system in CoRE and SCR.
In: Proc. 2nd ACM Workshop on Formal Methods in Software Practice (FMSP'98)
(1998)

21. Parnas, D.L., Asmis, G.J.K., and Madey, J.: Assessment of safety-critical software
in nuclear power plants.   Nuclear Safety , 32(2) (April–June 1991)

Modeling Software Tools Complex for Evaluation of Information Systems Operation Quality (CEISOQ)

Andrey Kostogryzov

Russian Academy of Rockets and Artillery Science
Postfach 117192, 135, 15 Winnitskaja ul., Moscow, Russia
akostogr@redline.ru

Abstract. The paper focuses on describing and using the unique software tools CEISOQ, supporting the mathematical models of information gathering, storage, processing and producing by computer-based systems [1-2]. An application of CEISOQ permits to evaluate a degree of achieving the information systems operation purpose: to meet users requirements for providing reliable, timely, complete, valid and confidential information. Software tools CEISOQ purposed to systems analysts (IS customers, designers, developers, users, experts of testing laboratories and certification bodies, as well as a staff of IS operation quality maintenance) for solving the next system problems: for substantiating framework and specifying quantified system requirements (to hardware, software, users, staff, technologies); for requirements analyzing, technical decisions evaluating, the bottle-necks exposuring; for investigating potential threats to information quality; for testing IS operation; for rational tuning IS technologies parameters.. CEISOQ may be also used in training and education in specialization "Information systems".

1   Introduction

A creature of modern systems in enterprise, manufacture, finance, customs, military, cosmic and other fields is by now inconceivable without wide computers use for providing information gathering, processing, storage and producing. Computer technologies implementations gave rise to a lot of new engineering problems which require practice computations in detail. On information technology market there is offered a wide choice of engineering solutions that are able to meet functional customer requirements. Admissible variant may be selected if to be guided by logical considerations. But whether will be this variant rational or not from the point of view of achieving integral system operation purpose? Very likely answer this question will be "rational" if one is intended for an enterprise system having the main operation purpose to derive the highest possible profits from goods manufacture and sale. Really, rationality criterion for a selection of some manufacture computer systems may be a criterion of a goods quality maximum (and, accordingly, a profits maximum) under an expenditures limitations. And if this question is applicable to information systems (IS), for which output is information...? One automatically makes the answer: it should be computed! Indeed, a rational ensuring of modern IS quality is impossible without a use of models and methods that allow to evaluate, investigate and optimize

V.I. Gorodetski et al. (Eds.): MMM-ACNS 2001, LNCS 2052, pp. 90–101, 2001.

processes of information gathering, processing, storage and producing. As a rule, in practice designers elaborate models considering the purposes and specificity of concrete IS. There exists a set of analytical and simulating models allowing to evaluate some separate characteristics of IS operation quality. Side by side with this the professional needs of systems analysts are to use complex models. These models are able to transform into day-to-day tools only if they possess a high level of universality and have convenient software support.

The analysis of different IS application shows that practically there may be formulated the next general purpose of IS operation from the users point of view: to meet users requirements for providing reliable, timely, complete, valid and confidential information for its following use . Of course, there may be a variations of this purpose (see Fig. 1.)

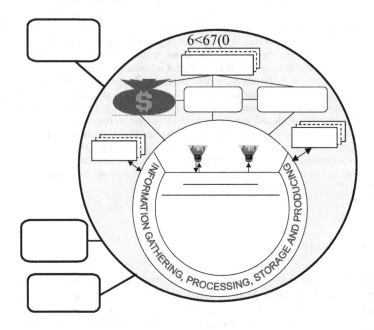

Fig. 1. The place and purpose of an information system in a SYSTEM

A totality of IS characteristics that bear on its ability to satisfy users needs in output information defines IS operation quality. Alongside with this, majority of IS are based on a use of quite standard information gathering, storage, processing and producing processes . Both of these considerations serve as the arguments of creating sufficiently universal models and implementing software tools to complex evaluate IS operation quality, to compare various IS engineering projects, to reveal "bottle-necks" and to optimize information gathering, storage, processing and production processes.

Proposed original mathematical models [1] are supported by created software tools CEISOQ (Complex for Evaluation of Information Systems Operation Quality, http://www.ceisoq.ru) [2].

Note. The models and software tools CEISOQ are not purposed for evaluating an IS mathematical support and software quality that bear on IS ability to be due to intolerable mistakes of processing.

An application of proposed models and software tools CEISOQ allows to answer the next system questions:

- what information operation processes should be duplicated?
- what calls flows processing technologies should be preferred?
- what calls flows and functional tasks operation time may be the main causes for bottle-necks?
- what productivity of preparation, transfer and input units and what gathering technologies should be preferred to provide information completeness and actuality?
- what about information check-up effectiveness?
- what qualification requirements should be due to staff and users?
- what about a danger of viruses influences and what anti-viruses security technologies should be chosen?
- what about a hazard of information security violation and under which conditions? and so on.

Connection between CEISOQ tools and characteristics of IS operation quality is reflected by the Table 1.

Table 1. Connection between CEISOQ tools and characteristics of IS operation quality

№	Threats realization results	Evaluated IS operation quality characteristics	The name of the model and corresponding software tool
1.	Non-produced information for using in consequence of system's unreliability	Reliability of information producing is the property causing capability of IS to operate required functions for information producing (or technological operations processing) when operated under specified conditions	The model of information producing considering hardware/software unreliability, «RELIABILITY»
2.	Untimely used information	Timeliness of information producing is the property causing capability of IS to operate timely required functions for information producing (or technological operations processing)	The models of calls processing, «TIMELINESS»
3.	Incomplete used information	Completeness of output information is the used information property to reflect conditions of all real objects of IS application domain according to specified purposes destination. One is composed of a completeness of IS function fulfillment, of a completeness of initial information filling and of IS filling by data as regards a new objects and processes of application domain during operation time.	The model of information system filling by data as regards a new objects of application domain, «COMPLETENESS»
4.	Deteriorated used information validity, including:	Validity of output information is the used information property to reflect real or estimated objects and processes of IS application domain conditions with the degree of approximation which is accept-	An evaluation of information processing accuracy and of data transmission validity is not supported by CEISOQ

№	Threats realization results	Evaluated IS operation quality characteristics	The name of the model and corresponding software tool
		able for effective using this information. One is composed of a validity of prepared and stored information, processed before composing, of a faultlessness and actuality of unprocessed input and stored composed information, of processing accuracy by IS software as well as of data transmission validity. Information faultlessness is the used information property to be without random errors and hidden distortions, including distortions as a result of unauthorized accesses and viruses influences.	
4.1	• non-actual used information;	• actuality of faultless information is the used faultless information property to correspond to the current conditions of IS application domain objects and processes with the degree of acceptable approximation. One is the property causing natural process when information is becoming antiquated in due course;	The models of information gathering from source, «Actuality»
4.2	• used information due to random errors missed during checking;	• information faultlessness after checking;	The model of input information check, «Faultlessness after check»
4.3	• used information due to random faults of users and staff;	• information faultlessness as a result of faultless users and staff actions;	The model of users and staff actions, «Faultlessness of staff and users actions»
4.4	• used information with hidden virus distortions;	• IS protection against viruses influences;	The model of computer viruses influences, «Viruses influences»
4.5	• used information with hidden distortions as a result of an unauthorized access	• IS protection against an authorized access.	The model of an unauthorized access to information and software resources, «Unauthorized access»
5.	Non-confidential used information	Confidentiality of used information is the used information property to be protected within required period from an unauthorized scanning.	The model of information confidentiality maintenance, «CONFIDENTIALITY »

The development of the models was founded on the using of limited theorems for regenerating processes [3], the proof of mathematical expressions see in [4].

CEISOQ models and software tools are the methodological and implementation foundation of the certification body and test laboratory of information systems and software in Russia. Since 2000 CEISOQ is subject to obligatory study in education (in Moscow State University of Oil and Gas, in Taganrog State Radio-Engineering University etc), CEISOQ has already found application in analysis special Space control system, the Russian State Customs Committee System, in certification of operation quality for special IS etc. CEISOQ application in Russia allowed to avoid unwarranted expenses 2000 year about several millions US dollars.

2   CEISOQ Models, Input, and Output

Model of information producing considering hardware/software unreliability ("Reliability").

Designations of CEISOQ input: the mean time between hardware/software failures for the k-th unit of the n-th subsystem; the mean time to system repair.

Designations of customer and users requirements: the required permanent time of reliable IS operation; the admissible probability of reliable information producing by IS during the required time.

Evaluated values by CEISOQ: the mean time between failures for the n-th subsystem; the probability of the reliable n-th subsystem performance during required IS operation time; the mean time between failures for complex composed by 1,...,n subsystems and for IS; the probability of reliable performance of complex composed by 1,...,n subsystems and IS during required operation time.

Models of calls processing ("Timeliness").
IS operation is modeled as calls processing by the priority and unpriority queueing systems $M/G/1/\mp$. The software tools CEISOQ allow to evaluate and to compare efficiency of the next dispatcher technologies:

- technology 1 for calls processing without priorities:
  - in consecutive order for the unitasking processing mode;
  - in time sharing for the multitasking processing mode;
- priority technologies 2-5 in the consecutive processing order:
  - technology 2 for the calls processing with relative priorities in the order FIFO ("first in - first out");
  - technology 3 for the calls processing with absolute priorities in the order FIFO;
  - technology 4 for the calls batching processing with relative priorities and the order FIFO within the batch;
  - technology 5 combined by technologies 2, 3, 4 for calls processing.

According to technology 1 (characteristic R="unitasking") calls are served in the consecutive order FIFO.

According to technology 1 (characteristic R="multitasking") if there exist n calls for processing, they all are served simultaneously, but every call is served in n times more slowly in comparison with one is served in the off-line mode.

According to technology 2 the queueing calls with higher priority will be served earlier than the queing calls with lower priority, there is used the order FIFO for serving the queueing calls with the equal priority. It is not interruption of begun serving.

Technology 3 allows the absolute interruption of begun serving in comparison with technology 2 if the priority of the coming calls is higher than priority of the serving calls. Processing of the interrupted calls continues from the interrupted place.

According to technology 4 the first call coming for processing at the free server forms the first batch. A next batch is formed by the calls coming for processing during the serving time of all calls of the previous batch. The next batch will begin to serve at once after all calls of the previous batch will be served. Within a serving batch a calls of this batch serve in full according to technology 2 without the dependence of a coming calls. A serving batch can not be interrupted for processing any other calls.

According to technology 5 all calls are distributed between n groups . A calls of g-th group have the higher priority than a calls of the e-th group if g<e (e, g = 1,n). A calls of every group have relative priority within this group.

Designations of CEISOQ input: the calls flow intensity (of the r-th type) for processing; the first (mean), the second and the third moments for calls run time in off-line mode; the number of group for combined technology 5; the mark of presence or absence of or absolute calls priority of the n_1–th group over calls of the n_2-th group; the dispatcher technology within group (it may be technology 2 or 4).

There exists the possibility to choose the timeliness criterion:
- the criterion of mean full time of calls processing;
- the criterion of probability of timely processing.

Designations of customer requirements: the admissible maximum time for the full calls processing time, including the wait queueing time; the admissible probability of the timely calls (of the r-th type) processing against the admissible time.

Evaluated values by CEISOQ: the mean wait queueing time for calls; the mean full processing time for the calls, including wait queueing time; the probability of timely processing for calls against the admissible time; the relative part of timely processed calls of all types; the relative part of timely processed calls of those types for which the customer requirements are met.

Model of IS filling by database new objects of application domain ("Completeness").

Designations of CEISOQ input: the frequency of appearing a new data about real objects and events in application domain; the mean time of information revealing and preparing; the mean time of information transfer; the mean time of information inputting.

Designation of customer requirements: the admissible probability of information completeness maintenance by new appeared data.

Designations of evaluated values: the probability of information completeness maintenance by new appeared data; the probability that storage information is not complete by new appeared data about conditions of K new real objects and events.

Model of information gathering from sources ("Actuality").

Definition. Used information is actual for deciding concrete task, if essential changes of objects conditions have updated and kept in IS to the moment of information using.

Designations of CEISOQ input: the frequency of essential changes of real objects and events conditions for effective SYSTEM performance; the mean time information collection and preparing; the mean time information transfer; the mean time of input into updated; the discipline of information gathering and updating (information gathering from sources may be carried out "at once after essential change" of real object and event conditions mirrored in database or beyond open dependence from changes of conditions).

Designation of customer requirements: the admissible probability of information actuality maintenance.

Designations of evaluated values: the probability of information actuality mainte-nance.

Model of input information check ("Faultlessness after check").

Definition. Information after data checking verification is faultless if all data er-rors have been detected and corrected and new errors have not brought in until as-signed check deadline.

Designations of CEISOQ input: the mean volume of checked information; the part of errors before information checking; the mean check speed; the frequency of check errors of the first kind (errors of checker); the mean time of checker attention concen-tration keeping; the mean permanent time during checking.

Designation of customer requirements: the admissible mean time of checking; the admissible probability of information faultlessness after checking.

Model of users and system staff random faults emergence and elimination ("Faultlessness of staff and users actions").

Designations of CEISOQ input: the frequency of failures for the k-th unit of the n-th subsystem; the mean time to system repair.

Designations of customer requirements: the required permanent time of reliable IS operation; the admissible probability of faultless staff and users actions during re-quired time.

Designations of evaluated values: the mean time between failures for the n-th sub-system; the probability of reliable the n-th subsystem performance during required time; the mean time between failures for complex composed by 1,...,n subsystems during required time; the probability of faultless performance of complex composed by 1,...,n subsystems during required time; the probability of faultlessness mainte-nance of IS staff and users actions during required time.

Model of computer viruses influences ("Viruses influences").

Definition. Information system is protected against potential computer viruses if viruses have not influenced during given period.

There are cheated the models for protection based on antivirus diagnostic, security monitoring and their combination.

Designations of input for protection based on antivirus diagnostic: the frequency of virus penetration into system; the mean virus activation time for influencing; the frequency of antivirus diagnostics; the mean diagnostic time.

Note. These considerations are right: used diagnostic tools allow to detect all vi-ruses or tracks of their influences and repair tools allow to provide the restoration of broken system integrity.

Customer requirements: the given period for IS security operation; the admissible probability of IS security operation during the given period.

Output: the probability of IS security operation during the given period.

Model of information and software resources unauthorized access ("Unauth-orized access").

Designations of CEISOQ input: the mean possible time for overcoming the m-th barrier; the frequency of regulated parameter value of the m-th barrier; the mean time between failures (or change) of protection devices of the m-th barrier (for non-regulated parameters).

Designations of customer requirements: the admissible probability of IS security maintenance against an unauthorized access.

Designations of evaluated values: the probability of the m-th barrier overcoming by violator; the probability of security maintenance by the 1,2,…m barriers; the probability of IS security maintenance against an unauthorized access.

Model of information confidentiality maintenance ("Confidentiality").

Designations of CEISOQ input: the mean confidentiality objective period (additionally to the input for the previous model).

Designations of customer requirements: the admissible probability of information confidentiality maintenance.

Designations of evaluated values: the probability of the m-th barrier overcoming prior to the confidentiality objective period has ended; the probability of information confidentiality maintenance by 1,…,m barriers prior to the confidentiality objective period has ended; the probability of IS information confidentiality maintenance prior to the objective confidence period has ended.

3   Examples

Example 1. Specialists know a man (as user or staff) is the most weak place for majority IS, because it is beyond human physical strength to act absolute correct all time. However few people conceive how much is human weakness in quantitative values. The computation by CEISOQ showed that for the file of 4-5 users (deciding one task in consecutive order and mistaking one per several hours) the probability of faultlessness users actions during day is practically zero. Remember your activities, please, there were untimely information, a great number of errors in documents prepared by inexperienced workers and so on! May be the CEISOQ model and computed zero allow to find the explanation of these matters. Indeed, for this reason experienced customers require to develop the special software "protection against fool".

Nobody will be accessed to work on a building without training in "safety measures". It needs to safe the workers life and to stave off the potential material damage. At the same time there are not compulsory qualification course for providing the information quality. And there are not one because the practical evaluations of exposing to danger for IS operation are often absent… Really, there are not dangers for human life and health. But the difficulty evaluated damage for output information is caused with probability approaching to unity. It exists natural principal question: what about the admissible level of users and staff training? Understandable that the answer for each IS will be different and may be clear only after modeling of users and staff actions and after evaluating the weight of their faults in the damage set for IS operation quality. The offered complex CEISOQ permits to get required system evaluations. Its application allows to evaluate more deeply and other man possibilities too.

Example 2. Side by side with this the same user or staff-man is the valuable IS component. Really, solely a man is able to reveal the semantic part of information, including in checking process. In practice there may be strong requirements to the information check deadline, abnormal conditions for labor and relaxation (for instance, in IS for control in emergence situations, in military systems etc). Could checker reveal all existing errors? Moreover, could he commits no own errors? So, the messages of one special real time system were been from 250 to 500 symbols in size, and their intensity was been about once per 20-30 minutes. And check time was been required less than 1-2 minutes for deadline. The developers had a doubt about necessity of special software for check support. They supposed that visual check will be quite sufficient. The customers required additional guarantee of faultlessness, because the check entrusted to users. But the users were not the checkers, they were only the specialists in making decisions. The computation by CEISOQ confirmed the medium-level users is quite enough for providing information faultlessness with the probability more than 0.95 for the messages no less than 250 symbols in size. At the same time for the checked messages about 500 symbols this probability decreased to 0.83-0.86 (as a result of strong requirements to possible check deadline), that was been inadmissible for reaching the purposes of the system operation. On the base of CEISOQ investigation results there were been forbidden the messages more than 250 symbols in size. And developers evaded wasted expenses. Moreover, there was been foreseen only plan training (no special training) for users.

Example 3. It is considered today that computer viruses present the most danger not only through suddenness and incomprehensibility of its influence but also mainly on account of one insufficient study. Really, full protection is guaranteed against any viruses only for the primordially "pure" (i.e. without "bugs") and closed systems into which some viruses penetrations are practically excluded (for instance into cash-apparatuses). And what about protection effectiveness for the systems with the possible channels for viruses penetration? There needs already to compute... Indeed any protection means the waste of resources and, as result, deterioration of the probabilistic and temporal characteristics of IS operation. To date there are two well-known modes of technological protection: "antiviruses diagnostic" and "monitoring of security". When are using mode "antiviruses diagnostic" integrity of information and software resources is the subject checking. Viruses may penetrate and influence between the neighboring diagnostics. The disturbed integrity is restored by suitable tools. When is using mode "monitoring of security" all suspicious integrity changes of information and software resources are due to be analyzed (by the automatic or automatized modes). The offered version of CEISOQ allows to evaluate the "antivirus diagnostic" effectiveness. So according to computations the daily diagnostics may provide system protection with the probability 0.90-0.96 against the accidental viruses befallen more seldom than once per week and made active about a several hours after penetration. At that time, if the frequency of viruses befalling will draw near once per day the probability of the antiviruses protection will be no more than 0.76. One might as well to think that for reaching the system purpose under the viruses attacks conditions it may be expediently to implement on duty system performance mode...without using infected IS The conception of viruses attacks should be defined and evaluated though. Indeed, if we can evaluate the potential threats, the IS protection possibilities than quite really to substantiate such viruses penetration frequency for which the probability of the IS secure operation will be less than admissible. The computed frequency will give an idea about the dangerous viruses attack characteristic. Software tools CEISOQ allow to cope with this task.

Example 4. Let a hypothetical SYSTEM of environment monitoring and control providing in emergency (including natural calamities, acts of terrorism and other) is due to be developed. And IS will be an important brain component of such SYSTEM. It is clear that effectiveness of taking measures in emergency, people safety and environment security are depended on IS operation quality (on information producing reliability and timeliness and on information completeness, validity and confidentiality). There is considered the system purposed for deciding the task set during a several days under the conditions of emergency. Objectively indeed for this period there is required to provide information confidentiality in the mobile command-staff car of the system. After task fulfillment the car should come back into the restricted hangar for keeping without the power supply until the next tasks appearing.

Traditional approach for information confidentiality providing consists in engineering fulfillment the standards requirements (for instance "Trusted Computer Security Evaluation Criteria"). At the same time under the cost limitation the expenses for neutralization of threats to confidentiality should be substantiately commensurable with the expenses for neutralization of other threats to IS operation quality. So, let customer requirements for IS operations during an emergency period are the next (in the part of confidentiality): IS protection technology should provide information and software resources integrity, confidentiality and authorized accessibility. In particular:

the probability of an unauthorized access prevention should be more than 0.999;

the probability of the information confidentiality maintenance should be more than 0.9999 during any permanent period no longer than seven days.

Let there are considered the next barriers of IS protection (see Table 2). An analysis of computed dependencies (see Fig.2 for barrier 10) shows:

- the strict maintenance of periodic changes regulated parameters value provides a higher level of IS protection against unauthorized access (for instance, Pprot.1...6 =0.87) in comparison with aperiodic changes (Pprot.1...6 =0.78). At that when Pprot.fi 1 this advantage is being inessential (0.997 against 0.991 for m=1,...,10). But it is true if change period and a time of changes are keeping as confidential. Therefore for m=1,2 periodic changes are more effective against inexperienced violators. And changes for other m‡2 should be aperiodic;
- the use of changes passwords allows to increase the protection probability from 0.21÷0.25 to 0.78÷0.87. At that the level of IS security by 1-6 barriers is low;
- the introducing of 7,8,9 barriers is practically useless;
- the use of cryptography allows to increase the probability to 0.991, but it does not satisfy high customer requirements.

In comparison with unauthorized access confidentiality maintenance should be provided during any 7 days. An analysis of computed dependencies (see Fig. 3) shows:

- in comparison with the result of previous subchapter the use of 6 barriers provides $P_{conf.1...6}$=0.992. It is more than $P_{prot.}$ for 10 barriers(!);
- the use of all 10 barriers provides required confidentiality, $P_{conf.}$ =0.99996. It decreases practically customer risk in IS security providing. This unexpected results should be understood by customer!

Table 2. Input for CEISOQ

Barrier	The frequency of changes	The mean time for barrier overcoming	Possible way overcoming
1. Guarded territory	Once per 2 hour	30 min.	Concealed penetration through guards
2. Parole for coming into office	Once per day	10 min.	Documents forgery
3. Block key for computer switching	1 one per 5 years (MTBF = 5 years)	1 week	Theft, collusion, forced confiscating
4. Password for input into system	Once per month	1 month	Collusion, forced extortion, spying, password decoding
5. Password for access to a program devices	Once per month	10 days	Collusion, forced extortion, spying, password decoding
6. Password for access to an information resources	Once per month	10 days	Collusion, forced extortion, spying, password decoding
7. Registered device for information recording	Once per year	1 day	Theft, collusion, forced confiscating
8. Confirmance of user authenticity during a computer seance	Once per month	1 day	Collusion, forced extortion, spying
9. Television monitoring	Once per 5 years (MTBF = 5 years)	2 days	Collusion, disrepair imitation, force roller
10. Cryptography	1 key per month	2 years	Collusion, deciphering

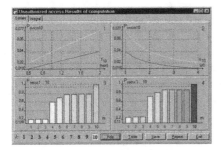

Fig. 2. Results of computation

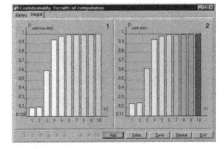

Fig. 3. Results of computation

Resume

An application of modeling software tools CEISOQ appears principal new potentiality for system analysts on the base of prompt analyzing thousands computed dependencies. It may reduce essentially hazards and unwarranted expenses and increase a scientific substantiation of customers, developers and users engineering decisions. At that expenses for investigation reduce by tens.

References

1. Kostogryzov, A.I.: Mathemetical: Models for Complex Evaluation of Information Systems Operation Quality. Proceeding of the 10th International Conference on Computing and Information, Kuwait (2000). Published by LNCS (2000)
2. Bezkorovainy, M.M., Kostogryzov, A.I., Lvov, V.M.: Modeling Software Tools Complex for Evaluation of Information Systems Operation Quality. "Sinteg", Moscow (2000) 114 http://www.ceisoq.ru
3. Klimov, G.P.: Probability theory and mathematical statistics. SU, Moscow (1983) (in Russian)
4. Kostogryzov, A.I., Petuhov, A.V., Scherbina, A.M.: Output information quality evaluation, providing and increasing foundations for automatized system. "Armament, Policy, Conversion ", Moscow (1994) (in Russian)

Analyzing Separation of Duties in Petri Net Workflows

Konstantin Knorr and Harald Weidner

Department of Information Technology
University of Zurich
CH-8057 Zurich
{knorr,weidner}@ifi.unizh.ch

Abstract. With the rise of global networks like the Internet the importance of workflow systems is growing. However, security questions in such environments often only address secure communication. Another important topic that is often ignored is the separation of duties to prevent fraud within an organization. This paper introduces a model for separation of duties in workflows that have been specified with Petri nets. Rules will be given as facts of a logic program and expressed in propositional logic. The program allows for simulating and analyzing workflows and their security rules during build time.

Keywords. Logical programming, Petri net, separation of duties, workflow

1   Introduction

It is well known that many computer related criminal activities are performed by insiders (Anonymous 1985, CSI 1999). One of the most prominent threats is fraud that is particularly difficult to detect in computerized environments such as workflow systems. Therefore it is of great importance to implement mechanisms to prevent such illegal activities. Separation of duties (SoD) has been identified by many authors as an efficient mechanism to prevent fraud within organizations (Ahn and Sandhu 1999, Bertino et al. 1999, Bussler 1995, Sandhu 1990, Stormer et al. 2000). It is in particular useful when applied to dynamic processes such as workflows. The physical and logical separation of tasks and their subjects can improve the prevention of fraudulent activities.

We present in this paper a simple logical representation for SoD rules that will restrict the execution of tasks by subjects, thus implementing a dynamic separation of duties. The SoD analysis proposed will be done in two steps: in a first step, logical programming (Prolog) will be used to calculate all valid execution chains of a workflow. In a second step, the result of the first step is used for further analysis (e.g. workload). A Petri net (PN) will express syntax and semantics of the business processes under consideration. The formal approach has the advantage of being analyzable before implementing a given workflow in a business environment. The specification can be searched for potential flaws like deadlocks, non-reachable end-states, and contradicting SoD rules.

V.I. Gorodetski et al. (Eds.): MMM-ACNS 2001, LNCS 2052, pp. 102–114, 2001.

The remainder of the paper has the following structure: Section 2 gives background information on workflows, roles, SoD, PNs, and PN workflows. Section 3 introduces an example that is used for illustration purposes throughout the paper. A formal SoD model is described in Section 4. Section 5 uses logical programming to analyze SoD rules of a business process. Section 6 discusses the findings of the paper, contrasts them to related work, and mentions further research topics.

2   Background

2.1   Workflows

Workflow management is an essential research area in computer science. It is an emerging technology used to automate and 'streamline' frequent business processes. This approach is especially useful for processes that rely on electronic documents because the workflow management system (WfMS) can administer and process the data objects. A workflow is an executable and computer-understandable business process whose administration, modeling, and execution is supported by a software system called WfMS. Before a workflow can be executed, it has to be described in a way, the WfMS is able to understand. This description is called a workflow specification. The definition is made during build time, i.e. before a workflow can be executed. During run time of the system many instances of the workflow are generated according to the specification. The main elements of a workflow specification are
- tasks,
- subjects,
- roles,
- and the control flow.

A workflow consists of several tasks whose chronological and logical order is given by the control flow. To describe a task, it has to be specified which subjects/roles are allowed to execute the task. Subjects can be associated with persons but also with machines and computer programs. For more comprehensive information see Cichocki et al. 1998 and Georgakopoulos et al. 1995.

2.2   Roles

In a workflow environment, usually tasks are not linked directly to subjects. The concept of roles forms a middle layer between subjects and tasks. Roles allow for tying the execution of a task to a specific group of persons with specific skills or qualifications. When a representative of a role leaves the company, task definitions can remain unchanged. Just the role has to be 'untied' from that person (Lawrence 1993).

An important field of applying the role concept is security. Access rights are enforced on roles and not on subjects which simplifies and cheapens security administration (cf. Proceedings of the 5th ACM Workshop on Role-based Access Control).

2.3   Separation of Duties

"SoD is a policy to ensure that failures of omission or commission within an organization are caused only by collusion among individuals and, therefore, are riskier and less likely, and that chances of collusion are minimized by assigning individuals of different skills or divergent interests to separate tasks" (Gligor et al. 1998).

SoD therefore is an important security mechanism to prevent fraud within an organization. Clark and Wilson (1987) stress the importance of SoD mechanisms in commercial settings. Within a business process context SoD rules express task dependencies and should be part of a company's security policy. Only recently the combination of workflow management and SoD has found considerable interest in the research community (Bussler 1995, Bertino et al. 1998, Stormer et al. 2000). In a workflow context, SoD has to be divided and extended into static SoD (SSoD) and dynamic SoD (DSoD). SSoD enforces certain rules during build time of the workflow. SSoD rules are therefore part of the workflow specification. In contrast, DSoD is enforced during run time. Examples: In a workflow different roles are specified for different tasks (SSoD). In a travel expenses workflow a manager should not be allowed to apply for the reimbursement of (task 1) and approve (task 2) his own travel expenses (DSoD). For the remainder of this paper we will concentrate on DSoD and call it SoD if not otherwise stated.

2.4   Petri Nets

Petri nets (PN) originated with Carl Adam Petri (Petri 1962). Nowadays, their theory and application is a vast research area with many publications. This section gives the basic definitions of Place/Transition Nets with arc weights equal to 1 (Reisig 1985).

Definition 1 (Petri net).   A Petri net N is a triple $N = (P, T, F)$. P is the finite set of the places, T the finite set of the transitions with $P \cap T = \emptyset$. The flow relation F is defined by $F \subseteq (P \times T) \cup (T \times P)$. Let $y \in P \cup T$. $\bullet y$ is called the preset of y and is defined by $\bullet y := \{x \in P \cup T \mid (x, y) \in F\}$. $y \bullet$ is called the postset of y and is defined by $y \bullet := \{x \in P \cup T \mid (y, x) \in F\}$. An element of $\bullet y$ and $y \bullet$ is called an input place/transition and output place/transition, respectively.

Figure 1 shows an example of a Petri net consisting of the places $p_1, \dots, p_6$ and the transitions ay, a1, a2, tf. The flow relation F equals to $F = \{(p_1, ay), (ay, p_2), (ay, p4), (p2, a1), (a1, p3), (p3, tf), (p4, a2), (a2, p5), (p5, tf), (tf, p6)\}$. The graphical interpretation of a Petri net is a bipartite graph, i.e. places can only be connected to transitions, transitions can only be connected to places. Places are represented graphically as circles, transitions as rectangles. The graphical interpretation of an element (x, y) ? F is an arrow from x to y. The preset and postset of a transition are sets of places — possibly empty. The preset and postset of a place are sets of transitions — possibly empty, too. Some presets and postsets from the example are $\bullet tf = \{p3, p5\}$, $\bullet p_1 = ?$, and $tf \bullet = \{p6\}$.

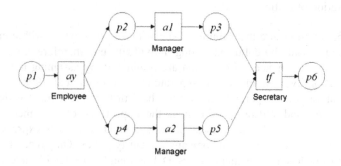

Fig. 1. Example of a Petri net

Definition 2 (Behavior of Petri nets)   A marking M is a mapping M:P → N_0 that associates with each place a number of tokens. A transition t is called activated under marking M, if M(p)>0 for all p ∈ t. An activated transition can fire. If a transition t fires, M changes to M^* such that $M^*(p)=M(p)- (p,t)+ (t,p)$ for every place of the net, where (x,y)=1 if (x,y) ∈ F and (x,y)=0 otherwise.

The behavior of Petri nets shall be illustrated following the example in Figure 1. Let the start marking be M(p1)=1 and M(p)=0 for all other places. A token is put on place p1. Transition ay is activated. If ay fires, the token from p1 duplicates and moves to p2 and p4. Transitions a1 and a2 are activated now. If a1 fires, the token from p2 moves to p3. Similarly, if a2 fires, the token from p4 moves to p5. tf can fire now, merging the tokens from p3 and p5 into a single token in p6. Now, there are no more activated transitions.

Definitions 1 and 2 give the 'classical' definitions for Petri nets. During the last years so called High Level Petri nets have been introduced. The major differences are that the tokens can be distinguished (Jensen 1992).

2.4   Petri Net Workflows

The fundamental idea behind the connection of PNs with workflows is to associate activities in the workflow with transitions in the PN. Through this association, the execution of an activity can be interpreted as the firing of a transition. The flow relation of the PN gives the control flow of the workflow. The marking of a PN can be associated with the state of the workflow. Start and end activities are derived from the start and end marking of the net. Subjects and roles are interpreted as attributes of transitions (Knorr 2000).

Definition 3 (Petri net workflow).   A Petri net workflow is a workflow whose tasks and control flow are specified through a Petri net where T is the set of tasks/transitions. Additionally, R is the set of roles and S the set of subjects. Function f_{TR}: T $\quad 2^R$ assigns a set of roles to each task, f_{SR}: S $\quad 2^R$ a set of roles to each role [1].

Petri net workflows offer the following advantages: The 'correctness' of their static and dynamic properties can be proven. There are various publications on this topic (Adam et al. 1998, Kindler and van der Aalst 1999, van der Aalst 1997). Note that it is possible to perform a validity check of the specification before run time, preventing any inconsistencies during execution resulting from an erroneous workflow specification. Next to their formal definition, PN workflows have an intuitive graphical interpretation that is well suited to express the dynamic nature of a workflow. Furthermore, PN could be used to standardize workflow specifications, an ongoing project led by the Workflow Management Coalition[2].

3   Sample Business Process

The last section introduced several concepts on an abstract level. This section gives an example of a business process that will be used for illustration purposes throughout the remainder of the paper. The business process deals with travel expenses reimbursement. Figure 1 shows the associated PN workflow. The workflow consists of four tasks: in a first task (ay), an employee applies for the reimbursement of his travel expenses by filling out an application form. Two managers have to approve this report (tasks a1 and a2). Finally, based on the approval of the managers a secretary will transfer the money to the employee's bank account (tf). Note that an instance of this workflow is created for every travel of an employee.

Formally, the set T of all transitions is T={ay, a1, a2, tf} and the set of roles R encompasses the manager, secretary and employee role, thus R={emp, man, sec} if the roles are abbreviated. The role associated with each task in Figure 1 is written under the corresponding transition (e.g. f_{TR}(ay)={emp}). Let the subjects of the company be Carpenter, Butcher, Snyder, Fisher, and the brothers A. Smith and B. Smith, thus S = {car, but, sny, fis, sma, smb}. All six subjects are employees, Carpenter, Butcher and B. Smith are managers, Fisher and Snyder are secretaries (therefore f_{SR}(fis) = {emp,sec}). Note that every manager (or secretary) is an employee, too. The partial order of roles builds up a so called role hierarchy.

Now we are ready to give SoD examples. The process definition in Figure 1 requires that different tasks are performed by different roles (the tf task by a secretary and the a1 task by a manager). That corresponds to SSoD rules. However, DSoD is of greater interest in a workflow context. The following rules are reasonable for this specific business process:

- A manager should not be allowed to approve his/her own travel claim.
- B. Smith should not be allowed to approve the claim of his brother A. Smith.
- A secretary should not be allowed to transfer the refund of his/her own travel expenses.

---

[1] In this definition 2^A denotes the power set of A – the set of all subsets of A.
[2] http://www.wfmc.org

- A manager should not be allowed to perform both approval tasks in the same workflow instance.

These examples clearly state the need for a formal model for SoD that takes into account the state of the underlying business process, which is the topic of the following section.

4   Formalizing Separation of Duties

This section introduces a SoD model based on PNs. For this purpose, the rules of a business process are stored in a rule base (RB).

Definition 4 (SoD rule base). RB is a set of entries (rules) of the form

$$(s1, t1) \qquad (s2, t2) \tag{1}$$

where $(s1,t1),(s2,t2) \quad S \quad T$ (S is the set of subjects and T is the set of tasks). A rule says that if subject s1 has done task t1 then s2 must not do t2.

Next to this separation of duties, a delegation of duties is also imaginable: $(s1, t1)$ $(s2, t2)$. If subject s has done a first task it might be reasonable that s performs a second task later on. Nevertheless, we restrict ourselves to separation since a delegation can be expressed by excluding all other subjects from performing the second task:

$$(s1, t1) \quad (s2, t2) \qquad [\, (s1, t1) \qquad \neg\, (x, t2) \quad x \quad S \setminus \{s2\}\,] \tag{2}$$

This equivalence will generate a large number of SoD rules. For a discussion of the complexity of the SoD analysis see Section 6.

   Not all rules given through (1) are meaningful. Therefore, the notion of soundness is introduced.

Definition 5 (Soundness). A rule in RB is sound if and only if
1. $f_{TR}(t) \quad f_{SR}(s)$ holds for both tuples in equation (1), i.e. the subject matches the role of the task.
2. $t1 < t2$ holds in equation (1), where '<' indicates the partial order given by the PN[3]. SoD rules can only be enforced based on the 'history' of the workflow.

RB is sound if all its rules are sound.

A sound RB of the example could contain the following rules (cf. the sample rules at the end of Section 3):
- (smb,ay) $\neg(smb,a1)$ and (smb,ay) $\neg(smb,a2)$ plus the same rules for the two other managers Carpenter and Butcher.
- (sma,ay) $\neg(smb,a1)$ and (sma,ay) $\neg(smb,a2)$.

---

[3]   The second postulation in Definition 5 requires a partial ordering of the tasks. This only holds for a special class of PNs — loop-free nets. For PNs with loops the 'firing history' of the net has to be taken into account which makes the SoD rules more complex. We therefore refrain from this approach.

- (fis,ay)   ¬(fis,tf) and (sny,ay)   ¬ (sny,tf).
- (smb,a1)   ¬(smb,a2) and (smb,a2)   ¬ (smb,a1) plus the same rules for the other managers.

Definition 6 (Execution chain). Given a start and an end marking. An execution chain of a PN workflow is a list of subject/task pairs whose execution (firing of the associated transitions by a legitimate subject) transforms the start marking into the end marking. If there is no contradiction to the SoD RB, the execution chain is called SoD valid or just valid.

Examples from our sample: Let the start marking be $M(p1)=1$ (the marking of all other places 0), the end marking be $M(p6)=1$ (the marking of all other places 0), and L1, L2, L3 be three lists with L1=[(fis,ay), (sma,a1), (but,a2), (sny,tf)], L2=[(fis,ay), (but,a1), (but,a2), (fis,tf)], and L3=[(fis,ay), (car,a1), (but,a2), (sny,tf)]. L1 is no execution chain since A. Smith is no legitimate subject for one of the approve tasks (a manager is needed). L2 is an execution chain but not SoD valid since Fisher transfers his/her own travel expenses and Butcher does both approval tasks. L3 is SoD valid.

5   Analyzing SoD Rules

In this section, we use logical reasoning to analyze the SoD rules that have been specified as part of a business process' security policy. The purpose is to find all SoD valid execution chains of a PN workflow given a sound SoD RB (cf. Definitions 3-6).

We make use of Prolog (<u>Pro</u>gramming in <u>log</u>ic) as the most popular logical programming language, to be specific GNU Prolog Version 1.1.2 by Daniel Diaz (Diaz 1999). For a comprehensive introduction to logical programming see O'Keefe (1990) and Hogger (1990).

Prolog allows for representing knowledge in a declarative, explicit, and machine independent way. A logical program is compact and flexible because it resembles more a specification than a traditional computer program. The result of a Prolog program is a logical consequence of its input (facts) using resolution with its strict mathematical background. A drawback in the use of Prolog is efficiency. Also, most implementations of Prolog mix declarative and procedural structures. Nevertheless, we will use Prolog in the first processing step due to the advantages stated above.

Figure 2 shows that our approach is divided into two processing steps. The first step uses two different input sets: the PN workflow specification and a sound SoD rule base. Subsections 5.1, 5.2 and 5.3 show how to represent these sets in Prolog in case of our example. The output of this first processing step are all valid execution chains (Section 5.4). In a second processing step, these chains are used as input for further analysis such as workload management. Nevertheless, the focus of this paper is on processing step 1.

To clarify the following program listings, some words about Prolog notation: The number of arguments of a predicate is called arity and is added with a slash at the end of the predicate (e.g. fire/3   ). Names of variables start with capital letters, constants with lowercase letters. Lists are written in square brackets and we use commas to separate the individual elements (L=[p2,p4]   ). Queries start with ?- . For more information see the GNU Prolog Manual (Diaz 1999).

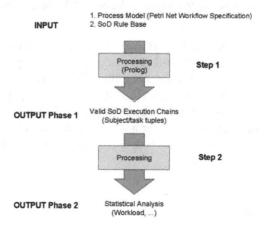

Fig. 2. SoD analysis during build time

5.1   Petri Nets in Prolog

According to Definition 1 a PN consists of places, transitions, and the flow relation. Those are represented through the predicates transition/1   , place/1   , and flow/2   . Listing 1 shows the facts of our sample business case. The two predicates preset/2   and postset/2   give the preset and postset of a transition and return the corresponding marking (bagof/3   is a build-in predicate that stores all possible unifications of a variable of a given expression in a list). A marking is represented as a multi set of places that in Prolog is represented as list L. Example: ?- post-set(ay,L)   results in L=[p2,p4]   . The flow/2   -facts give the flow relation, i.e. define which places are connected to which transitions and vice versa. Example: (p4,a2) is an element of the flow relation, therefore flow(p4,a2)   is in the listing.

The dynamic properties of a PN are expressed through the rule fire/3   . In fire(T,L1,L2)   it is checked if the firing of transition T changes marking L1 to L2. This is done in several steps:

- First of all, T has to be a transition.
- For T to be activated, the preset of T has to be a sublist (subset) of L1. The sub-list/2   predicate (not shown in the listing) is used which returns true if L is a sublist of L1.

- If T fires, the marking changes: The preset of T is 'subtracted' (setminus/3   )
  and the postset of T is added (append/3   ). The new marking is returned in the
  variable L3.

Examples :   ?-fire(ay,[p1],L)   results in L=[p2,p4]   , ?- fire(T,
[p2,p4],[p3,p4])   results in T=[a1]   , while ?-fire(ay,[p2,p4],[p6])
results in failure.

```
transition(ay). transition(a1). transition(a2). transition(tf).
place(p1). place(p2). place(p3).place(p4). place(p5). place(p6).
flow(p1,ay). flow(ay,p2). flow(ay,p4). flow(p2,a1). flow(p4,a2).
flow(a1,p3). flow(a2,p5). flow(p3,tf). flow(p5,tf). flow(tf,p6).

preset(T,L):- bagof(X,flow(X,T),L). postset(T,L):- bagof(X,flow(T,X),L).
fire(T,L1,L2):- transition(T), preset(T,L), sublist(L,L1),
   setminus(L,L1,L3),postset(T,M), append(M,L3,L4),sort0(L4,L2).
```

Listing 1: Petri nets in Prolog

5.2   Organizational Facts

Roles and subjects are important parts of a workflow specification. Existing roles are
stored in the predicate role/1   , subject information in the predicate subject/1   .
The predicate play/2   allows for defining which subject can activate which role.
Note that three letter abbreviations are used (man stands for the role manager, fis   for
the subject Fisher). Finally, roles need to be tied to tasks (transitions) which is done
via the predicate execute/2   (cf. Listing 2). Note that play/2   and execute/2
correspond to the functions f_{SR} and f_{TR} defined in Definition 3.

```
role(man). role(sec). role(emp).
subject(sma). subject(smb). subject(but).
subject(car). subject(fis). subject(sny).
play(sma,emp). play(smb,emp). play(but,emp). play(car,emp).
play(fis,emp). play(sny,emp). play(smb,man). play(car,man).
play(but,man). play(sny,sec). play(fis,sec).
execute(man,a1). execute(man,a2).
execute(emp,ay). execute(sec,tf).
```

Listing 2: Organizational facts of the sample process

5.3   SoD Data Base

The predicate separate/2   defines which task/subject tuples should be separated
from others. Listing 3 shows the SoD RB for our sample process. The first six facts
state that managers must not approve their own claim. The next two facts prevent B.
Smith from approving the claim of his brother. Subsequently, the following two facts
make sure that secretaries do not transfer the money for their own travel expenses.
Finally, the last six facts state that no manager can perform both approval tasks within
the same workflow instance.

```
% Managers must not approve their own claim
separate(perform(smb,ay),perform(smb,a1)).
```

```
separate(perform(smb,ay),perform(smb,a2)).
separate(perform(car,ay),perform(car,a1)).
separate(perform(car,ay),perform(car,a2)).
separate(perform(but,ay),perform(but,a1)).
separate(perform(but,ay),perform(but,a2)).
% Brother example
separate(perform(sma,ay),perform(smb,a1)).
separate(perform(sma,ay),perform(smb,a2)).
% Secretaries must not transfer their own money
separate(perform(fis,ay),perform(fis,tf)).
separate(perform(sny,ay),perform(sny,tf)).
% Managers must not do both approves
separate(perform(smb,a1),perform(smb,a2)).
separate(perform(smb,a2),perform(smb,a1)).
separate(perform(car,a1),perform(car,a2)).
separate(perform(car,a2),perform(car,a1)).
separate(perform(but,a1),perform(but,a2)).
separate(perform(but,a2),perform(but,a1)).
```

Listing 3: SoD database for the sample business process

5.4   Analysis Predicates

The basis of the SoD analysis are the Prolog predicates generate_chain/3   and
check_chain/1   . Listing 4 shows the corresponding Prolog code. The first predicate generates all possible execution chains. The definition of generate_chain/3
uses recursion — a very powerful programming technique in Prolog. The initialization
utilizes the empty list if the markings L and M are permutations of each other. In the
body of the recursion it is checked whether subject S is able to activate the correct role
and the start marking L_s activates transition T.

The predicate check_chain/1   checks if an execution chain is SoD valid. The
predicate is initialized with a single list entry perform(S,T)   where S has to be a
subject and T a transition. Then — starting with the first element — all possible separation tuples (second tuple in the separation facts) are stored in the list L. If this list is
disjunct (checked by the predicate disjunct/2   not shown in Listing 4) from the
rest of the execution chain (Tail   ), the predicate succeeds, otherwise fails. Finally, the
predicate valid_chains/3   returns all valid SoD chains in the variable Valid-
Chain .

```
generate_chain(L,M,[]):- permutation(L,M).
generate_chain(L_s,L_e,[perform(S,T)|L_v]):-
   play(S,R), execute(R,T),
   fire(T,L_s,L_t), generate_chain(L_t,L_e,L_v).
check_chain([perform(S,T)]):- subject(S), transition(T).
check_chain([H|Tail]):- bagof(X,separate(H,X),L),
   disjunct(L,Tail), check_chain(Tail).
valid_chains(StartM,EndM,ValidChain):-
   generate_chain(StartM,EndM,L),
   check_chain(L), ValidChain is L.
```

Listing 4:   Prolog predicates used for SoD analysis

6   Discussion and Further Work

6.1   Discussion

The Prolog program discussed in Section 5 generates 28 possible execution chains:

1. [perform(sma,ay),perform(car,a1), perform(but,a2),perform(fis,tf)]
2. [perform(sma,ay),perform(car,a1), perform(but,a2),perform(sny,tf)]
...
28. [perform(sny,ay),perform(car,a1), perform(but,a2),perform(fis,tf)]

A first analysis yields the following:

- For the initial marking $M(p1)=1$ and final marking $M(p6)=1$ (all other 0) there are no deadlocks. For every employee in the example, the travel expenses workflow can be finished. No workflow has to be canceled due to too restrictive SoD rules. Although this statement may seem obvious for the example, in a more complex setting it is not.
- Every workflow requires four different persons for its execution. Without the SoD rules two persons would be sufficient (e.g. first three tasks by a manager, last one by a secretary).
- A further analysis of the valid execution chains can be used to do a 'workload management'. In the example, the managers Carpenter and Butcher have more work because of the 'brother rule' (cf. Table 1).
- The analysis shows that RB of the example contains no contradicting rules.

Some words about the complexity of the analysis: The Prolog program performs a depth first search for all SoD valid execution chains of the workflow. The complexity of the program grows with the number of places/transitions of the PN, SoD rules, subjects, and roles. Consequently, the calculation for larger business processes can be very demanding concerning time and resources. We clearly understand the need for reduction techniques. Nevertheless, the analysis is designed to be done during build time of the business process, so it is not time critical. The proposed Prolog program uses the generate/test paradigm (cf. the vaild_chains/3   predicate in Listing 4). On the one hand this makes the code more comprehensible, on the other hand it is very inefficient because the generate and check predicates could be combined to cause an earlier back-tracking.

With minor modifications, the Prolog code can be used during run time, too. In most WfMSs, when a subject logs into the system, a work list is generated that shows all pending tasks for this subject. The generation of the work list can be supported by our Prolog code. With the predicates defined in Section 5, a new predicate complete_chain/1   could be defined to complete partial execution chains. Example:

?-complete_chain([perform(fis,ay),perform(car,a1),X1,X2])

would result in X1=perform(but,a2), X2=perform(sny,tf)   plus all other possible completions. Note that usually a WfMS manages numerous business processes and multiple instances of every process. Therefore an extension of the tuples in the execution chains with an instance and process identification number is necessary. Furthermore, in this case the computation would be time critical.

Table 1. Statistical analysis of the sample workflow. N(s,t) is the number of valid execution chains in which subject s performs task t

N(s,t)	sma	smb	car	but	sny	fis
ay	4	4	4	4	4	4
a1	0	8	10	10	0	0
a2	0	8	10	10	0	0
tf	0	0	0	0	14	14

6.2 Related Work

This paper introduced a SoD model based on Petri net workflows that was analyzed through a logic program. There are more complex SoD models. Ahn and Sandhu (1999) introduced RSL99 (role based separation of duties language) and Bertino et al. (1999) proposed a comprehensive language to specify and enforce authorization constraints. Going beyond our model, these languages can express constraints such as 'a manager can only execute a specific task five times'. Furthermore, use permissions for operations on objects are an important issue.

Nevertheless, our rules are sufficient to 'catch' many of the more prominent SoD rules in today's business processes, our model is on a comprehensible level, and — most important — the other languages are not based on Petri nets.

6.3 Future Work

Future work will focus on the following issues:
- Knorr and Stormer (2001) introduce a tool that allows for graphically modeling SoD rules on top of an existing process definition and analyzing them as proposed in this paper. Interfaces will be provided for existing process definitions. Security officers of a company can then edit, model, simulate and analyze a security policy prior to run time of their processes.
- Our SoD model can be extended in several directions: (1) Within a task that requires access to different data items, certain privileges may be prohibited by the security policy of a company. E.g. a subject could not be allowed to change a document that he/she created in an earlier task. (2) As indicated earlier, our SoD model depends on loop-free PNs. For general PNs, the SoD model has to take into account the history of task activations. (3) More complex SoD rules are imaginable: A subject could be eligible for a third task if he/she performed a first and not a second task. Furthermore, role hierarchies and inheritance of privileges are important features in commercial settings. A second task could only be performed by a role higher (or lower) in a hierarchy than a first task.
- An empirical study — preferably in a large company — has to show if our model is sufficient for real business processes and security policies.

References

1. van der Aalst, W.M.P.: Verification of Workflow Nets. In: Proc. of Application and Theory of Petri Nets, LNCS 1248, Springer. (1997) 407–426
2. Anonymous: Internal Security, PC Week, 18(2) (May 1985) 89–91
3. Adam, N.R., Atluri, V. and Huang, W.-K.: Modeling and Analysis of Workflows Using Petri Nets. Journal of Intelligent Information Systems (10:2), (March 1998) 131–158
4. Ahn, G.-J. and Sandhu, R.: The RSL99 Language for Role-based Separation of Duty Constraints. In: Proc. of the Fourth ACM Workshop on Role-Based Access Control, Fairfax, VA, (October 28–29, 1999)
5. Bertino, E., Ferrari, E., and Atluri, V.: The Specification and Enforcement of Authorization Constraints in Workflow Management Systems. ACM Trans. on Inf. and Sys. Sec., 2(1):65-104, (Feb. 1999)
6. Bussler, C.: Policy Resolution in Workflow Management Systems. Dig. Tech. J., 6(4) (1995)
7. Cichocki, A., Helal, A., Rusinkiewicz, M. and Woelk, D.: Workflow and Process Automation — Concepts and Technology, Kluwer Academic (1998)
8. Clark, D. and Wilson, D.: A Comparison of Commercial and Military Computer Security Policies. In: Proc. of the IEEE Sym. on Sec. and Privacy, Oakland, CA, (1987) 184–194
9. CSI (Computer Security Institute): Issues and Trends – 1999 CSI/FBI Computer Crime and Security Survey, http://www.gocsi.com/summary.htm
10. Diaz, D.: GNU Prolog (Version 1.1.2) Manual, Edition 1.1 (November 29, 1999)
11. Georgakopoulos, D., Hornick, M. and Sheth, A.: An Overview of Workflow Management, Distributed and Parallel Databases (3) (1995) 119–153
12. Gligor, V., Gavilla, S., and Ferraiolo, D.: On the Formal Definition of Separation-of-Duty Policies and their Composition. In: Proc. of the IEEE Sym. on Sec. and Priv. (1998)
13. Hogger, C.J.: Essentials of Logic Programming, Clarendon Press (1990)
14. Jensen, K.: Coloured Petri Nets — Basic Concepts, Analysis Methods and Practical Use, Volume 1, EATCS Monographs on Theoretical Computer Science, Springer (1992)
15. Kindler, E. and van der Aalst, W.M.P.: Liveness, Fairness, and Recurrence in Petri Nets, Information Processing Letters (70), (1999) 269–274
16. Knorr, K. and Stormer, H.: Modeling and Analyzing Separation of Duties in Workflow Environments, in: Proc. of 16th IFIP/SEC, Paris, France (June 11–13 2001)
17. Knorr, K.: WWW Workflows Based on Petri Nets, in: Proc. of the 9th Intl. Conf. on Information Systems Development, Kristiansand, Norway (2000)
18. Lawrence, L. G.: The Role of Roles, Computers & Security, (12) (1993) 15–21
19. R. O'Keefe: The Craft of Prolog, MIT Press (1990)
20. C.A. Petri: Kommunikation mit Automaten, PhD Thesis, Universität Bonn (1962)
21. Proceedings of 5th ACM Workshop on Role-Based Access Control, Berlin (July 2000)
22. Reisig, W.: Petri Nets — An Introduction, Springer (1985)
23. Sandhu, R.: Separation of Duties in Computerized Information Systems. In: Proc. of the IFIP WG 11.3 Workshop on Database Security, Halifax, UK, Sep. 1990
24. Stormer, H., Knorr, K. and Eloff, J.: A Model for Security in Agent-based Workflows. INFORMATIK / INFORMATIQUE. 6 (Dec. 2000) 24–29

Information Security with Formal Immune Networks

Alexander O. Tarakanov

St. Petersburg Institute for Informatics and Automation
39, 14th Liniya, St. Petersburg, 199178, Russia
tarakanov@togetherlab.nw.ru

Abstract. We propose a biological approach to information security based on a rigorous mathematical notion of formal immune network. According to our previous developments, such networks possess all the main capabilities of artificial intelligence system, and could be considered as an alternative to the wide spread artificial neural networks or intelligent agents. We consider also the main distinctions of our approach from the modern information security by agent-based modeling and artificial immune systems.

1   Introduction

Nowadays the natural immune system is treated by specialists as "the second brain of vertebrates" [3]. In fact, the immune system possesses all the main features of Artificial Intelligence (AI) systems: memory, ability to learn, to recognize and to make decision how to treat any macromolecule (antigen) even if the latter has never existed before on the Earth. Of especial interest for computer science is the widespread theory of immune networks, formed by the interactions between specific proteins (antibodies) of the immune system. The existence of such networks is established now beyond all doubts, because their fragments and interactions have been detected experimentally by molecular immunology. It is worth to note that almost the similar networks under the name of molecular circuits have been even proposed as a possible molecular basis of neuronal memory in the human brain [1].

Based on biological principles of immune system, there arises a new and rapidly growing field of Artificial Immune Systems (AIS), offering powerful and robust information processing capabilities for solving complex problems [4]. Like Artificial Neural Networks (ANN), AIS can learn new information, recall previously learned information, and perform pattern recognition in a highly decentralized fashion. AIS have already been applied in several specific problems, including information security, faults detection, vaccine design, control of robots, data mining, etc.

Among these applications, information security becomes increasingly important for everyday life. The matter is that the growing scale of computer networks and sophisticated software codes make them more and more vulnerable to alien intrusions, such as computer viruses, non-authorized access, intentional corruption, etc. Such intrusions could cause rather serious failures of computer-based information and control systems. The example of the well-known Y2K problem shows how deeply such failures could affect our society.

V.I. Gorodetski et al. (Eds.): MMM-ACNS 2001, LNCS 2052, pp. 115–126, 2001.

In the same time, currently used computer security systems show insufficient speed, reliability, flexibility and modularity to satisfy the modern requirements [11]. That is why AIS seem to be the most perspective way to accept the challenge of modern information security on the basis of the highly appropriate biological prototype.

In fact, computer viruses could be inferenced from J.von Neumann's studies of self-replicating mathematical automata in the 1940s. Although the idea of programs that could infect computers dates to the 1970s, the analogy between information security and biological processes was recognized in 1987, when the term "computer virus" was introduced by Adelman [7]. The idea of using immunological principles in information security started since 1994 when S.Forrest and her team have been working on a research project with a long-term goal to build AIS for computers. Nowadays several of such AIS are being under development, but all of them represent a set of heuristic algorithms, using ideas from genetic algorithms, ANN, agent-based modeling, etc.

However, there exists a strong need for a proper mathematical basis of AIS in general, and, especially, of AIS designed for information security. The problem is caused by very specific objects and interactions of immune networks, which differ remarkably from any of genetic algorithm, cellular automata, ANN, or intelligent agent. On the other hand, such mathematical basis could raise AIS up to the level of the widely spread ANN, and even allow to speak about hardware implementation of AIS in a new kind of computer – immunocomputer [15].

Thus, our paper is intended to fulfill the existing gap. Our general goal is a rigorous mathematical basis of immune networks intended for information security assurance. This goal can be accomplished by developing the novel mathematical notion of formal immune network [15] and its application to the field of information security. We consider also main distinctions of immune networks from modern information security approaches by agent-based modeling and AIS.

2   Modern Information Security with AIS

Though there are many security-related products and technologies, yet there exist no detection system that can catch all types of different violations in networked computer systems and the potential threats and vulnerabilities remain intractable. An influx of new approaches is needed to enhance security measures. Researches have been exploring various AI-based approaches for intrusion detection. Among them agent-based modeling seems to become more and more promising, because Internet evolves towards an open, free-market information economy of automated agents buying and selling a rich variety of goods and services. Over time, agents will progress naturally from being mere facilitators of electronic commerce transactions to being financial decision-makers in their own right. Ultimately, inter-agent economic transactions may become an inseparable and perhaps dominant portion of the world economy.

Thus, in the agent-based systems, humans delegate some of their decision-making processes to programs that are in some sense intelligent, mobile, or both. "Intelligent" agents have reasoning capabilities, e.g., rule-based inferencing, probabilistic decision analysis, and/or learning. For example, an agent-based model of information security system is proposed in [8] based on ontology (a network with a sense of existence) where agents solve, jointly, the entire multitude of tasks of information security. The model introduces intelligent meta-agents that solve management and coordination of decisions of the subordinate security agents.

Such approach to information security, as well as any other, has its strength and weaknesses in real world applications. The matter is that the intent of information security system is to provide the least amount of impact to the network performance. But securing of a network by filling it with complicated intelligent agents and ontology hardly corresponds to the intent. Moreover, any intelligent coordinating center, such as meta-agent, becomes the most vulnerable object of the network itself.

Fortunately, we have the natural immune system, which solves the similar problems, but in the way that is radically different from those of traditional information security. The immune system involves many unreliable, short-lived, and imperfect components (mainly B- and T-cells), which circulate at various primary and secondary lymphoid organs of the body. There is no central organ or "meta-agent" that controls the functions of the immune system. The system is autonomous and self-regulatory by nature. It is not "correct", because it sometimes makes mistakes. However, in spite of these mistakes, it functions well enough to help keep most us alive for many years, even though we encounter potentially deadly parasites, bacteria, and viruses every day.

Up to date, related works on the field of immune-based information security are concentrated on isolated ideas and mechanisms of the immune system (e.g. negative selection algorithm [7]). But now there is a larger vision in terms of a set of organizing principles and possible architectures for implementation.

For example, the work [5] focuses on the investigating of immunological principles in designing a multi-agent system for intrusion/anomaly detection and response in networked computers. In this approach, the immunity-based agents roam around the machines (nodes or routers), and monitor the situation in the network (i.e. look for changes such as malfunctions, faults, abnormalities, misuse, intrusions, etc.).

The types of agents and the scope of each agent type are considered to be similar in function and purpose as that of immune cells: monitoring agents (correspond to B-cells), communicator agents (correspond to proteins secreted from T-cells to stimulate B-cells and antibodies), decision/action agents (correspond to helper-, killer-, and suppressor cells). The immune agents can simultaneously monitor networked computer's activities at different levels (such as user level, system level, process level and packet level) in order to determine intrusions and anomalies. They can mutually recognize each other's activities, learn and adapt to their environment dynamically, and detect both known and unknown intrusions.

The above example shows how fruitful it could be to translate the structure of the human immune system into information security. However, several biological solutions could not be directly applicable to our computers because of the serious differences in basic elements and mode of functioning. We also have a risk to overlook non-biological solutions that are more appropriate. So the success of the analogy will be ultimately based on our ability to identify the correct level of abstraction, preserving what is essential from an information security perspective and discarding what is not.

Therefore, we propose another level of abstraction where the core consists in a proper mathematical basis of immune networks. Our approach is somewhat analogous to the proper mathematical basis of neural networks, abstracted from the features of their biological prototype and leading to the wide spreading of the ANN [19].

3   Mathematical Basis of Information Security

Immunologists traditionally describe the problem solved by the immune system as the problem of distinguishing "self" from dangerous "other" (or "nonself") and eliminating other [3]. Self is taken to be the internal cells and molecules of the body, and nonself is any foreign material, particularly bacteria, parasites, and viruses, as well as degenerated self-cells. Distinguishing between self and nonself in natural immune systems is difficult for several reasons. But the main reason is that the components of the body are constructed from the same basic building blocks as nonself, particularly proteins. Proteins are important constituent of all cells, and the immune system processes them in various ways, including the processing in fragments called peptides, which are short sequences of amino acids.

The problem of protecting computer systems from malicious intrusions can similarly be viewed as the problem of distinguishing self from nonself. In this case nonself might be an unauthorized user, foreign code in the form of a computer virus or worm, unanticipated code in the form of a Trojan horse, or corrupted data, etc. In principle, information security could be completely specified based on the abstract representation of self and nonself as sets of bit strings, at that designated even as "proteins" and "peptides"[7].

For example, "protein" could be a sequence of viral bytes in a legitimate program, or a "signature" of computer virus. To preserve generality, in [9] it has been proposed to represent both the protected system (self) and infectious agents (nonself) as dynamically changing sets of bit strings, because in cells of the body the profile of expressed proteins (self) changes over time. In [7] "peptide" for a computer system is defined in terms of short sequences of system calls executed by privileged processes in a networked operating system. Preliminary experiments on a limited testbed of intrusions and other anomalous behavior show that short sequences of system calls (currently sequences of length 6) provide a compact signature for self that distinguishes normal from abnormal behavior. By this analogy proteins can be thought of as "the running code" of the body while peptides serve as indicators of its behavior [7].

More generally, from the viewpoint of computer science we can consider that natural proteins (and peptides) realize main functions of information processing and information security in the whole living Nature. In fact, namely the proteins recognize and execute programs (instructions) represented in the form of genetic code. Being the neuromediators and the receptors of neurons proteins control the electrical activity of the brain. Proteins also can be considered as the main components of the immune system: receptors of B-cells and T-cells, antibodies and messengers (factors, limphokynes). Apparently, proteins should play the key role both for immune and intellectual processes.

In spite of exceptional complexity of proteins' behavior there exist convincing evidence for the following principles:

- function of any protein depends on its spatial conformation;
- this conformation, in its own turn, is determined by the linear sequence (word) of amino acid's code of given protein.

Based on the above postulates a mathematical notion of formal protein, or formal peptide (FP), has been introduced in [14]. This notion abstracts a biophysical principle of the free energy dependence over the space conformation of protein's chain. According to [15], the model of FP demonstrates such important features of protein, as a self-organized reaching of stable state (self-assembly, or folding), and its dependence from the number and the order (non-commutativity) of the links.

The main condition for a protein to function is its binding with another protein (or molecule). Such binding is highly specific (selective), because it depends like "key and lock" on the existence of highly adjusted local shapes of interacting proteins. The proposed model also permits to determine in a natural way the free energy of interaction between FPs as a binding energy. As a result of interaction, a binding (recognizing) of FPs occurs, if binding energy is lower than some threshold; otherwise FPs do not bind.

As a result of binding, protein can change its spatial shape (the so-called allosteric effect). Furthermore, by this effect protein can receive an ability to bind with such molecule (antigen, antibody, messenger, transmitter, etc.), which it couldn't bind before. Thus, new proteins are able to become involved in such process of subsequent binding, forming networks of binding (or molecular circuits). Based on this fact we have introduced the notion of (formal) network of binding, which implies any subsequence of binding between FPs with allosteric effects.

For the modeling properties of immune networks we have supplied the networks of binding with the models of reproduction and death of cells. For this purpose we have introduced a notion of formal B-cell and defined a formal immune network (FIN) as a network of bindings, which includes B-cells [15]. Unlike cellular automata or artificial neural networks, with fixed elements and connections, FIN's elements (B-cells and FPs) are allowed to displace and to bind freely with each other.

Namely, formal B-cell is a 4-tuple

$$B = < P, Ip, Is, Im >,$$

which includes formal protein P as a cell receptor, receptor state indicator Ip, cell state indicator Is, and mutation indicator Im. A behavior of the B-cell is defined by the following conditions:

1. B-cell can be only in the states Is = {0, 1, 2};

2. State Is = 0 corresponds to death when B-cell is destroyed;

3. State Is = 1 corresponds to recognition when B-cell possesses the abilities of its receptor P;

4. Is = 2 corresponds to reproduction when B-cell is divided to the two copies with the cell states Is = 1 and the receptor states determined by the Im;

5. Transition from the state Ir=1 to the state Ir=2 occurs only as a result of binding between FPs.

For example, consider the simplest variant of FIN - an one-dimensional integer-valued network $1DN(n, n_h)$, which is defined by the following conditions:

1. Ip = {0, 1,..., n-1} for every B-cell. Accordingly, designate the states of receptors as P(0), P(1), ... , P(n-1), and cell states as B(0), B(1), ... , B(n-1);
2. An integer-valued threshold of binding n_h is given;
3. Energy of interaction between FPs is defined by the formula

$$w(P(i), P(j)) = \min \{ (i-j) \bmod(n), (j-i) \bmod(n) \} .$$

4. B-cells form one-dimensional sequence (population) without gaps, with beginning (left) and ending (right);
5. If cell B(j) reproduces, then one of its copy remains on the former place, and the other copy is added to the end of the population;
6. If cell B(j) dies, then the other cells shift to the left and fill the gap.

We have introduced and studied two kinds of 1DN: the so-called AB-networks and BB-networks.

AB-network AB(n, n_h) is defined as such 1DN, which possesses, apart from B-cells, also free FPs (antigens) of the n sorts: A(0), A(1), ... , A(n-1), with the following rules of displacement and interaction:

1. Population of antigens is displaced over the population of B-cells so, that to every B-cell no more than one antigen is corresponding.
2. Interaction is allowed only for the B-cell and the antigen over it.
3. B-cell dies, if there is no antigen over it, or if $w > n_h$.
4. If $w = 0$, then B-cell makes two precise copies of itself (without mutations).
5. If $0 < w £ n_h$, then B-cell makes two copies of its nearest sorts (with mutations).
6. The interaction brings no influence on the antigen.
7. Interactions are realized consequently from left to right.
8. When the end of population is achieved, interactions continue from the beginning.

The following result has been proved for such networks:

Theorem 1. If all antigens in a AB(n, n_h) network are of the same sort, and at least one B-cell binds an antigen, then after a finite number of steps, for every antigen a matching B-cell will correspond.

This result affirms, that even the simplest variant of FIN shows the mechanisms, by which FPs (antigens) control reproduction and death of B-cells. Besides, we have determined the conditions of arising and supporting of formal immune response, which implies the B-cells' intention for acceptation of antigen's sort [15].

We have studied also a case, when several sorts of B-cell are generated and stored by interactions between B-cells themselves, in the absence of any antigen. For this purpose we have defined a notion of BB-network BB(n, n_h), as 1DN with population of B-cells satisfying to the following rules:

1. Interactions are allowed only between the neighboring B-cells with the numbers 2k-1, 2k , where k = 1,2, ... , is a number of the pair of B-cells;
2. If the last B-cell in population is odd (without pair) then it dies;
3. If $w > n_h$, then the second B-cell in the pair dies and its place remains free;
4. If $0 < w £ n_h$, then the second B-cell in the pair reproduces with mutations, where the first copy remains at the former place, and the second copy is delayed;
5. After all pairs of the population have interacted once, B-cells are shifted to the left for filling gaps remaining from the died cells;
6. Then the delayed copies are added to the end of the population in the increasing order of their numbers.

Theorem 2. For any initial population of any BB(n, n_h) network only one of the three regimes is possible: 1) death of all B-cells, 2) unlimited reproduction of B-cells, and 3) cyclic reproduction of the initial population (formal immune memory).

Theorem 3. For any n there exists such threshold n_h that at least one cyclic regime is possible in BB(n, n_h) network.

In fact, there exists a number of cyclic regimes with several periods and dimensions of populations, including those, where the number of B-cells changes from population to population. Namely such regimes of FIN represent a mathematical model of self-maintaining immune memory, where several sorts of B-cell are generated and stored by interactions between B-cells themselves, in the absence of any external antigen [17].

The obtained results show that even the simplest variants of FIN demonstrate such important effects, as:

- immune response under the control of antigen;
- immune memory and generation of a new immune repertoire in the absence of outer antigen by means of the cyclic regimes of FIN.

We have introduced also a notion of formal T-cell, which synthesizes FP of the definite type when all receptors of the T-cell become bound by FPs. It has been shown also in [15], that a special set of such T-cells, called T-FIN, is equivalent to an inference engine for problem solving and decisions making.

In general, according to biological prototypes, the principal difference between the mathematical models of immune networks and the models of neural networks is determined by functions of their basic elements. If artificial neuron is considered as a summation with a threshold, then FP as the basic element of FIN ensures self-assembly (folding) of its stable states, as well as a free binding with any other element, as a function of their reciprocal states. Namely on the base of such interaction between FPs we have developed the mathematical concept of FIN. Theorems 1-3 demonstrate rigorously, that even the simplest variants of FIN possess the intrinsic properties of immune memory and immune response.

4   Information Security with FIN

Consider an arbitrary column vector $X = [\, x_1 \ldots x_n \,]^T$ where upper case $"^T"$ is a symbol of transposing and components $x_1, \ldots, x_n$ are real values and/or integers. Let such vector represent a set of information security indicators. For example, it can be a bit string of a legitimate program, a signature of computer virus, a coded sequence of system calls, statistics of current activity of the network, etc. Consider a space $\{X\}$ of such indicators, partitioned to k subspaces (classes) $\{X\}_1, \ldots, \{X\}_k$. For example, k = 2, where $\{X\}_1$ is normal behavior and $\{X\}_2$ is "infection". Then, having a concrete vector X, the task consists in determining it's class $c = \{X\}_c$ where c=1,...,k . Thus the problem is reduced to the well-known pattern recognition.

The main feature of the FIN approach to pattern recognition consists in treating an arbitrary pattern as a way of setting the binding energy between FPs [14]. The idea follows from the principles of associative recognition of antigen by proteins (antibodies and cells' receptors) of the natural immune system [3].

A mathematical basis of the approach was considered in a rather detailed way in our previous works [10, 15]. It is based essentially on the properties of Singular Value Decomposition (SVD) of an arbitrary matrix over the field of real numbers. According to the approach the task of pattern recognition is solved as follows.

4.1   Supervised Learning

4.1.1   Folding Vectors to Matrices

Fold vector X of dimension $n \times 1$ to a matrix A of dimension $n_i \times n_j = n$. It has been shown strictly in [10], that such folding increases the specificity of recognition.

4.1.2   Learning

Form matrices $A_1, ..., A_k$ for all classes $1, ..., k$, and compute singular vectors of the matrices by the SVD:

$$\{X_1, Y_1\} - \text{for } A_1 , ... , \{X_k, Y_k\} - \text{for } A_k .$$

4.1.3   Recognition

Compute k values of binding energy for every input pattern A:

$$w_1 = - X_1^T A Y_1 , ... , w_k = - X_k^T A Y_k .$$

Determine the class to be found by the minimal value of the energy:

$$c : w_c = \min_c \{w_1, ..., w_k\} .$$

4.2   Unsupervised Learning

Consider the matrix $A = [X_1 ... X_m]$ of dimension $n \times m$ formed by m input vectors. Compute the SVD of this matrix:

$$A = s_1 \begin{pmatrix} w1_1 \\ ... \\ w1_n \end{pmatrix} Y_1^T + s_2 \begin{pmatrix} w2_1 \\ ... \\ w2_n \end{pmatrix} Y_2^T + ..., \tag{1}$$

where s_1, s_2 are the first two singular values, and Y_1, Y_2 are right singular vectors.

According to [10], there exists a rigorous correspondence between vectors and FPs. Thus, consider two FPs: {FP1, FP2} as antibodies, which correspond to the vectors Y_1, Y_2 . Consider also n FPs: {FP$_1$,..., FP$_n$}, which correspond to the strings of the matrix A . Then every string A_i , which represents the values of the indicator number i: $i = 1, \dots, n$, is mapped to the two values $\{w1_i, w2_i\}$ of binding energy between FP_i and antibodies:

$$w1_i = w(FP1, FP_i), \quad w2_i = w(FP2, FP_i).$$

Therefore, every vector with n components can be represented and viewed as a point in two-dimensional space of binding energies $\{w1, w2\}$. This plane could be treated also as a shape space of FIN, according to [6]. Such representation of initial data allows to classify vectors in a rigorous and visual way.

The results obtained in [10, 15] show, that this approach to pattern recognition is rather effective. It is able to give fine classification and sharply focus attention on the most dangerous situations. It is worth to note also, that the approach was successfully used for processing indicators of the natural infections. Namely, it has allowed to detect nontrivial similarities in the dynamics of infectional morbidity and to predict a risk of the plague epizooty.

According to [9], information security is supposed to address five issues: confidentiality, integrity, availability, accountability, and correctness. In the immune system, however, there is really only one important issue, survival, which can be thought as a combination of integrity and availability. Likewise, the immune system is not concerned with protecting secrets, privacy, or other issues of confidentiality. This is probably the most important limitation of the analogy, and one that we should keep in mind when thinking about how to apply our knowledge of immunology to problems of computer security.

Nevertheless, being a mathematical abstraction, FIN could be also applied to the other issues of information security. Consider, for example, data hiding and encryption.

According to [2], data hiding, a form of steganography, embeds data into digital media for the purpose of identification, annotation and copyright. It represents a class of processes used to embed data, such as copyright information, into various forms of media such as image, audio, or text with a minimum amount of perceivable degradation to the "host" signal; i.e., the embedded data should be invisible and inaudible to a human observer. Note that data hiding, while similar to compression, is distinct from encryption. Its goal is not to restrict or regulate access to the host signal, but rather to ensure that embedded data remain inviolate and recoverable.

Let an arbitrary matrix A represent the initial data array. It could be an image, a folded audio signal, etc. Consider the SVD of the matrix in the form (1). Let us add to this sum a FP in the form $s_{r+1} W_{r+1} Y^T_{r+1}$, where r is a rank of the matrix, $W^T_{r+1} W_{r+1} = Y^T_{r+1} Y_{r+1} = 1$, $s_r > s_{r+1}$, and s_r is a minimal singular value of the matrix. According to the mathematical properties of SVD, such FP only slightly disturbs the matrix. Although such disturbance is invisible or inaudible to a human observer, the presence of the "hidden" FP can be surely detected in the shape space of FIN. So FIN functions like the natural immune system, which verifies identity by the presence of peptides, or protein fragments.

Consider now data encryption. In modern cryptography, the secret of keeping encrypted information is based upon a widely known algorithm and a string of numbers that is kept secret called a key. The key is used as a parameter to the algorithm to encrypt and decrypt the data. Decryption with the key is simple, but without the key is very difficult and in some cases nearly impossible. Therefore the "fundamental rule of cryptography" is that both sides of the message transfer know the method of encryption used [13].

As an example of encryption, consider a $BB(n,n_h)$ network from the previous section. According to Theorem 3, such network possesses a cyclic regime for any n . Specifically, in the network BB(10,2) for any sort i = 0, ... , 9 of B-cells the following populations repeating with the period 4 :

$$(i+2)\ (i)\ (i-2)\ (i)\ .$$

For example,

1979 fi   187800 fi   1770991 fi   17980 fi   1979 fi   ... .

Consider now the numbers {10, 2} as a key, which define the network BB(10,2). Then the string 1979 could encrypt the string 1770991. Knowing the key, the data could be decrypted, say, as the string of the maximal length, generated by the network BB(10,2) from the given string 1979. Although the example seems rather simple, it shows the principal possibility of using FIN in cryptography.

5   Conclusion

The developing of the FIN theory has already appeared to be useful in solving a number of important real world tasks, including detection dangerous ballistic situations in near-Earth space, complex evaluation of ecological and medical indicators in Russia, and prediction danger by space-time dynamics of the plague infection in Central Asia [10, 15, 18]. In addition, FIN could be successfully applied for synchronization of events in computer networks [15] and even for online virtual clothing in Internet [16].

The obtained results show, that FIN is rather powerful, robust and flexible approach to pattern recognition, problem solving, and modeling of natural systems dynamics. Thus, FIN could be effectively applied also for information security assurance. An advantage of FIN in this field could be seen as a sharp and surely focusing attention on the most dangerous situations, especially in the cases that are beyond the power of traditional statistics or AI (e.g. see [18]).

Therefore, we should like to highlight three features, which determine perspectives of FIN approach to information security:

• highly appropriate biological prototype of immune networks;
• rigorous mathematical basis of FIN;
• possibility of hardware implementation of FIN by special immune chips.

It is worth to note, that the theory of FIN gives a mathematical basis for developing special immune chips proposed to be called also as immunocomputers (IC). Besides, the properties of the biological immune networks admit to hope, that IC would be able to overcome the main deficiencies that block the wide application of neurocomputers [19] in those fields, where a cost of a single error could be too high. An important example of such field gives us information security. Thus, IC could raise the information security issues to a new level of reliability, flexibility and operating speed.

Acknowledgement. This work is supported by the EU in the frame of the project IST-2000-26016 "Immunocomputing".

References

1.   Agnati, L.F.: Human brain in science and culture (in Italian). Casa Editrice Ambrociana, Milano (1998)
2.   Bender, W., Gruhl, D., Morimoto, N., Lu, A.: Techniques for data hiding. IBM Systems J. Vol. 35. 3–4 (1996) 313–336
3.   Coutinho, A.: Immunology: the heritage of the past. Letters of the L.Pasteur Institute of Paris (in French). 8 (1994) 26–29
4.   Dasgupta, D. (ed.): Artificial immune systems and their applications. Springer-Verlag, Berlin Heidelberg New York (1999)
5.   Dasgupta, D.: Immunity based intrusion detection system: a general framework. In: Proc. of the 22th National Information Security Conference. Arlington, Virginia, USA (1999)
6.   DeBoer, R.J., Segel, L.A., Perelson, A.S.: Pattern formation in one and two-dimensional shape space models of the immune system. J. Theoret. Biol. 155 (1992) 295–333
7.   Forrest, S., Hofmeyer, S., Somayaji, A.: Computer immunology. Communication of the ACM, Vol. 40. 10 (1997) 88–96
8.   Gorodetsky, V.I., Kotenko, I.V., Popyack, L.J., Skormin, V.A.: Agent based model of information security system: architecture and framework for behavoir coordination. In: Proc. of the 1st Int. Workshop of Central and Eastern Europe on Multi-Agent Systems (CEEMAS'99). St.Petersburg, Russia (1999) 323–331
9.   Hofmeyr, S., Forrest, S.: Immunity by design: an artificial immune system. In: Proc. of the Genetic and Evolutionary Computation Conference (GECCO-99). (1999) 1289–1296
10. Kuznetsov, V.I., Milyaev, V.B., Tarakanov, A.O.: Mathematical basis of complex ecological evaluation. St. Petersburg University Press (1999)
11. Scormin, V.A., Delgado-Frias, J.G.: Biological Approach to System Information Security (BASIS), A White Paper. Air Force Research Lab., Rome, NY (2000)
12. Somayaji, A., Hofmeyr, S., Forrest, S.: Principles of a computer immune system. In: New Security Paradigms Workshop, ACM (1998) 75–82
13. Tannenbaum, A.S.: Computer networks. 3rd edn. Prentice Hall (1996)
14. Tarakanov, A.O.: Mathematical models of biomolecular information processing: formal peptide instead of formal neuron (in Russian). In: Problems of Informatization J. 1 (1998) 46–51
15. Tarakanov, A.: Formal peptide as a basic agent of immune networks: from natural prototype to mathematical theory and applications. In: Proc. of the 1st Int. Workshop of Central and Eastern Europe on Multi-Agent Systems (CEEMAS'99). St. Petersburg, Russia (1999) 281–292
16. Tarakanov, A., Adamatzky, A.: Virtual clothing in hybrid cellular automata. (2000) http://www.ias.uwe.ac.uk/~a-adamat/clothing/cloth_06.htm
17. Tarakanov, A., Dasgupta, D.: A formal model of an artificial immune system. In: BioSystems J. Vol. 55, 1–3 (2000) 151–158

18. Tarakanov, A., Sokolova, S., Abramov, B., Aikimbayev, A.: Immunocomputing of the natural plague foci. In: Proc. of Int. Genetic and Evolutionary Computation Conference (GECCO-2000), Workshop on Artificial Immune Systems. Las Vegas, USA (2000) 38–39

19. Wasserman, P.: Neural computing. Theory and practice. Van Nostrand Reihold, New York (1990)

BASIS: A Biological Approach to System Information Security

Victor A. Skormin[1], Jose G. Delgado-Frias[1], Dennis L. McGee[1],
Joseph V. Giordano[2], Leonard J. Popyack[2],
Vladimir I. Gorodetski[3], and Alexander O. Tarakanov[3]

[1]Binghamton University, Binghamton NY, USA
{vskormin, delgado, mcgee}@binghamton.edu
[2]Air Force Research Laboratory at Rome NY, USA
Leonard.Popyack@rl.af.mil
[3]St.Petersburg Institute for Informatics and Automation, St. Petersburg, Russia
{gor, tar}@mail.iias.spb.su

Abstract. Advanced information security systems (ISS) play an ever-increasing role in the information assurance in global computer networks. Dependability of ISS is being achieved by the enormous amount of data processing that adversely affects the overall network performance. Modern ISS architecture is viewed as a multi-agent system comprising a number of semi-autonomous software agents designated to prevent particular kinds of threats and suppress specific types of attacks without burdening the network. The high efficiency of such a system is achieved by establishing the principles of successful individual and cooperative operation of particular agents. Such principles, evolved during evolution, are known to be implemented in biological immune systems. The aim of this paper is the exploration of the basic principles that govern an immune system and the potential implementation of these principles in a multi-agent ISS of a heterogeneous computer network.

1   Introduction

With the increase of size, interconnectivity, and number of points of access, computer networks have become vulnerable to various forms of information attacks, especially to new, sophisticated ones. It should be pointed out that biological organisms are also complex and interconnected systems that have many points of access; these systems are vulnerable to sabotage by alien microorganisms. During evolution, biological organisms have developed very successful immune systems for detecting, identifying, and destroying most alien intruders. In this paper we intend to establish a connection between the basic principles that govern the immune system and potential uses of these principles in the implementation of information security systems (ISS) for computer networks. In order to be dependable, existing ISS require an enormous amount of data processing that adversely affects the network performance. A modern ISS must include a number of semi-autonomous software agents designated to prevent particular kinds of threats and suppress specific types of attacks. The ability of such an ISS to provide the required level of information security without burdening the

V.I. Gorodetski et al. (Eds.): MMM-ACNS 2001, LNCS 2052, pp. 127–142, 2001.

network resources could be achieved only by adopting the most advanced principles of interaction between particular agents. Such principles have been used in biological immune systems. These systems have distributed cells (agents) of various types that attack anything suspected to be alien. Cells interact by sharing information about the type and location of an intruder, utilize the feedback principle for engaging only "as many cells as necessary," and are capable of learning about intruders that results in immunity to repeated attacks.

When necessary functions of the ISS agents are established, the multi-agent system theory can be utilized as a mathematical apparatus to facilitate the formalization of the complex interaction between particular agents in the fashion that is observed in biological immune systems. The individual and collective behavior of particular immune cells need to be investigated to establish a synthetic immune system operating in computer networks.

This paper has been organized as follows. In Section 2 a brief outline of the approach is provided. Some basic principles of a biological immune system are described in detail in Section 3. The major components of the proposed Biological Approach to System Information Security (BASIS) are presented in Section 4. Some concluding remarks have been drawn and presented in Section 5.

2 Proposed Approach Outline

BASIS, proposed in this paper, reflects the similarities between the computer network security problem and the task of protecting a biological system from invading microorganisms. We propose to synthesize an ISS of a computer network that follows the basic principles of operation of the biological immune system. The utilization of biological defensive mechanisms developed by evolution has a great potential for the assurance of information security in large-scale computer networks.

With the complexity of modern information security systems, an ISS must be considered as a number of independent, largely autonomous, network-based, specialized software agents operating in a coordinated and cooperative fashion designated to prevent particular kinds of threats and suppressing specific types of attacks. This approach provides the required level of general security of information according to a global criterion. A biological immune system of an advanced organism already has been considered as a good and clear example of a modern agent-based ISS [1, 2]. The immune system consists of distributed white blood cells, which attack anything that they consider alien. By having as many cells as necessary, the animal body is able to defend itself in a very efficient way. If the animal body is infected in one area, then cells move to that area and defend it. Modern multi-agent system technology presents a valuable approach for the development of an ISS that is expected to have very promising advantages when implemented in a distributed large scale multi-purpose information system. A consideration of an agent-based model of an ISS, consistent with the BASIS concept, is given in [3, 4].

Our ISS approach can be viewed as a set of semi-autonomous distributed agents capable of detecting, recognizing, pursuing, and learning about the attackers. The algorithms of agents' individual behavior need be consistent with those established in immunology. While the implementation of particular ISS agents may be computationally intensive, a feasible ISS system requires a high degree of interaction, coordination, and cooperation between agents. It is believed that these, almost intelligent, interactions constitute the centerpiece of a biological defensive mechanism and assure its high efficiency. Until recently, quantitative description of the cooperative behavior of semi-autonomous agents presented a problem that could not be solved within a general framework. Such a framework has been provided by the multi-agent system approach. This framework facilitates the development of the rules and procedures of the interaction between ISS agents thus leading to the development of a feasible ISS operating as an artificial immune system. The feasibility of such an ISS could be assured only by the implementation of the rules of agents' collective behavior observed in a biological immune system that could be formalized using the multi-agent system theory.

Research in immunology has established various mechanisms of individual behavior of cells resulting in their ability to detect, identify, pursue and destroy an alien entity; to accumulate knowledge on attackers, to adopt behavior to a new situation; and to determine the proper response. These mechanisms, developed by evolution, are highly efficient and successful. In addition to individual cell operation, immunology presents numerous examples of collective, almost intelligent, "unselfish" behavior of various types of defensive cells. This collective cell behavior allows the achievement of high efficiency and minimum response time of the immune system, as well as maximum utilization of its limited resources. The major difference in the targets between the immune system and an ISS is: the immune system treats as an "enemy" any foreign entity within the organism, while an ISS must recognize and treat as an "enemy" any illegitimate entry or software. Table 1 shows some of the similarities between the two systems.

A number of mathematical models of individual behavior of defensive cells utilizing methods of statistics, discrete mathematics, and numerical simulation, have been successfully implemented. The attempts to develop an artificial immune system that could be applied for such a practical problem as information security assurance are much less successful, primarily because of the necessity to describe mathematically cooperative cell behavior. However, modern multi-agent system technology presents the most plausible approach for solving this problem in the framework of the development of an ISS. The resultant ISS would operate as a synthetic immune system and is expected to have very promising advantages when implemented in a distributed large-scale multi-purpose computer information network.

We propose the following steps to achieve ISS by means of biologically inspired schemes:

a) Analysis and Qualitative Description of Immune Systems. An analysis of the recent biological research and development of a comprehensive qualitative description of the operation of a biological immune system are needed to capture the behavior immune system as it may relate to ISS-related problems.

b) Algorithmic Description Immune Cells. The next step is to develop a mathematical/algorithmic description of the individual behavior of immune cells utilizing already established methods and models, and their cooperative operation using the multi-agent system theory.

Table 1. Similarities Between Biological and Computer Systems

Biological Systems	Computer Networks
High complexity, high connectivity, extensive interaction between components, numerous entry points.	High complexity, high connectivity, extensive interaction between components, numerous entry points.
Vulnerability to intentionally or unintentionally introduced alien microorganisms that can quickly contaminate the system resulting in its performance degradation and collapse.	Vulnerability to malicious codes (including computer viruses) that being introduced in the system result in unauthorized access to information and services and/or denial of service.
Alien microorganisms as well as cells of a biological system are composed of the same building blocks - basic amino acids.	Malicious codes as well as the operational software of a computer network are composed of the same building blocks - basic macro commands.
The difference between alien microorganisms and the healthy cells of a biological system is in the (gene) sequencing of their building blocks.	The difference between malicious codes and the operational software of a computer network is in the sequencing of their building blocks.
Biological immune systems are capable of detecting, recognizing and neutralizing most alien microorganisms in a biological system.	Information security systems should be capable of detecting, recognizing and neutralizing most attacks on a computer network.

c)   Software Implementation of Models. A software implementation of the established mathematical models, rules and algorithms resulting in an ISS operating as an artificial immune system is necessary to test and check these models.

d)   Simulation Environment. A simulation environment suitable for the representation of a computer network with a resident ISS and various forms of threats and attacks needs be developed to prove the model.

e)   System Analysis. Simulations need to be analyzed to fine-tuning of the resultant ISS, This analysis should include assessment of its impact on the network vulnerability.

The BASIS approach requires a multidisciplinary effort; the disciplines included are: advanced control theory, mathematics, computer science, computer engineering, biology, information warfare, and computer programming.

The immune system, like computer network systems, is extremely complex. It comprises a massive whole body response mechanism involving multiple cell types and specialized tissues. For the purpose of feasibility, the immune system needs to be represented/modeled in its basic components and then one could consider the interaction of these components resulting in a complete body response to three generalized foreign agents: a toxin, a bacteria, and a virus. The toxin represents a foreign agent presented to the system in large amounts (e.g. injected or swallowed) or produced in large amounts by an infectious agent which causes harm to the host system. In a computer network environment, effects of a toxin could be paralleled with an illegal entry which results in the destruction of significant amounts of resident data. The bacteria would be an agent which can replicate itself independent of the host and which causes harm to the host system. A corresponding attack on a computer network would comprise an illegal entry facilitating future illegal utilization of the network facilities including the ability to manipulate confidential information. Finally, the virus would represent an agent which replicates itself through the host system and which then would cause harm to the host. Unsurprisingly, growing and replicating itself using the host facilities at the expense of resident software, is exactly what a computer virus does rendering the network useless.

3   Information Security Tasks in a Biological Immune

A detailed description of the biological immune system is provided in this section. This description includes the basic immune system response, the major players in this system, and the interaction of these components.

3.1   The Basic Immune Response

A basic immune response can consist of one or more of three components [5, 6]. The first component is the innate or non-specific immune response. It consists of anatomic or physiological barriers (e.g. the skin and the acidity of the stomach) and the inflammatory response that is responsible for the redness and swelling at a wound site and the influx of cells such as neutrophils and macrophages which phagocytize. This innate immune response does not show specificity for any particular foreign agent and does not show an enhanced response, or memory, with the second encounter. For all practical purposes, the innate immune response could be paralleled with a system of passwords, intended to prevent unauthorized access to the network. This can be considered as the "skin" of a computer system. Its response is prompted by reading a password -an "external label" that has nothing to do with the internal nature of the attacker. At the computer network level, innate responses could include firewalls that do not allow an intruder to get into a closed sub-network. In this case the ID domain would prevent any other user from a different domain (non-self) to get access to any computer in the present domain (self).

The other two responses show specific acquired or adaptive immune responses. These responses show the exquisite capacity to specifically recognize unique foreign substances (or antigens as they are called), distinguish self from non-self, and show a heightened and rapid memory response with each subsequent encounter with the antigen. The humeral immune response has classically been defined as a response of the body to foreign antigens by producing large amounts of antibodies; serum proteins which have a binding site that binds with high specificity to the antigen (or a defined portion of the antigen) to inactivate the antigen or allow for the removal of the antigen. The B lymphocytes cells (or B cells) produce these antibodies. Antibodies generally work only on antigens which are found exposed in the body such as floating in the blood or body fluids (e.g. toxins) or on bacteria, viruses, etc. which are exposed and not sequestered inside of an infected cell. These antibodies are useful in clearing the infectious agents or toxins from the body fluids and tissues. The computer equivalent of this response is the ability to distinguish between a legitimate user and an illegal intruder that successfully penetrated the system. In general, messages/programs containing system calls are considered suspicious, thus the internal composition of these messages/programs can be used to determine self and non-self. Non-self messages/programs need be discarded by specialized software (antibody) that deals with this type of attackers. This response is facilitated by the ability to recognize an attacker because of its foreign internal nature as well as behavior.

The cell-mediated immune system has evolved to attack extracellular bacteria and viruses along with those infectious agents which may be hidden inside of cells. These infected cells become factories that produce the viruses or bacteria, yet in many instances the antibodies cannot get to the infectious agents sequestered inside of the cell -therefore, the infection persists. This cell-mediated immune system then uses either specialized killer cells (cytotoxic T lymphocytes or CTLs, see below) with highly specific cell surface receptors to recognize and kill virus infected cells or other antigen-specific T lymphocytes (T_{DTH} cells) to direct the action of non-specific types of cells such as the phagocytic macrophages to destroy bacteria, protozoa or fungi. The computer equivalent of this response is based on the ability to detect a hidden attacker disguised as or within a legitimate piece of software. For instance, an email may contain a malicious virus; this virus is hidden within an apparently legitimate piece of software. Antivirus programs are usually utilized to deal with these problems. However, these viruses need be recognized by an immune-like system before these cause any major harm.

Table 2. Immune and computer system responses

Immune Response	Immune System	Computer/Network System
Innate or non-specific	Anatomic or physiological barriers (e.g. the skin and the acidity of the stomach) and the inflammatory response.	System's passwords and firewalls that do not allow an intruder to get into a closed sub-network.
Humeral	Antibodies identify antigens and help to clear the body from infectious agents.	Specialized software help to identify system calls in messages/programs. "Non-self" programs are discarded.
Cell-mediated	CTL and T_{DTH} cells are used to destroy bacteria, protozoa or fungi.	Antivirus programs specialized on a particular computer virus.

Of importance to note, the mechanisms which regulate the activation and function of these immune cells are very stringent and complex. Without proper regulation, the immune system would over-respond to foreign agents resulting in potential harm to the host or the system may allow self reactive cells to function resulting in deadly autoimmune responses. By its nature, this effect is nothing but a sophisticated feedback mechanism that could be implemented in a computer network. Table 2 presents a summary of the immune system responses along with some parallel responses of computer/network systems.

3.2   Immune Response Major Players

Antigens may be composed of several types of compounds (protein, sugars, lipids, etc.), however protein antigens are the type that induce the most vigorous response. Proteins are comprised of chains of simpler compounds called amino acids. Since there are 20 different amino acids, the combinations of the amino acids can yield an incredible number of possible distinctly different proteins.

The antibodies are proteins themselves which, by the nature of their amino acids at the binding sites, can bind strongly to specific short sequences of amino acids. These specific amino acid sequences may then appear within the longer sequence of a certain protein and therefore the antibody can "recognize" the protein via this shorter sequence and then bind. One important property of this antigen-antibody recognition system is that it is extremely specific. However, antibodies may also bind, with lesser strength, to amino acid sequences which are almost identical to the recognized sequence allowing for "cross-reactivity" of the antibodies. This could allow the antibodies to recognize a slight variation of the original infectious agent and therefore confer some immunity. Yet the more different the sequence is from the recognized sequence, the greater the chance for non-recognition. Another important consideration of this system is that in general, a specific antibody must exist for essentially all of the antigens which one could possibly encounter. Still, the immune system has the capacity to randomly generate more than 10^{11} different antigen binding antibodies.

The antibody proteins are produced by the B cells. These cells randomly generate the capacity to produce an antibody with a single antigen recognition site such that one B cell produces an antibody which recognizes only one antigen (to have a cell which produces several antibodies which recognize different antigens would be a regulation nightmare!). This B cell then produces the antibody only as a cell surface receptor and remains in a resting state, waiting to encounter the specific antigen. When the antigen is encountered, it binds to the antibody on the B cell surface and stimulates the B cell to awake and get ready to function. However, in most instances, the B cell cannot begin to undergo cell division (to amplify the number of antigen specific cells and therefore amplify the response) or begin secreting antibodies until it obtains a second signal from a Helper T lymphocyte (Th cell). This prevents the B cell from producing potentially harmful antibodies without a confirmation that the response is needed. Once the B cell is activated, it then begins to differentiate into either a Plasma Cell, which produces large amounts of antibodies and then dies, or a Memory Cell which eventually reverts back to a resting state and waits for a second encounter with the antigen. These Memory Cells are the basis of the greatly elevated and rapid memory response to the same antigen, as there are now greater numbers of these cells present which now have a less stringent requirement for activation.

As mentioned above, the B cell requires a second signal from a Th cell in order to continue its activation sequence. This Th cell is also an antigen specific cell with a specialized receptor, called the T cell receptor, for a very specific antigen amino acid sequence. The T cell receptor is similar to the antibody molecule, yet it is limited in its ability to recognize and antigen. During development of the Th cell (indeed all T cells including the CTLs and T_{DTH} described below), the cells pass through a specialized tissue, the thymus, in which cells with T cell receptors that recognize self-antigens are killed. This eliminates the majority of the self reactive cells and does an excellent job of preventing autoimmune responses. Also in the thymus, only the T cells with T cell receptors which can recognize antigen segments which are "presented" to them by accessory cells with specialized cell surface antigen-presentation receptors, the Major Histocompatability Complex class II receptors (MHC II), are allowed to survive. Indeed, it has been estimated that greater than 90% of the T cells in the thymus never survive these stringent regulatory requirements to leave the thymus.

Once a Th cell leaves the thymus, it is fully capable of functioning, yet, like the B cell, it too is in a resting state. The T cell receptors of these Th cells cannot recognize antigens alone so they cannot be activated directly by antigen. In the case of the humoral immune response, the B cell binds the antigen via its cell surface antibody which gives the B cell its first activation signal. This antibody bound antigen is then taken internally by the B cell and "processed" into short amino acid segments which are then "loaded" onto the MHC II receptors. The B cell then places these MHC receptors loaded with antigen segments on its surface and is now ready to interact with a Th cell. This interaction then consists of the B cell "presenting" the antigen to the Th cell to activate the Th cell. This presentation of the antigen-MHC II to the Th cell provides an activation signal for the Th cell to begin cell division (amplification of the response) and to differentiate into a mature helper T cell capable of helping the B cell. The mature, activated Th cell then produces signals (via cell surface receptors or small secreted factors) which tell the B cell to continue on its activation sequence to cell division and antibody secretion.

Of interest, this B cell-Th cell interaction presents another site for amplification of the immune response. One B cell can "process" a large antigen into several small different segments which could be used to activate several Th cells with different antigen specificities. This would increase the probability that the B cell would get a second signal from a Th cell even if the Th cell did not recognize the exact same segment of the antigen amino acid sequence that the B cell recognized. Also, a single activated Th cell could then interact with several B cells to allow the production of several different types of antibodies (one specific type from each different B cell). However, only those B cells which have encountered the antigen for the first activation step would be sensitive to the Th cell help. The overall response would be a more complete activation of several B cells and T cells with different antigen specificities - essentially responding to several different segments of a single antigen.

Before continuing, another important group of cells must be considered. These are the accessory cells which are not antigen specific but play a very important role in the immune response. The accessory cells consist mainly of macrophages and dendritic cells which function to engulf or phagocytize cells, bacteria, viruses, or even cellular debris and proteins. These engulfed cells or substances are then enzymatically destroyed and "processed" much like the B cell processes antigens bound to their cell surface antibody receptors to yield short segments of amino acids. As in the B cells, these antigen segments are also loaded onto MHC II receptors for the accessory cells to "present" to any nearby Th cells. Indeed, the initial activation of most Th cells usually occurs via the presentation of foreign antigens by these accessory cells. This system then allows for the constant sampling of the body's environment via macrophage phagocytosis (for bacteria, viruses, and etc.) or dendritic cells (for cell debris and individual proteins). Therefore, antigen sampling (and hence Th cell activation) is not limited to antigen specific B cells only, but also certain non-specific accessory cells.

In the cell-mediated immune response, one of the most potent killer cells is the CTL, a T cell with a T cell receptor which recognizes foreign antigens present inside of an infected cell. Like the B cell, this CTL is normally in a resting state and needs two independent signals for activation. The first signal comes when the resting CTL encounters an infected cell. Almost all cells of the body have an internal system which constantly destroys old proteins as new ones are produced by the cell. The destruction of old proteins results in the production of short amino acid segments which may then be loaded onto a different type of MHC receptor called the MHC class I receptor (MHC I) which is then placed on the surface of the cell. Therefore, most cells of the body constantly display a variety of "self" amino acid segments in conjunction with the MHC I receptor on the surface of the cell. However, when a cell becomes infected with a virus, the virus uses the cells machinery to replicate itself. Yet this replication of the virus inside of the cell allows the cell's internal system to sample some of the virus proteins by destroying them and placing the short virus amino acid segments onto the MHC I receptors (again, random sampling as with the accessory cells above). Subsequently, when the viral antigen loaded MHC I receptors are on the surface of the cell, the cell is now labeled as an infected cell even though the immune system cannot directly get at the virus inside of the cell.

Once a CTL comes in contact with a virus infected cell, if its T cell receptor can recognize the virus segment, then the CTL obtains its first activation signal. However, the CTL cannot be completely activated until it receives help from a Th cell, which has also been activated. The activated Th cell (usually activated by antigens from the same virus as presented by accessory cells) then produces a T cell growth factor (interleukin-2) necessary for the CTL to begin cell division (once again, an amplification step) and mature into a functional CTL. As with the B cells, memory CTLs are also produced to produce a memory response in subsequent encounters with the virus. When the mature CTL encounters the virus infected cell again (via the T cell receptor binding to the virus antigen segment and the MHC I receptor), the CTL kills the virus infected cell. Of importance, the mature CTL can kill many virus infected cells over its life span.

The final player in the cell-mediated immune response is the T_{DTH} cell. Once again, this T cell has an antigen specific T cell receptor and must have the antigen presented by an accessory cell along with MHC II. They are essentially helper T cells, which have their first encounter with the antigen in the lymph nodes. They then undergo cell division (amplification of the response) and maturation to competent T_{DTH} cells. These cells then leave the lymph nodes and actively seek out the areas of infection (see below how they are recruited into areas of infection). They migrate into the area where they receive the second encounter with the antigen (presented by resident macrophages actively phagocytizing the bacteria or viruses or dendritic cells sampling the infection debris) and then produce large amounts of inflammation inducing factors or cytokines which activate the phagocytic macrophages, neutrophils and other cells in the area.

3.3   The Interaction of the Immune System Components

The immune system components interact in a particular way depending on the type of foreign agent that the biological organism is exposed to. In this subsection we describe the immune system respond against toxins, bacteria, viruses, and infections.

3.3.1   Immune System Respond against Toxins

In the case of a toxin, the best defense is the antibody molecule. A toxin by itself usually can cause harm to the cells of the body and therefore must be neutralized. This is usually effectively done by the binding of the antibody molecule (usually by blocking the toxin from entering into a cell or blocking the toxin function). Therefore, the major players in the anti-toxin response would be the B cell which produces an antibody to neutralize the toxin and the Th cell which provides help for the B cells. However, macrophages can also have a role in these responses by providing another cell type to present the toxin antigens to Th cells. Macrophages also have receptors on their cell surface which can bind to antibodies which have bound to an antigen (these receptors often do not bind well to antibody which has not bound to an antigen). This then provides a way for the macrophage to attach to the antigen and engulf or phagocytize the toxin to remove it from the body. Finally, some toxins may also activate macrophages and induce them to secrete soluble factors which can enhance B cell division, antibody production, and Th cell responses. On the second encounter with the toxin, the body already has antibody present to neutralize the toxin and the greater number of antigen specific B cells and Th cells (memory cells) are rapidly activated to produce extremely high levels of anti-toxin antibody (which is the basis of the booster shots in vaccines).

3.3.2   Immune System Responds against Bacteria

For a bacteria not sequestered inside of a cell, the first line of defense is the innate immune response: the inflammatory response and macrophage phagocytosis of the bacteria. The purpose of this innate immune response is to hold the infection at bay until the immune response can be activated. Macrophages which have phagocytized bacteria or dendritic cells which have picked up bacterial debris begin to present bacterial antigen segments with MHC II. These cells travel to nearby lymph nodes where they then can present the antigens to the Th cells to begin their activation. Meanwhile, bacterial debris or even whole bacteria present in the lymph (the fluid surrounding the cells of the body) are carried via the lymphatic system to the lymph nodes. This allows for B cells in the lymph nodes to become stimulated and even resident macrophages in the lymph nodes to pick up antigen for presentation to Th cells. The activated Th cells then interact with the activated B cells and eventually the B cells begin to produce massive levels of antibody. This antibody then gets into the blood circulatory system and is carried to the infection site where it can have a number of effects. Antibody binding directly to bacteria can allow macrophages and neutrophils to attach to the antibody to enhance phagocytosis and killing of the bacteria. Antibody bound to the bacteria can also activate the inflammatory system which eventually results in the activation of macrophages to become better bacteria killers and to cause the release of signals which recruit more macrophages, neutrophils, and even T cells from the blood to the site of infection.

In some instances, T_{DTH} cells in the lymph nodes may also be activated by the accessory cells bringing in antigen. These T_{DTH} cells then leave the lymph nodes to seek out the area of infection. Once in the infection area, the macrophages present more antigen to the T_{DTH} cells to induce them to release several powerful inflammation inducing factors. These include factors that recruit more macrophages and neutrophils from the blood into the area, factors that activate the neutrophils and macrophages to become master killers of microbes (this in addition to the virus-specific antibody greatly enhances the macrophage function), factors that provide help for other Th and T_{DTH} cells in the area, and they can help B cell to enhance antibody production.

The end result of these responses is a massive influx and activation of killer macrophages and neutrophils which phagocytize the bacteria, the influx of antibodies which neutralize the bacteria and enhance their phagocytosis, and the activation of the inflammatory response. The invading bacteria is usually destroyed, however host tissue damage may occur in cases of massive infection. Of note, often the induction of the immune and inflammatory response results in the secretion of high levels of activation factors by the macrophages and T cells. As the levels of these activation factors increase, they often induce the production of inflammation suppressing factors by the immune cells and resident cells of the tissues. This, along with the reduction in the levels of antigens or bacteria for stimulation, allows for the down-regulation of the response and the beginning of wound healing.

3.3.3   Immune System Respond against Viruses

Viruses present an interesting challenge to the immune response in that these agents have an intracellular phase, in which they are not available to many of the immune response elements, and often an extracellular phase when the virus is shed from an infected cell to spread to and infect nearby cells. Most of the above immune mechanisms (antibodies and T_{DTH}-macrophage responses) can effectively handle the extracellular phase of the virus. Antibodies bind to the extracellular viruses and prevent their binding to or entering other cells, enhance their destruction by allowing macrophages a handle to bind to the bound antibody and induce phagocytosis, and bound antibody can induce the inflammatory response. T_{DTH} type helper T cells can migrate to the site of infection and direct the activation of macrophages to become master killers with a greater phagocytic capacity, produce factors or cytokines which recruit other T cells, macrophages, and neutrophils into the area of infection, help the activation and maturation of CTLs (see below), and, since the T_{DTH} cells are specialized Th cells, they can help B cells to enhance antibody production. In addition, the T_{DTH} cells can secrete a very potent factor, interferon, which induces all nearby cells to turn on their own internal antiviral defense mechanisms to prevent viral replication and help in preventing the spread of the infection.

Yet the above mechanisms generally have no effect on the viruses hidden within infected cells. The result is that the infection continues because the source (the virus infected cell) has not been destroyed and in some cases the infection can spread via direct cell-to-cell transfer of the virus without an extracellular phase. The destruction of the virus infected cells requires the action of the antigen specific CTLs. CTLs activated at the site of the virus infection can receive immediate help from the T_{DTH} type Th cells in the area to become mature, active CTLs to kill the virus infected cells. This also releases any internal viruses to be exposed to antibody and macrophages for destruction. Finally, CTLs also can produce interferon which induces more nearby cells to turn on their internal antiviral mechanisms.

The overall effect is that the virus spread and source of infection is stopped. Of course large numbers of memory B cells, T_{DTH} type Th cells, and CTLs are also produced so that in subsequent encounters with the same virus, the specific immune response is very rapid and much stronger; hence, immunity.

3.3.4   Whole Body Responses to Infections

In addition to the above described immune responses to infectious agents, several other mechanisms are induced which can help in preventing the spread of the infectious agents to different parts of the body.

One of the most striking features of the immune system is that the immune response cells are not centralized, but are spread out in strategically placed lymph nodes throughout the body. The fluids collected from around the cells in only a defined section of the body pass through any single lymph node (e.g. the lymph nodes of the groin area filter fluids from various sections of the legs). These lymph nodes provide a staging area for the interactions which are required for the immune response to occur, interactions which could not occur in the rapidly flowing blood or most normal tissues where the immune cell numbers would be too low. To ensure that the antigens of the infectious agents get to the lymph nodes (often well before the antigens or viruses and bacteria actually reach the lymph node on their own), the macrophages and dendritic cells at the infection site specifically migrate to the local nodes carrying samples of any infection in the tissues. This way, the immune system does not have to initially seek out the infection - it is brought to the immune system. The lymph nodes also provide a filter where lymph node macrophages remove many of the bacteria and viruses from the fluids to prevent the spread of the infection. Indeed, several lymph nodes may be strung in succession to ensure the filtering of infectious agents. Therefore, the immune response is localized and direct for a specific area of the body.

However, the results of the localized lymph node immune response is disseminated throughout the body. Antibodies and infection-seeking activated T_{DTH} cells and CTLs quickly reach the blood circulatory system and are spread throughout the body to prevent the spread of the infection. As mentioned above, these cells are actively recruited to the areas of infection by the factors produced as a result of the inflammatory response and activated immune cells. After the close of the immune response, the memory B and T cells continue to migrate throughout the body, spending varying amounts of time in each lymph node on the way. This insures that the memory cells will then be (or soon will be) at the appropriate lymph node to respond to a second encounter with the antigen wherever it may occur.

4   Biological Approach to System Information Security (BASIS)

A modern information security system (ISS) is considered as a number of independent, largely autonomous, network-based, specialized software agents operating in a coordinated and cooperative fashion designated to prevent particular kinds of threats and suppressing specific types of attacks. The modern multi-agent system technology presents a valuable approach for the development of an ISS that, when implemented in a distributed large scale multi-purpose information system, is expected to have important advantages over existing computer security technologies.

There are several principles of the immune system [1, 2] that can be applicable to information security. These principles can make ISS more robust and reliable.

Distributability. There is no central or master cell/organ that is in charge of diagnostic of foreign cells, distribution and reproduction of antibodies, and immune system memory. This in turn implies that there is no single point of failure. This is a very desirable feature for ISS, since it not only avoids bottlenecks and vulnerability but also provides faster response and robustness. A multi-agent approach is able to accomplish this feature/principle for ISS.

Multi-barrier/mechanism. The immune system has multiple barriers or mechanisms to prevent an intruder cell to get in the body and cause harm. A foreign cell will face multiple barriers to penetrate and damage a body. ISS should have different mechanisms to deal with an undesirable piece of software. A combination of these mechanisms could render a highly effective security system.

Diverse mechanisms. Having different mechanisms greatly help the vulnerability of the immune system. These mechanisms may react to a similar antigen in a different way. Each mechanism has its weak points; however, the immune system as a whole is robust. A multi-agent ISS should have diversity in its agents to reduce vulnerability.

Self-rule (autonomous) cells. Most of the cells in the immune system require no management from other places. Each cell with its own mechanism determines the proper reaction to a foreign cell or request from other cells. This feature helps a great deal in providing a fast reaction to an attack and finding a proper response. As mobile code becomes more common practice, autonomous agents will be required to have a more effective treatment of undesirable software codes.

Adaptability and memory. The immune system is capable of recognizing new antigens and figure out the proper response. The immune system is also capable of remembering antigens that it has dealt with before. The immune system constantly makes a space/time tradeoff in its detector set; at a given time the system maintains a sample of its detector set. In ISS, it is difficult to recognize a new threat. An ISS should be able to learn how to detect new threats based on previous experiences. New threats may be recognized by their abnormal behavior. The agents that have been more successful in combating attacks should be kept and update while agents with no success should be either set in resting place (store) or mutated to have new agents.

4.1   Main BASIS Components

Conceptually, a modern multi-agent ISS is viewed as a well coordinated and highly-cooperative team of the following types of agents distributed both across the network and on the host computer itself [3, 4]. The basic agents include [7, 8, 9, 10]:

(1) Access control agents that constrain access to the information according to the legal rights of particular users by realization of discretionary access control rules (ACR) specifying to each pair "subject - object" the authorized kinds of messages. Various access control agents cooperate for the purpose of maintaining the compliance with discretionary ACR on various sites of network. These agents supervise the flows of confidential information by realization of mandatory ACR not admitting an interception of confidential information. These agents act as part of the computer network's skin or innate system response.

(2) Audit and intrusion detection agents detecting non-authorized access and alerting the responsible system (agent) about potential occurrence of a security violation. As a result of statistical processing of the messages formed in the information system, these agents can stop data transmission processes, inform the security manager, and specify the discretionary ACR. A statistical learning process, crucial for the successful operation of these agents, is implemented. It utilizes available

information about normal system operation, possible anomalies, non-authorized access channels and probable scripts of attacks. These agents are used as part of the computer network humeral response. Here agents that have proved to be successful are kept active while the others are either discarded or stored.

(3) Anti-intrusion agents responsible for pursuing, identifying and rendering harmless the attacker. Anti-intrusion agents are the parallel of antibodies in the immune system. These agents use knowledge about potential attackers in a similar fashion that an antibody can recognize an antigen. Once the attacker is identified the agent neutralizes it. It should be pointed out that there is no buddy system in our approach as in the immune system with a Helper T lymphocyte (Th) cell.

(4) Diagnostic and information recovery agents assessing the damage of unauthorized access. These agents can be seen as part of the cell-mediated response of the computer network system. Here the agents assess the damages (or potential damages) and prescribe an appropriate response.

(5) Cryptographic and steganography agents providing safe data exchange channels between the computer network sites. These can be seen as part of the way the immune system communicates with different cells (and organs). Here is important to stress the importance of reliable communication between nodes in the network, since agents are distributed all over the network. Thus, access to the proper agent depends greatly on a reliable communication channel.

(6) Authentication agents responsible for the identification of the source of information, and whether its security was provided during the data transmission that provides the identity verification. They assure the conformity between the functional processes implemented and the subjects initiated by these processes. While receiving a message from a functional process, these agents determine the identifier of the subject for this process and transfer it to access control agents for realization of discretionary ACR. In is extremely important to set apart self and non-self messages; authentication agents need to use immune system techniques to achieve this task.

(7) Meta-agents that carry out the management of information security processes, facilitate coordinated and cooperated behavior of the above agents and assure the required level of general security according to some global criteria.

It could be observed that functions of a number of ISS agents are consistent with the specific functions performed by the components of the biological immune system. Since verbal definitions of the above problems are well established, the BASIS team will utilize its expertise in modern immunology to detect similar tasks performed by the immune system and to establish the qualitative and mathematical description of the relevant immune problems. When applicable, "immune" algorithms will be formalized, implemented in software and subjected to thorough investigation. While this paper presents only an outline of the proposed research, consequent publications will feature current and future effort in this direction.

4.2   Genetic Scheme

In our approach we intend to use other biologically inspired solutions such as genetic schemes. Each of the messages that enter in the network will be assigned a genetic print that is based on its message id, destination, type of commands (system calls), and sequence of commands. As a message enters into the network, its genetic print (chromosome) is generated by the host computer. This host computer uses a fitness function to message's chromosome to determine if the message is suited to enter into the network. If message is not fit to enter the network, this message is analyzed further to determine if it is indeed an undesired message. If the message is found to be an authorized one, the fitness function needs to be modified to allow evolution to take place; otherwise, this message is discarded. The host computer analyzes trends in the chromosomes to detect potential large number of "clones" in this population. The host computer collects a representative chromosome sample and sends it to other hosts in the network. This approach will allow the network to identify a large number of clones that come from different parts of the network. This in turn could help to prevent denial of service problems.

The fitness function is modified to allow new applications to be part of the system. As new applications become dominant in the network, the system should learn to let these applications to pass messages in the network. Thus, learning is be accomplished by allowing the fitness function be modified. Since the population (messages) is constantly changing, it is extremely important to include a flexible fitness function to allow changes in different generations. If a message does not meet the fitness function requirements, this is analyzed by a meta-agent to determine if indeed this message is a dangerous one. If the message is determined as a non-dangerous the fitness function needs be modified.

Having a genetic scheme could be seen as a buddy system (along with an immune system scheme) to information security. This buddy system could add robustness to network security, since due to its diversity more potentially attacks can be detected.

5   Conclusion

In this paper, we have presented a biological approach to deal with information security on a heterogeneous network of computers. This approach involves a distributed multi-agent scheme along and a genetic approach. This scheme provides implements a buddy system where the two approaches complement each other and provide a better information security system.

Acknowledgement. The authors are grateful to the Air Force Research Laboratory at Rome NY for funding this research.

References

1.   Forrest, S., Hofmeyer, S.A., and Somayaji, A.: Computer Immunology. Communication of the ACM, Vol. 40, No. 10, (October 1997) 88–96
2.   Somayaji, A., Hofmeyr, S., and Forrest, S.: Principles of a Computer Immune System. 1997 New Security Paradigms Workshop, Langdale, Cumbria, UK (1997) 75–82
3.   Crosbie, M., Spafford, E.: Active Defending of a Computer System using Autonomous Agents. Technical Report No. 95-008. COAST Group, Purdue University, (1995) 1–15

4. Balasubramaniyan, J., Garcia-Fernandez, J., Isakoff, D., Spafford, E., and Zamboni, D.: An Architecture for Intrusion Detection using Autonomous Agents. In Proceedings of the 14th Annual Computer Security Applications Conference, Phoenix, Arizona. (December 7-11, 1998)
5. Kuby, J.: Immunology. 3rd Edition. W.H. Freeman and Co., New York (1997)
6. Janeway, C.A., Travers, P., Walport, W., and Capra, J.D.: Immunobiology. The immune system in health and disease. Garland Publishing, New York (1999)
7. Stolfo, S.J., Prodromidis, A.L., Tselepis, S., Lee, W., Fan, D.W., and Chan, P.K.: Jam: Java agents for meta-learning over distributed databases. In Proceedings of the 3rd International Conference on Knowledge Discovery and Data Mining, Newport Beach, CA, (1997) 74–81
8. White, G., Fish, E., and Pooch, U.: Cooperating Security Managers: A Peer-Based Intrusion Detection System. IEEE Network (January/February 1996) 20–23
9. Stillman, M., Marceau, C., and Stillman, M.: Intrusion Detection for Distributed Applications. Communications of the ACM, Vol. 42, No. 7, (July 99) 63–69
10. Warrender, C., Forrest, S., and Pearlmutter, B.: Detecting Intrusions Using System Calls: Alternative Data Models. IEEE Symp. on Security and Privacy, (1999) 133–145

Learning Temporal Regularities of User Behavior for Anomaly Detection

Alexandr Selez nyov, Oleksiy Mazhelis, and Seppo Puuronen

Computer Science and Information Systems Department
University of Jyv" askyl'a
P.O. Box35, FIN-40351, Jyv" askyl'a, Finland
alexandr,mazhelis,sepi@it.jyu.fi

Abstract.   Fast expansion of inexpensive computers and computer networks has dramatically increased number of computer security incidents
during last years. While quite many computer systems are still vulnerable
to numerous attacks, intrusion detection has become vitally important
as a response to constantly increasing number of threats. In this paper
we discuss an approach to discover temporal and sequential regularities in user behavior. We present an algorithm that allows creating and
maintaining user pro les relying not only on sequential information but
taking into account temporal features, such as events' lengths and possible temporal relations between them. The constructed pro les represent
peculiarities of users' behavior and used to decide whether a behavior of
a certain user is normal or abnormal.
Keywords: Network Security, Intrusion Detection, Anomaly Detection,
Online Learning, User Pro ling, User Recognition

1   Introduction

During last decade we may witness an "explosion" of PC hardware and software development. As a result of this "explosion" inexpensive computers and
computer networks are getting involved in more areas of human life. However,
becoming cheaper and faster, computer systems, at the same time, tend to be
more and more complicated introducing new weaknesses that may be exploited in
order to penetrate systems' defenses. Therefore, there is a strong need to provide
means for detecting security breaches, i.e. identify intruders, detect intrusions,
and collect evidences. Intrusion detection is aimed to ful ll this role.

Intrusion detection systems use a number of generic methods for monitoring
of vulnerabilities exploitations. Basically, the intrusion detection approaches may
be divided into two main categories: misuse and anomaly detection [2]. Systems
based on misuse detection attempt to detect intrusions that follow well- known
patterns of attack (signatures) exploiting known software vulnerabilities.

Intrusion detection systems of the second category (for example IDES [1])
are detecting abnormal behavior or use of computer resources. They classify
usual or acceptable behavior and report other irregular behavior as potentially
intrusive. The techniques used in anomaly detection are varied. Some rely mainly

V.I. Gorodetski et al. (Eds.): MMM-ACNS 2001, LNCS 2052, pp. 143–152, 2001.

on statistical approaches and resulted in systems that have been used and tested extensively.

In this paper we focus on user-oriented anomaly detection by formulating an anomaly detection problem as one of classi cation of user behavior (by comparison with his pro le) in terms of incoming multiple discrete sequences. We develop an approach that allows creating and maintaining users' behavior pro les relying not only on sequential event information but taking into account events' lengths and possible relations between them. In particular, we develop a mathematical background that is used to nd deviations from normal behavior, then to decide whether it intrusion or normal behavior changes, and automatically update encoded patterns if necessary.

A de nition and description of basic concepts are given in Section 2. In Section 3 we present algorithm to extract behavioral patterns from a stream of events. Section 4 shows how to detect abnormal behavior by comparing a current one with user pro le. Finally, section 5 concludes this paper.

2   Building User Pro le

In this section we provide necessary de nitions and describe basic concepts of our approach. Here we consider a problem of user pro le building as a bringing to conformity events, provided by operating system log facilities, with notions used to build a user behavioral model. At the beginning we introduce   a layer structure of events, which is a three levels used for describing incoming information on di erent abstraction levels.

Layer is a concept generalization level of relations between occurrences, each of which represents a single event on a di erent abstraction level. The higher layer the more general and more descriptive the notions describing the user behavior are. Thus, on the highest layer we describe the user behavior using most general way not depending on a source where information is coming from.

At the lowest — instant layer all occurrences are represented as a time point (instant) on an underlying time axis. A single occurrence on this layer is called an event. Information on this layer is source-dependent (for example, it depends on operating system or logging facility used to collect it). On the next —   action layer, events with their simple relations are described and form actions. The action is considered as a temporal interval.

The last is an activity layer that is represented by actions with relations between them. It describes user behavior in a general source- independent way. A single occurrence on this level we call an activity . A relation between two actions (temporal intervals) is de ned as one of Allen's interval temporal relations [5].

These three layers are the way by which systems classify certain patterns of change. No one is more correct than other, although some may be more informative for certain circumstances. They are aimed to manage incoming information from multiple sources. For example, if system detects that a WWW browser is active and it exchanges information using HTTP protocol then it may conclude that user is browsing WWW pages in Internet. If the system observes network packets' headers it may come to the same conclusion when it detects connection establishment between user workstation and some server on port 80 followed by an information exchange. Therefore, our point is | by monitoring di erent sources (sequences of events) it is possible to come to the same conclusions or, in other words, build a stream of user activities, which are source and platform independent. Later it is used to obtain information about normal user behavior, encode it in a set of patterns that represents common sequences in user behavior, and nally use these patterns to detect attacks. The set of patterns forms a general model of user behavior that is presented by a probabilistic tree, where the patterns form branches of this tree. However, the sequential data is not the only information that is possible to get from a stream of discrete happenings. It also contains some hidden information that is usually neglected: time relations between events. Therefore, our approach is also aimed to extract and encode temporal patterns as well as sequential. The temporal information is added to every node and edge of the probabilistic tree and thus, we are going to call it as a temporal-probabilistic tree .

2.1   Temporal-Probabilistic Tree

As a general concept we de ne   action class — A, which describes one of the possible kinds of actions. It provides a formal description of an action without providing any speci c details. Action class contains descriptions of events that start and end that action and possible (or/and necessary) events between them. A number of all possible actions is limited by operation system and installed software; therefore, it is nite and known beforehand.

An instance — I^A describes a certain group of actions that belong to the same action class and have similar temporal characteristics. Since an action has a beginning and an ending, it can be described by a temporal interval   T. These lengths must be distributed normally for actions in order to be grouped into a same instance.

Each instance has information to which action class it belongs and has statistical information about actions forming this instance: characteristics of length distribution μ and   , and temporal interval [T_{min} , T_{max}] that limits the length of the actions in this instance. Also, a class may have "free" actions — n_{act} in it. The meaning of n_{act} is to register all actions of a class that do not belong to any pattern yet.

In order to build a temporal-probabilistic tree, sets of actions, with similar temporal characteristics, have to be formed. Then, it is necessary to nd sequences in these sets in a way that in each sequence corresponded pairs of adjacent instances would be connected by temporal relations that have similar temporal characteristics. We call temporal relations as   transitions and these sequences as patterns .

Here we assume that user typical behavior corresponds to individual temporal- probabilistic tree $S(G, E)$, where $G(I, n)$ is a set of nodes and $E(;)$ — set of edges. Every node $G_i^k \in G\}$ on level k represents a certain instance (level k is de ned as a shortest distance from this node to the root). In order to know how often a certain node is used,   n is stored for each node. It is a number of times node G_i^k has been involved into classi cation process.

A single edge is described as $E_i^k(\;;\;)$, and it represents a certain transition between connected nodes from consecutive levels. It connects node on level   k − 1 with G_i^k on level k. By analogy with a node, it has   μ and   parameters of transition lengths distribution. They are calculated using t_i time history of temporal distances between actions. Here t_i — temporal distance between two actions of corresponding instances.

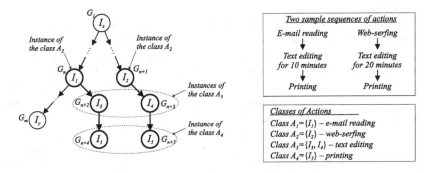

Fig. 1.   An example of a temporal-probabilistic tree

In gure 1, an example of a temporal-probabilistic tree is shown. Several crucial aspects may be identi ed in this gure:

- each branch of the tree represents one or several alternatives of user behavior and, correspondingly, one or several user behavior patterns;
- two nodes of the tree may represent a same instance;
- two actions of the same class may have di erent lengths and, as a result, belong to di erent instances.

Each branch of the tree or its part may represent a certain pattern(s) of user behavior. Therefore, a pattern can be described as $M_i(G^0, E^0, n)$ 2 M, where M — is a set of user's patterns, G^0 2 f G}, E^0 2 f E }, and n — number of times this pattern was used. Thus, it is also possible to de ne and consider a temporal-probabilistic tree as a set of patterns   S(G, E) f M}.

Actions grouped into instances and transitions between them are used as a building material for patterns (and trees) creation. Not all of them are taken into account. Below we consider criteria that a ect selection.

A set of actions of the same class may form an instance I_i that has temporal parameters:

$$
\mu = \frac{\overset{X^n}{\underset{i=1}{}} T}{n}; \qquad = \sqrt{\frac{\overset{X^n}{\underset{i=1}{}} (T - \mu)^2}{n^2}} \tag{1}
$$

This instance is taken into account by classi cation process if following conditions are met:

$$T_{max}, \quad \text{and} \quad n \quad n_{min}. \tag{2}$$

Here parameter n_{min} de nes a minimal amount of actions that may form a pattern — it has to be statistically big enough to minimize possible mistakes. This variable a ects the pattern extracting process, i.e. regulates its sensitivity. The less value n_{min} has the more instances (and hence patterns) of user behavior the process may nd. Therefore, it is possible to automatically vary the size of the tree by changing n_{min}.

Variable T_{max} de nes an upper limit for the deviation that is allowed for the action length to be included in a certain instance. This variable is chosen experimentally; it a ects the algorithm's selectivity, meaning that the less value T_{max} has the more precise description of user behavior algorithm creates. Too big and too small values of T_{max} lead to false negatives and false positives correspondingly. Thus, by choosing carefully T_{max} it is possible to minimize amount of false negatives as well as false positives.

Temporal parameters of each relation between actions may be calculated similarly. As we de ned above there are two time points describing each action: where it begins (T_{beg}) and ends (T_{end}). Let e represent a point of time, which corresponds to either starting or nishing action's point $e \ 2 \ f \ T_{beg}, T_{end}$ }. Allow $t(e_1, e_m)$ to represent the length of temporal interval between time point e_1 of the action O_1 (belong to instance I_j) and time point e_m of the action O_m (belong to instance I_{j+1}); and action O_1 is "before" O_m, i.e. O_1 began earlier than O_m did. These actions would form a pair, if they both happened during the same activity. Let n be amount of such pairs of actions. Then for transition between I_j and I_{j+1}, temporal-probabilistic parameters would be:

$$\mu = \frac{\sum_{i=1}^{n-1} t_i(e_1, e_m)}{n}; \quad = \min_{e_1, e_m \ 2 f \ T_{beg}, T_{end}\}} \sqrt{\frac{\sum_{i=1}^{n-1} (t_i(e_1, e_m) - \mu_t)^2}{n^2}} \tag{3}$$

Let a result of expression 3 is produced when variables e_1 and e_m are initialized with some values e_{prev} and e_{next} correspondingly. If conditions 2 are met we can say that there is a stable[1] relation between actions O_1 and O_m. If such relation between two sets of actions is discovered it means that these sets of actions and relation between them can be used to describe user behavior and thus, it corresponds to the transition $E (; ; e_{prev}, e_{next})$ between nodes that represent instances I_j and I_{j+1}.

Meaning of variable t_{max} is similar to T_{max}. It is an upper limit for the deviation of the transition length. However, since temporal distance between pairs of actions connected by a stable relation varies in signi cantly smaller range than action lengths, the value of T_{max} should be chosen bigger than t_{max}.

---

[1] The relation is stable if cases grouped by it form a normal distribution.

3   Pattern Extraction

Taking as an input a set of instances, system creates new patterns and optimizes the tree by trying to merge its branches. While merging, a current pattern M_i is concatenated with another one — M_j, when sequence of user actions, represented with the M_j, is the most likely continuation of the sequence of user actions represented by the pattern M_i. Consider in details how the system can automatically perform it.

Each newly created instance I^A is assigned to the new pattern M_i. At the same time usage frequency n of this pattern is assigned to amount of actions that belong to this instance. After that, a new node is created that represents the instance I^A. This new node forms a new branch of the tree that at the moment contains the only node and represents the only pattern.

This operation is performed over a set of newly created instances. Thus, as a result a set of new patterns appears that contains only instances (as a simplest cases of patterns). After that the system needs to optimize the tree. To perform this it is necessary to sequentially analyze patterns sorted by the eld n (usage frequency). For each pattern system looks for concatenation possibilities (i.e. another pattern(s) to concatenate with). For the current pattern M_j the search is performed by following steps:

- Last node G_i of the pattern M_j is selected and an instance I^A that is represented by the node G_i is identi ed.
- A set L_i of actions is created that contains only actions that belong, at the same time, to both: node G_i and to the pattern M_j.
- Then for the node G_i a vector of edges is created. The jth element of this vector corresponds to an edge, connecting set of actions L_i of the node G_i with the set of actions L_j of the rst node G_j of the pattern M_j.
- For each element of obtained vector, parameters $; ; e_{prev}, e_{next}$ are calculated using formulas 3. When an element with the smallest value of is found, the pattern M_j which contains this element is selected. After that the system concatenates two patterns M_i and M_j.

Before concatenation it is necessary to calculate the amount of actions that will be included into a new pattern. Let concatenate two patterns M_i and M_j, and let $n_[$ be a number of relations forming an edge E_j between G_i (last node of pattern M_i) and G_j (rst node of pattern M_j).

The process of patterns concatenation depends on following conditions:

- how many actions of the pattern M_i are not being included into concatenated pattern: $n_{\ i} = n_i - n_[$.
- how many actions of the pattern M_j are not being included into concatenated pattern: $n_{\ j} = n_j - n_[$.
- whether the node G_j is a common node for several patterns.

Consider all possible cases in details.

1. If for the pattern M_i the following condition is met $n_i < n_{min}$ then all nodes of the pattern M_i are selected. Knowing these nodes we are able to identify the instances represented by these nodes. For each found instance amount of "free" actions is increased by n_i.

2. If the condition is not met then the pattern M_i is duplicated. While duplicating a new pattern M_i^0 appears. Both M_i^0 and M_i patterns are represented by identical sets of nodes and edges.

3. If for the pattern M_j the $n_j < n_{min}$ condition is met then:

   - Node G_j is shared between patterns. For the pattern M_j its own branch in the tree is created so that all nodes of this branch would not be shared with any other pattern. Then for the pattern M_j all the nodes are selected. Knowing nodes, instances represented by these nodes are identi ed. For each instance amount of \free" actions is increased by n_j.

   - Node G_j is not shared between patterns. All nodes of the pattern M_j are selected. For each instance of every selected node a number of "free" actions is increased by n_j.

4. If for the pattern M_j the above condition is not met then the pattern is duplicated producing a new pattern M_j^0 with identical sets of nodes and the edges. In case if it is necessary to duplicate M_i or M_j, than only a new pattern M_i^0 or M_j^0 takes part in concatenation.

Finally, patterns M_i^0 (M_i) and M_j^0 (M_j) are concatenated. It means that a set of l nodes of the pattern M_j^0 (M_j) (with the edges connecting them) becomes the l last nodes of the pattern M_i^0 (M_i). After that each instance I^A, for which amount of "free" actions has been increased during concatenation process, is checked whether it satis es to a following condition: n_{act} n_{min}. If it does, the system looks for a pattern, containing the only node that represents this instance. If such pattern is found, its usage frequency n is increased by n_{act}. In case when the condition is not met a new pattern M_i, that includes this instance, is created. The system checks whether the tree has already a branch started by node G_i that represents an instance I^A. If it does then this new pattern is assigned to be represented by this single node. If no t — a new node for a new pattern is created that represents the instance I^A. This new node forms a new branch of the tree that at the moment contains the only node.

The operation of concatenation is repeated until all patterns present in the tree are checked for concatenation possibilities among each other. When it is not possible to nd a pattern to concatenate any more, the temporal- probabilistic tree creation process is over. This tree is stored in   user pro le database   [4] and it is used for online user behavior classi cation.

Application of pattern concatenation on the tree produces less number of branches in it. However, they are longer, which means that they re ect more strictly user behavior peculiarities, and hence lead to a less amount of errors in user behavior classi cation.

4   Detecting Abnormal Behavior

As a measure showing how much a current user behavior differs from profiled one a coefficient of reliability [3] is used. At the beginning of a pattern matching process, an intrusion detection system does not have yet any evidence of either trust or distrust, therefore it assigns average coefficient of reliability to it. Then the coefficient of reliability is being updated for every active user according to deviations between current behavior and predicted. Below we consider the process of matching in details.

In our intrusion detection architecture presented in [4] there is a classification module that takes an action stream as an input and calculates a coefficient of reliability change for every action. At the beginning, when the first action O_i comes to the classifier, it starts the matching process by creating a new record related to this action.

If a node, that represents an instance I^A, is not present on a level 1 in a tree then the record cannot be created. In this case, the amount of "free" actions of the instance I^A is incremented by '1'.

A new matching process is launched (and correspondingly its record is created) if a new action comes for which its instance does not match any anticipated actions of all matching processes. Similarly, when a new action arrives and there is not any matching processes active, then a new matching process is initiated.

When a new action appears in a stream of actions, a classifier should determine what instance this action belongs to. After that it decides what node of the tree this action may correspond. In order to accomplish this task system searches among possible nodes on a next level, stored in a record of each matching process. During searching for the next node a transition is defined between the current node and preceding one. If there are several nodes for which the transition may be defined, then the node with starting time closest to μ_t is chosen. When the next node is found, a corresponding record of the matching process is updated. In case if there is no match found, the classifier looks for possibility to launch a new matching process.

During search for a next node each matching process is constantly checking time elapsed after current action O_i of this process and compares it with T_{lim} that is a temporal interval during which next action related to O_i is expected to appear. The value T_{lim} is calculated as $\mu_t + 3\sigma_t$, where μ_t and σ_t are the parameters of the longest transition connecting G_i^k with any node on level $k+1$ (that is this edge has the biggest value μ_t among all the edges coming out of G_i^k). If elapsed time interval is longer than T_{lim} then it is necessary to check whether there exists a pattern with last node G_i that refers to action O_i. For this pattern a value of usage frequency n is incremented by '1'. It may probably happen that such pattern does not exist. In this case, all nodes, that represent this pattern, are selected. For instances of selected nodes the amount of "free" actions is incremented by '1'. Regardless of the search result the current matching process has to be finished.

For every node G_i or edge E_i its mean and standard deviation recalculated each time it being used for classification according:

$$\mu_{T_i} = \frac{\mu_{T_{i-1}} \times 1 - T_{i-1} + T_i}{1}, \tag{4}$$

$$T_i = \frac{{}^s_{T_{i-1}}{}^2 \times 1 - (\mu_{i-1} - T_{i-1})^2 + (\mu_i - T_i)^2}{1^2}. \tag{5}$$

where T | time length of being classified action/transition;
1 — volume of time lengths' pool. That is a limit of temporal lengths history based on which temporal-probabilistic characteristics for nodes and edges are calculated.

On each step of classification a coefficient of reliability   r is updated accordingly: $r = r0 + c$ $_r$, where $r0$ | the value of the coefficient on a previous step, $_r$ | coefficient of reliability change on this step, and   c | coefficient that determines system's sensitivity to deviations.

The value of coefficient of reliability change should depend on temporal-probabilistic characteristics of an instance and a transition that connect previous node with current one. Besides, for the normal user behavior, average value of this coefficient should aspire to zero. Thus, a value of change for the coefficient of reliability may be calculated as:

$$_r = {}^T_r + {}^t_r,$$
$${}^T_r = f(T) - f(\mu_T + 0.67 {}_T), \tag{6}$$
$${}^t_r = f(t) - f(\mu_t + 0.67 {}_t),$$

where $f(x) = {}^p 2 {}^2 \times \exp((x - \mu)^2 / (2 {}^2))$ — probability density function of the instance's time distribution.

In our work we use an assumption that normal user behavior should lead to no average punishment. To satisfy this requirement shifted probability density function is chosen. We define this function accordingly to our assumption and suppose that for normal user behavior, the action (transition) times are lying within time interval [$\mu - 0.67$; $+0.67$] with probability 0,5. Therefore, probability density function is shifted to the value of $f(\mu + 0.67)$.

As can be seen from above, values of ${}^T_r$ and ${}^t_r$ can be positive as well as negative. The more action's (transition's) time differs from the mean of its instance the less value of the coefficient of reliability change has. If deviation of current action (transition) time differs from the mean more than 0   .67 , the user behavior is considered as abnormal rather than normal. Therefore, the coefficient of reliability is "punished" by negative value of ${}^T_r$ (${}^t_r$) and as a result the value of coefficient of reliability   r is reduced. If current action's (transition's) time differs from the mean less than 0   .67 , the user behavior is considered as normal. Therefore, the coefficient of reliability is "encouraged" by positive value of ${}^T_r$ (${}^t_r$) and thus, the value of coefficient of reliability   r is increased. At the end of matching process the coefficient of reliability   r is analyzed and if it is lower than a certain threshold the alarm is fired.

There are two cases when it is impossible to calculate coefficient of reliability change using the method described above:

- a new action cannot be assigned to any of possible nodes as well as to any of rst states in a pattern set;
- after appearance of the current action G_i^k in a stream of actions, time more than T_{lim} has passed; and there is no any pattern existing that has G_i^k as a nal state.

It is obvious, that in these cases the temporal-probabilistic tree does not contain such relation and, therefore, the coe cient of reliability punishment is calculated as: $_r = f(\mu_T + 3 _T) - f(\mu_T + 0.67 _T)$. As can be seen from this formula, in spite of the probability that this action may belong to the interval $[\mu_T - 0.67 _T, \mu_T + 0.67 _T]$, the value of $_r$ is negative. Hence, the coe cient of reliability is decreased showing that the current behavior does not correspond to the user pro le.

5   Conclusions

In this paper we proposed a new approach for anomaly detection. It introduces features of arti cial intelligence in anomaly intrusion detection and employs online learning. The main assumption behind this approach is that the behavior of users follows regularities that may be discovered and presented using the temporal-probabilistic networks.

The presented approach allows detecting anomalies in user behavior by comparing events caused by user in operating system with his pro le. The pro le consists of extracted temporal and sequential patterns grouped into temporal-probabilistic network in a form of tree. Using these pro les a monitoring system evaluates every user action according to its length and temporal position (relation) relatively to others actions. According to the evaluation a coe cient of reliability is changed. Basing on it a decision is made whether the current behavior is normal or anomalous.

References

1. Lunt, T. and Tamaru, A. and Gilham, F. and Jagannathan, R. and Neumann, P. and Javitz, H. and Valdes, A. and Garvey, T.: A Real-Time Intrusion detection Expert System (IDES) — Final Technical Report. Computer Science Laboratory, SRI International, Technical (1992)
2. Smaha, S.: Tools for Misuse Detection. ISSA'93, Crystal City, VA (1993)
3. Seleznyov, A. and Puuronen, S.: Anomaly Intrusion Detection Systems: Handling Temporal Relations between Events. 2nd International Workshop on Recent Advances in Intrusion Detection, Lafayette, Indiana, USA (1999)
4. Seleznyov, A. and Puuronen, S.: HIDSUR: A Hybrid Intrusion Detection System based on Real-time User Recognition. 11th International Workshop on Database and Expert Systems Applications, Greenwich-London, England (2000)
5. Allen, J.: Maintaining Knowledge About Temporal Intervals. Communications of the ACM, 26, (1983) 832–843

Investigati ng and Eval uati ng Behavioural Profiling and Intrusion Detection Using Data Mining

Harjit Singh, Steven Furnell, Benn Lines, and Paul Dowland

Network Research Group, Department of Communication and Electronic Engineering,
University of Plymouth, Drake Circus, PLA 8AA, United Kingdom
{hsingh, sfurnell, blines, pdowland}@plymouth.ac.uk

Abstr act. The continuous growth of computer networks, coupled with the increasing number of people relying upon information technology, has inevitably attracted both mischievous and malicious abusers. Such abuse may originate from both outside an organisation and from within, and will not necessarily be prevented by traditional authentication and access control mechanisms. Intrusion Detection Systems aim to overcome these weaknesses by continuously monitoring for signs of unauthorised activity. The techniques employed often involve the collection of vast amounts of auditing data to identify abnormalities against historical user behaviour profiles and known intrusion scenarios. The approach may be optimised using domain expertise to extract only the relevant information from the wealth available, but this can be time consuming and knowledge intensive. This paper examines the potential of Data Mining algorithms and techniques to automate the data analysis process and aid in the identification of system features and latent trends that could be used to profile user behaviour. It presents the results of a preliminary analysis and discusses the strategies used to capture and profile behavioural charac-teristics using data mining in the context of a conceptual Intrusion Monitoring System framework.

Keywords. Data Mining, Intrusion Detection Systems, Knowledge Discovery, Behavioural Profiling, Intelligent Data Analysis.

1 Introduction

The increasing reliance upon IT and networked systems in modern organisations can have a calamitous impact if someone deliberately sets out to misuse or abuse the system. Systems may be affected by internal and external categories of abuser, as a result of both mischief and malice, leading to a range of undesirable consequences for the affected organisations (e.g. disruption to activities, financial loss, legal liability and loss of business goodwill). A recent study conducted by the US Computer Security Institute (CSI), in collaboration with the FBI, reported that 70% of respondent organisations had detected unauthorised use of their computer systems in the previous 12 months [1] – which represented an 8% increase on previous findings from 1999. The level of reported incidents highlights the paucity of security measures in current systems and, hence, the need for more comprehensive and reliable approaches. In particular, it can be suggested that traditional user authentication and access controls (e.g. passwords and user/group-based file permissions) are not sufficient to prevent determined cases of abuse or re-occurrence, in the case of successfully breached account(s), and misuse occurring from a legitimate user.

V.I. Gorodetski et al. (Eds.): MMM-ACNS 2001, LNCS 2052, pp. 153–158, 2001.
© Springer-Verlag Berlin Heidelberg 2001

Having passed the frontline controls and having the appropriate access privileges, the user may be in the position to do virtually anything without being further challenged. However, appropriate monitoring and analysis of user activity within an active session may potentially reveal patterns that appear abnormal in relation to their typical behaviour, or which are compatible with the sign of recognised intrusion scenarios. It is from this perspective that many Intrusion Detection Systems (IDS) have been conceived. Various IDSs [2, 3] have been proposed, which generally can be categorised based on the data source, audit trails or network traffic data, and intrusion model employed, anomaly detection or misuse detection model. The approaches used are generally focused on providing continuous monitoring and involve analysing vast amounts of audit trails, which in an eight-hour period can amount to 3-35MB [4] of data generated.

There is an increasing need for a more coherent paradigm for audit processing in terms of automating the data analysis stages. The current trend of network components providing audit trail or audit logs provides the foundation for IDSs to explore database automated match and retrieval technologies. This can be seen in audit processor components for instance the SecureView in the Firewall-1 using Data Mart to store the audit trails [5]. This available information could be used for security audit trail analysis in IDSs by utilising the technology in the data analysis stages. The need to eliminate the manual and ad-hoc approaches in the data analysis stages in IDSs is attracting interest in applying Intelligent Data Analysis (IDA) techniques. In this paper is discussed the potential of Data Mining (DM) algorithms and techniques as an IDA tool. We use DM to automate the data analysis process in identifying system features and latent trends for classifying user behaviour from the collected audit trails. DM is a rapidly expanding field which, has been exploited in lucrative domains such as the financial [6] and communications [7] sectors. Although some reported work has been carried out to analyse network traffic data [8, 9], none has been carried out in analysing host-based audit trails using DM for the purpose of user authentication, which is the focus of this research work.

2   Data Mining

Data Mining can be described as a collection of techniques and methodologies used to explore vast amounts of data in order to find potentially useful, ultimately understandable patterns [10] and to discover relationships. DM is an iterative and interactive process, involving numerous steps with many decisions being made by the user. The fundamental goals of data mining are finding latent trends in data, which enable prediction and description [11] of the analysis phases. Different algorithms are optimised based on the predefined DM task. This involves deciding whether the goals of the DM process are classification, association, or sequential [10]. Classification has two distinct meanings. We may aim to classify new observations into classes from established rules or establishing the existence of classes, or clusters in data [12]. Association attempts to generate rules or discover correlation in data and is expressed: $X \Rightarrow Y$, where X and Y are sets of items. This means that an event or a transaction of a database that contains X tends to contain Y. Sequential looks at

events occurring in a sequence over time or time-ordered sequences. This could be
H[SUHVVHG WKURXJK WKH IRO⊕RZ⟨L⟨Q IRU⟨H⟩W RI HYHQW W\SHV DQG DQ H
SDLU $ W   ZKHUH⟨(⟨LV DQ HYHQW W\SH DQG W UHSUHVHQWV WKH WLP
occurrence of an event. This is followed by predefined sets of possible intrusion
FODVVHV ZKHUH & LV D VHW RI LQWUX⟨V⟨⟨RQ⟨FOD⟩VVQW⟨UX⟨QLRQ W\SH KH
example: 90% of the time, if the event (A, t) occurs, it is followed by intrusion type I.
The subsequent process, once the DM task is defined can be derived from the four
main activities; selection, pre-processing, data mining and interpretation , also known
as post-processing [12].

3   Classification of User Behaviour

Distinguishing user behaviour patterns and classifying it as normel or intrusive is a
subtle task. Furthermore exploring the vast amount of audit trail data often yields a
small fraction of intrusion or misuse. Besides managing these tasks, IDSs have to
limit the errors that could occur from misclassification of user behaviour such as false

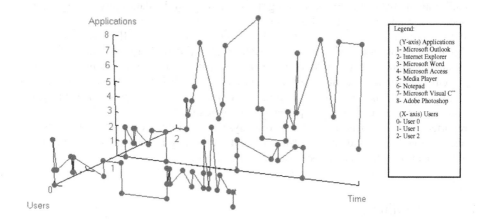

Fig. 1. Graphical representation of applications run by users

positive or false negative errors. Therefore, it is essential to ensure that accurate
profiles of users are established in order to improve the accuracy of intrusion
classification. Hence the need to gather as much information as possible pertinent to a
user's interaction with the system in order to distinguish between similar behavioural
patterns of users that could occur. Auditing the applications that users run could for
instance provide a distinctive pattern of the user's interaction with the system as
depicted in Figure 1 . Users patterns once identified could be incorporated into an
anomaly detector framework in conjunction with other key indicators of user
behaviour in order to identify unauthorised access when compared against this
distinctive usage patterns. Hence it is essential to collect as much information as
possible regarding such behavioural indicators in order to correlate the possibility of
intrusion.

4   Methodology

We use DM to extract latent patterns or models of user behaviour from the collected audit trail. This is then reflected in the DM algorithm classifiers (e.g. through rule induction) to recognise deviation, if it occurs, from normal use. This approach is based on the assumption that a user's behaviour has regularity and that using the classifiers this behaviour can be modelled. Using this analogy, anomalous behaviours can then be categorised as a possible unauthorised user or use of that system. The audit trail data analysed was collected from networked computers on a participating local area network (LAN) using an independent agent installed locally in order to audit user interaction with the system. This is based on the assumption that users performing their regular tasks will impose similarly regular demands upon system resources. Hence system features involved for continuous monitoring of user interaction with the system such as resource usage, process-related information such as creation, activation and termination, etc, is audited. Similar system features have been used in other published work [2]. However, previous work was focused on statistical and neural network analysis. A user's behaviour profile can be uniquely identified by: <user name, absolute time, date, hostname, $event_{1,...,}$ $event_n$ >, which is the semantic used for the audit trail where, $events_n$ denotes the system features being monitored.

4.1   Data Mining Audit Trail

The methodology used is derived from the four main activities of DM: selection, pre-processing, data mining and interpretation, and is as depicted in Figure 2. The

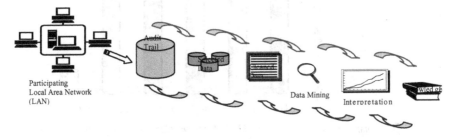

Fig. 2. Methodology for Behaviour Profiling

collected audit trail is split into various sample sizes. These subsets form the target data sets, which will undergo the analysis to identify patterns and to test specific hypotheses. The cleaned data, containing both categorical and numerical data, is then subjected to analysis by the DM algorithms. There are a wide variety of DM techniques available, each of which performs more accurately over certain characteristic data sets (e.g. numerical or categorical) and is also relative to the number of variables or attributes and classes. The Intelligent Data Analysis (IDA) Data Mining Tool [13] is used to analyse the sample data sets which incorporates algorithms from the fields of Statistical, Machine Learning and Neural Networks. Six algorithms, k-NN, COG, C4.5, CN2, OC1 and RBF were chosen for this investigative work. For the purpose of this work, the data sets were split into ratios of 9:1, 8:2 and 7:3, hence into two parts, which is a commonly used technique known as train and test. The algorithm or classifier is subjected initially with the training set and then the classification accuracy is tested using the unseen data set or testing set. The results give an indication of the error rate (or false positives) and the overall classification accuracy of the trained algorithms.

5   Results

The initial results obtained from the analysis as depicted in Figure 3, suggest that Machine Learning and Statistical-based algorithms are better for these types of data sets. C4.5 and OC1 decision tree based algorithms in particular, out performed the CN2 rule-based and RBF algorithms. The classification accuracy obtained, using k-NN in comparison to C4.5, shows some significance for further investigative work despite the slower classification times observed. Amongst the statistical algorithms, k-NN faired better then COG but is slower in comparison to the classification times observed. The classification accuracy obtained overall depicts RBF classification accuracy as inverse proportional to the sample sizes. These results support other reported work [12]. In addition to the consistency in classifying the data sets and the overall average classification accuracy, our initial investigations also identified that C4.5 has overall quicker train and test time and outputs explicit rules.

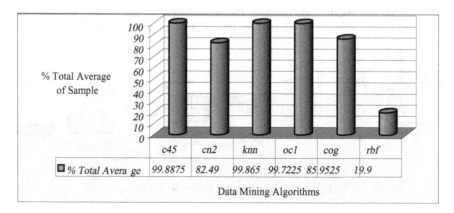

Fig. 3. Total percentage average classification accuracy of selected Data Mining algorithms

6   Discussion and Conclusion

The classification accuracy obtained suggests that DM techniques could be integrated into an IDS framework in order to provide a mechanism to detect intrusions. The approach used in these initial trials has shown the potential that DM techniques can be used to detect anomalies or intrusions through the behaviour model generated by the DM algorithm's classifiers. The high classification accuracy obtained and fast response time exhibited in classifying the user behaviour by some of the DM algorithms further demonstrate the potential of applying DM techniques within a real-time application for identifying intrusions [14]. Another important element identified is the interpreted rules obtained from the data mining process. The systems features outlined by the classifiers to detect anomalous behaviour can be used to detect known intrusions. The results so far have been based around the classifiers used that are optimised to classify either new observed user behaviour into classes from established rules or establishing the existence of classes using the DM algorithms. While this has been the fundamental goal in our approach, another important aspect of identifying user behaviour from frequent patterns developing over time has yet to be addressed.

References

1. Computer Security Institute, "2000 CSI/FBI Computer Crime and Security Survey", Vol. 6, No.1, SPRING-2000
2. Lunt, T.F.: IDES: an intelligent system for detecting intruders. Proc. of the Computer Security, Threat and Countermeasures Symposium, Rome, Italy (November 1990)
3. Mukherjee, B., Herberlein, L.T. and Levitt, K.N.: Network Intrusion Detection. IEEE Network-1994, Vol. 8. 3 26–41
4. Frank, J.: Artificial Intelligence and Intrusion Detection: current and future direction. Proc. of the 17th National Computer Security Conference (October 1994)
5. Amoroso, E.G.: Intrusion Detection: an introduction to internet surveillance, correlation, traps, trace back, and response. Intrusion.Net-1999, ISBN 0-9666700-7-8
6. Westphal, C. and Blaxton, T.: Data Mining Solution, Methods and Tools for Solving Real-World Problems. Wiley-1998, ISBN 0-471-25384-7, 531–585
7. Sasisekharan, R. and Seshadri, V.: Data Mining and Forecasting in Large-Scale Telecommunications Networks. IEEE Expert Intelligent Systems and Their Applications- 1996, Vol. 11. 1 37–43
8. Lee, W. and Stolfo, S.: Data Mining Approaches for Intrusion detection. Proc. 7th USENIX Security Symposium (1998)
9. Warrender, C., Forrest, S., Pearlmutter, B.: Detecting Intrusion Using Calls: alternative data models. Symposium on Security and Privacy (1999)
10. Fayyad, U.M.: Data Mining and Knowledge Discovery: making sense out of data. IEEE Expert-1996, Vol. 11. 6 20–25
11. Adriaans, P. and Zantinge, D.: Data Mining. Addison-Wesley-1998 , ISBN 0-201-40380-3
12. Michie, D., Spiegelhalter, D.J. and Taylor C.C.: Machine Learning, Neural and Statistical Classification. Ellis Horwood-1994, ISBN 0-13-106360-X, 136–141
13. Singh, H., Burn-Thornton, K.E. and Bull, P.D.: Classification of Network State Using Data Mining. Proc. of the 4th IEEE MICC & ISCE '99,Malacca, Malaysia, Vol. 1. 183–187
14. Furnell, S.M. and Dowland, P.S.: A Conceptual Architecture for Real-time Intrusion Monitoring. Information Management & Computer Security-2000, Vol. 8. 2 65–74

Typed MSR : Syntax and Examples

Iliano Cervesato[?]

Advanced Engineering and Sciences Division, ITT Industries, Inc.,
2560 Huntington Avenue, Alexandria, VA 22303 — USA
iliano@itd.nrl.navy.mil

Abstract. Many design flaws and incorrect analyses of cryptographic protocols can be traced to inadequate specication language s for message components, environment assumptions, and goals. In this paper, we present MSR, a strongly typed specication language for security protocols, which is intended to address the rst two issues. Its typing infrastructure, based on the theory of dependent types with subsorting, yields elegant and precise formalizations, and supports a useful array of static check that include type-checking and access control validation. It uses multiset rewriting rules to express the actions of the protocol. The availability of memory predicates enable it to faithfully encode systems consisting of a collection of coordinated subprotocols, and constraints allow tackling objects belonging to complex interpretation domains, e.g. time stamps, in an abstract and modular way. We apply MSR to the specication of several examples.

1   Introduction

The design and analysis of cryptographic protocols are notoriously complex and error-prone activities. Part of the diculty derives from subtleties of the cryptographic primitives. Another portion is due to their deployment in distributed environments plagued by powerful and opportunistic attackers. We claim that a third major source of problems arises from the use of ambiguous, complex or inexpressive languages for the specication of protocols, of the assumptions on their operating environment, and of their goals. The Dolev-Yao model of security [19], [14] tackles the rst problem by promoting an abstraction that has the eect of separating the analysis of the message flow from the validation of the underlying cryptographic operations. It assumes that elementary data such as principal names, keys and nonces are atomic rather than bit strings, and views the message formation operations (e.g. concatenation and encryption) as symbolic combinators. The cryptographic operations are therefore assumed to be flawless. This model is generally reasonable for authentication protocols and underlies most systems designed for protocol analysis, e.g. [5], [18], [16], [1], [13]. Within the Dolev-Yao model, the capabilities of the intruder are circumscribed.

[?] Partially supported by NRL under contract N00173-00-C-2086.

V.I. Gorodetski et al. (Eds.) : MMM-ACNS 2001, LNCS 2052, pp. 159–177, 2001.

They can be in many respects neutralized by relying on appropriate message formats [2], [22]. However, practical reasons, such as limited bandwidth, sometimes make such architectures inviable.

We claim that a signi cant source of faulty designs and contradictory analyses can be traced to shortcomings in the languages used to specify protocols. The popular "usual notation" relies on the Dolev-Yao model and describes a protocol as the sequence of the messages transmitted during an expected run. Besides distracting the attention from the more dangerous unexpected runs, this description expresses fundamental assumptions and requirements about message components, the operating environment and the protocol's goals as side remarks in natural language. This is clearly ambiguous and error-prone. Strand formalizations [16], like most modern languages, represent protocols as a collection of independent roles that communicate by exchanging message. Their reliance on a fair amount of natural language still makes it potentially ambiguous.

In [9], [15], we proposed MSR, a language based on multiset rewriting, as a formalism for unambiguously representing authentication protocols, with the aim of studying properties such as the decidability of attack detection. The actions within a role were formulated as multiset rewrite rules, threaded together by dedicated role state predicates . The nature and properties of message components was expressed in a relational manner by means of   persistent information predicates and to a minor extent by typing declarations. In particular, variables that ought to be instantiated to "fresh" objects during execution were marked with an existential quanti er. In [11], [10], we proved the substantial equivalence between MSR and extensions of popular formalisms such as strand spaces. Nonetheless, the resulting speci cations were not completely satisfactory for two reasons: persistent information proved di cult to reason about, and the rigid structure of MSR rules limited its applicability to basic authentication protocols.

This paper proposes a thorough redesign of   MSR and establishes this formalism as a usable speci cation language for security protocols. The major innovations include the adoption of a ﬂexible yet powerful typing methodology that subsumes persistent information predicates, and the introduction of memory predicates and of constraints on interpreted domains that signi cantly widen the range of applicability of this language.

The type annotations of our new language, drawn from the theory of dependent types with subsorting, enable precise object classi cations for example by distinguishing keys on the basis of the principals they belong to, or in function of their intended use. Therefore, the public key of any two principals can be assigned a di erent type, in turn distinct from their digital signature keys. Protocol speci cations, called protocol theories in   MSR, are strongly typed, and we have devised algorithms for statically catching type violations,   e.g. the use of a shared key to perform public-key encryption [7]. Our typing infrastructure can point to more subtle access control errors, such as a principal trying to encrypt a message with a key that does not belong to him [6].

Memory predicates allow a principal to remember information across role executions. Their presence opens the doors to the speci cation of protocols structured as a collection of coordinated subprotocols. In this paper, we exemplify this possibility by formalizing the Neuman-Stubblebine repeated authentication protocol [20], which lies outside the reaches of our previous version of   MSR. In [8], we use this technique to give a speci cation of the Dolev-Yao intruder that lies fully within the syntax of   MSR roles.

Constraints are another novelty of the language presented in this paper. They permit referring to objects belonging to complex interpretation domains in an abstract and modular way. Our speci cation of the Neuman-Stubblebine protocol [20] relies on constraints to verify the validity of timestamps: how these objects and their operations are implemented is invisible (and irrelevant) to the resulting protocol theory.

This presentation is organized as follow. In Section 2, we introduce the syntax of MSR. The next three sections formalize as many popular case studies: Section 3 implements the abridged version of the Needham-Schroeder public-key authentication protocol; Section 4 extends this speci cation to the full protocol, inclusive of the server activity; and Section 5 formalizes the Neuman-Stubblebine protocol. Section 6 summarizes the ideas discussed in this paper and hints at directions of future work.

2   Typed MSR

In the past, cryptoprotocols have often been presented as the temporal sequence of messages being transmitted during a "normal" run. Recent proposals champion a view that places the involved parties in the foreground. A protocol is then a collection of independent   roles that communicate by exchanging messages, without any reference to runs of any kind. A role has an owner, the principal that executes it, and speci es the sequence of messages that he/she will send, possibly in response to receiving messages of some expected form.   MSR adopts and formalizes this perspective. A role is given as a parameterized collection of multiset rewrite rules that encode the expected message receptions and the corresponding transmissions. Rule ring emulates receiving (and accepting) a message and/or sending a message, the smallest execution steps. The messages in transit, the actions and information available to the roles, and other data constitute the state of execution of a protocol. Rules implement partial transformations between states. Their applicability is constrained by the contents of the current state and by the satisfaction of guards. Execution is preceded by static type-checking [7] and access control validation [6] which limits the number of run-time checks and allows catching common speci cation errors early. This section describes the form of an   MSR speci cation. More speci cally, in Section 2.1, we de ne our notion of messages. In Section 2.2, we present the predicates that appear in a state, in turn de ned in Section 2.3. In Section 2.4, we introduce the typing infrastructure that allows us to make sense of these objects. In Section 2.5, we discuss rules and their constituents. Roles and protocol theories are de ned in Section 2.6.

2.1   Messages

Messages are obtained by applying a number of message forming constructs, discussed below, to a variety of   atomic messages . The atomic messages we will consider in this paper are principal identi ers   A, keys k, nonces n, timestamps T, and raw data m (i.e. pieces of data that have no other function in a protocol than to be transmitted). We formalize our notion of atomic message in the following grammatical productions:

$$a ::= A \mid k \mid n \mid T \mid m$$

We will also use B to denote a principal while we reserve the letter   S for servers. Although we limit the discussion in this paper to these kinds of atomic mes- sages, it should be noted that others can be accommodated by extending the appropriate de nitions.

The message constructors   we will consider consist of concatenation ($t_1 t_2$), shared-key encryption $\{t\}_k$, public-key encryption $\{\!\{t\}\!\}_k$, and digital signature $[t]_k$. Altogether, they give rise to the following de nition of a   message , or more properly a   term .

$$t ::= a \mid x \mid t_1 t_2 \mid \{t\}_k \mid \{\!\{t\}\!\}_k \mid [t]_k$$

Observe that we use a di erent syntax for shared-key and public-key encryption. We could have identi ed them, as done in many approaches. We choose instead to distinguish them to show the flexibility and precision of our technique. Similarly, we de ne digital signatures as an independent primitive operation rather than as asymmetric key encryption with a private key. As usual, [$t]_k$ denotes the term   t being signed, together with the signer's certi cate cryptographically constructed from t using the key k.

Again, other constructors, for example hash functions, can easily be accom- modated by extending the appropriate de nitions. We refrain from doing so since their inclusion would lengthen the discussion without introducing substantially new concepts.

A parametric message   allows variables x wherever terms could appear. We use a sans-seri ed font to denote possibly parametric principals   A (or B), keys k, nonces n, timestamps T and raw data m. Constants and variables constituted the class of elementary terms , denoted with the letter   e.

2.2   Message Predicates

Message predicates are the basic ingredient of states, de ned in Section 2.3. They are atomic rst-order formulas with zero or more terms as their arguments. More precisely, they are applied to ordered sequences of terms called   message tuples and denoted t̄.

The predicates that can enter a state or a rewrite rule are of three kinds:

- First, the predicate   N(_) implements the contents of the   public network in
  a distributed fashion: for each (ground) message   t currently in transit, the
  state will contain a component of the form   N(t).
- Second, active roles rely on a number of   role state predicates , generally one
  for each rule in them, of the form $L_1(\_,\ldots,\_)$, where l is a unique identifying
  label. The arguments of this predicate record the value of known parameters
  of the execution of the role up to the current point.
- Third, a principal   A can store data in private memory predicates of the
  form $M_A(\_,\ldots,\_)$ that survives role termination and can be used across the
  execution of di erent roles, as long as the principal stays the same.

The reader familiar with our previous work on   MSR will have noticed a number
of di erences with respect to the de nitions given in [9], [10]. Memory predicates
are indeed new. They are intended to model situations that need to maintain
data private across role executions: for example, this allows a principal to remember his Kerberos ticket, or the trusted-third-party of a fair exchange protocol
to avoid fraudulent recoveries from aborted transactions. Memory predicates
can further be used to represent such entities as local clocks, as we will see in
Section 5. Another di erence with respect to our earlier work is the absence of
a dedicated predicate retaining the intruder's knowledge. This can however be
easily implemented using memory predicates, as described in [8].

2.3   States

States are a fundamental concept in   MSR . They are the objects transformed by
rewrite rules to simulate message exchange and information update and, together
with execution traces, they are the hypothetical scenarios on which protocol
analysis is based. A   state S is a nite collection of ground state predicates:

$$S ::= \cdot \mid S, N(t) \mid S, L_1(\bar{t}) \mid S, M_A(\bar{t})$$

Protocol rules transform states. They do so by identifying a number of predicates, removing them from the state, and adding other, usually related, state
elements. The antecedent and consequent of a rewrite rule embed therefore sub-
states. However, in order to be applicable to a wide array of states, rules usually
contain variables that are instantiated at application time. This calls for a parametric notion of states. For the most part, this reduces to admitting variables in
embedded terms. However, role state predicates need to be created on the spot
in order to avoid interferences between concurrently executing role instances. We
achieve this by introducing variables, denoted   L, that are instantiated to actual
role state predicates during execution.

2.4   Types

While types played a very modest role in the original de nition of   MSR [9],
[10], they stand at the core of the extension presented in this paper. Through

typing, we can enforce basic well-formedness conditions (e.g. that only keys be used for encrypting a message), as described in detail in [7]. Types also provide a statically checkable way to ascertain complex desiderata such as, for example, that no principal may grab a key he/she is not entitled to access. This aspect is thoroughly analyzed in [6]. The central role of types in our present approach is witnessed by the fact that they subsume and integrally replace the "persistent information" of the original   MSR [10].

The typing machinery that best ts our goals is based on the type-theoretic notion of dependent product types with subsorting   [3], [21]. Rather than delving into the depth of the de nitions and properties of this formalism, we introduce only the facets that we will use, and only to the extent we will need them.

Types are syntactic constructions that are used to classify other syntactic expression, such as terms. By doing so, they give them a   meaning , saying for example that an object we interpret as a key is not a nonce. Whenever a key is used where a nonce is expected, something has gone wrong since the meaning of this term has been violated. The types we will use in this paper are summarized in the following grammar:

::= principal | nonce | shK A B | pubK A | privK k | sigK A | verK k | time | msg

Needless to say, the types " principal" and " nonce" are used to classify principals and nonces respectively. The next three productions allow distinguishing between shared keys, public keys and private keys. Dependent types o er a simple and flexible way to express the relations that hold between keys and their owner or other keys. Given principals " A" and " B", a shared key " k" between " A" and " B" will have type " shK A B". Here, the type of the key   depends on the speci c principals \ A" and " B". Similarly, a constant " k" is given type " pubK A" to indicate that it is a public key belonging to "   A". We use dependent types again to express the relation between a public key and its inverse. Continuing with the last example, the inverse of " k" will have type " privK k", from which it is easy to establish that it belongs to principal "   A". A similar design principle applies in the case of digital signatures: the signature key "   k" of principal " A" has type " sigK A" while its inverse, the veri cation key \   k⁰", has type " verK k". Timestamps are assigned type " time".

We use the type msg to classify generic messages. Clearly raw data have type msg. This is however not su cient since nonce, keys, timestamps, and principal identi ers are routinely part of messages. We solve this problem by imposing a subsorting relation between types. In this paper, each of the types discussed above, with the exception of signature keys and their inverses, will be a subtype of msg. With the appropriate array of typing rules (see [7]), not de ning signature and signature veri cation keys as subsorts of \   msg" has the e ect of banning these keys from well-typed messages, except as the unrecoverable indices of signed messages: any attempt at transmitting a signature key will be statically marked as violating the typing policy.

Again, the types and the subsorting rules above should be thought of as a reasonable instance of our approach rather than the approach itself. Other schemas can be speci ed by de ning appropriate types and how they relate to each other. For example, an application may nd it convenient to see each of the above types related to encryption or decryption as a subtype of a universal key type, say, key, in turn a subsort of msg. Alternatively, we may want to de ne distinct types for long-term keys and have them not be a subsort of   msg, prohibiting in this way the transmission of long-term secrets as parts of messages. We are already handling signature keys in this way.

Predicate symbols are assigned a type by listing the type of their arguments. It is tempting to de ne the type of a tuple as the sequence of the types of its components. Therefore, if A is a principal name and k_A is a public key for A, the tuple (A, k_A) would have type " principal× pubK A" (the Cartesian product   symbol "×" is the standard constructor for type tuples). This construction allows us to associate a generic principal with A's public key: if B is another principal, then (B, k_A) will have this type as well. We will often need stricter associations, such as between a principal and his own public key. In order to achieve this, we will rely on the notion of dependent type tuple . In this example, the tuple (A, k_A) will be attributed type " principal$^{(A)}$ × pubK A", where the variable A in "principal$^{(A)}$" records the name of the principal at hands and forces the type of the key to be "pubK A" for this particular A: therefore (A, k_A) is a valid object of this type, but (B, k_A) is now ill-typed since k_A has type " pubK A" rather than the expected "pubK B".[1]

We attribute a type to a term tuple by collecting the type of each constituent message, but we label these objects with variables to be used in later types that may depend on them. Thus, a   dependent type tuple is an ordered sequence of parameterized types:

$$\tau^- ::= \cdot \mid \tau^{(x)} \times \tau^-$$

Given a dependent tuple type $\tau^{(x)} \times \tau^-$, we will drop the label $^{(x)}$ whenever the variable x does not occur (free) in τ^-. The resulting simpli ed notation, $\tau \times \tau^-$, will help writing more legible speci cations when possible. As for term tuples, we will omit the leading " ·" whenever convenient.

2.5   Rules

The core of a rule has the form " lhs ! rhs". Rules are the basic mechanism that enables the transformation of a state into another, and therefore the simulation of protocol execution: whenever the antecedent "   lhs" matches part of the current state, this portion may be substituted with the consequent "   rhs" (after some processing).

It is convenient to make protocol rules parametric so that the same rule can be used in a number of slightly di erent scenarios (e.g. without xing inter- locutors or nonces). A typical rule will therefore mention variables that will be

---

[1] Our dependent type tuples are usually called strong dependent sums in the type theoretic community, and the standard notation for the dependent type tuple we have written as " principal$^{(A)}$ × pubK A" is " A   : principal. pubK A". We believe that our syntax is likely to be more clear to the target audience of this paper.

instantiated to actual terms during execution. Typed universal quantiers can conveniently express this fact. This idea is captured by the following grammar:

$$r ::= lhs \rightarrow rhs \mid \forall x : \tau . r$$

Both the right-hand side and the left-hand side of a rule embed a nite collection of parametric message predicates , some ground instance of which execution will respectively add to and retract from the current state when the rule is applied:

$$\bar{P} ::= \cdot \mid \bar{P}, N(t) \mid \bar{P}, L(\bar{e}) \mid \bar{P}, M_A(\bar{t})$$

Observe that predicate sequences dier from states (see Section 2.3) mainly by the limited instantiation of role state predicates: in a rule, these objects consist of a role state predicate variable applied to as many elementary terms as dictated by its type (this is enforced by the typing rules in [7]). Recall that elementary terms are either variables or atomic message constants. Network and memory predicates will in general contain parametric terms, although not necessarily raw variables as arguments.

The Dolev-Yao model [14] champions a symbolic interpretation of cryptographic primitives that reduces messages to expressions in an initial algebra. Some of the components of a message, such as timestamps, are however subject to operations or tests that are not conveniently expressed in this way. We reconcile the simplicity of the Dolev-Yao model and the necessity to accommodate objects drawn from complex interpretation domains by treating the latter as atomic constants when embedded in messages, but by relying on dedicated constraint handlers to perform operations and resolve tests. These invocations enter the syntax of MSR as constraints in the left-hand side of a rule. We use the letter to denote them. Constraints are not part of the state, but should rather be thought of as guards to the applicability of a rule.

In this paper, the only message constituents that require a constraint handler are timestamps. We need to check their validity against the current time, which is modeled by arithmetic constraints involving the usual ordering relations; a possible such constraint could be $(\ T < T_{now}\ )$ where T and T_{now} are respectively the timestamp and the current time. We will also need to set alarms T_{alarm} by adding predetermined temporal values, say T_{val}, to the current time T_{now}. This operation is expressed as the constraint $(\ T_{alarm} = T_{now} + T_{val}\ )$. The domain of timestamps will correspond to the real numbers or any suciently precise approximation supported by the implementation at hand.

The use of constraints allows for an abstract architecture since it isolates the specication of interpretation domains away from the formalization of the security protocols that use them. The interface is limited to a few type declarations and a syntax for the operations and tests that can enter a constraint. Constraint handlers are then external and interchangeable modules that can be plugged to the protocol specication on demand.

The left-hand side , or antecedent , of a rule is a nite collection of parametric message predicates guarded by nitely many constraints on interpreted data:

$$\text{lhs} ::= \bar{P} \mid \text{lhs};$$

The right-hand side , or consequent , of a rule consists of a predicate sequence possibly pre xed by a nite string of fresh data declarations such as nonces or short-term keys. We rely on the existential quanti cation symbol to express data generation:

$$\text{rhs} ::= \bar{P} \mid 9x : : \text{rhs}$$

2.6   Roles and Protocol Theories

Role state predicates record the information accessed by a rule. They are also the mechanism by which a rule can enable the execution of another rule in the same role. Relying on a xed protocol-wide set of role state predicates is dangerous since it could cause unexpected interferences between di erent instances of a role executing at the same time. Instead, we make role state predicates local to a role by requiring that fresh names be used each time a new instance of a role is executed. As in the case of rule consequents, we achieve this e ect by using existential quanti ers: we pre x a collection of rules   that should share the same role state predicate   L by a declaration of the form " 9L : ¯ ", where the typed existential quanti er indicates that   L should be instantiated with a fresh role state predicate name of type ¯ .

With this insight, the following grammar de nes the notion of   rule collection :

$$::= \quad j \ 9L : \bar{\ } : \quad \mid r;$$

It should be observed that this de nition allows for role state predicate parame-ters declarations and rules to be interleaved in a rule collection. We will however generally divide a collection in a   preamble where all roles state parameters are declared, and a body that lists the rules that constitute a role.

A role is given as the association between a   role owner A and a collection of rules . Some roles, such as those implementing a server or an intruder, are intrinsically bound to a few speci c principals, often just one. We call them anchored roles and denote them as ^{A}. Here, the role owner A is an actual principal name, a constant. Other roles can be executed by any principal. In these cases A must be kept as a parameter bound to the role. These   generic roles are denoted ^{8A}, where the implicitly typed universal quanti cation symbol implies that A should be instantiated to a principal before any rule in   is executed, and sets the scope of the binding to   . Observe that in this case   A is a variable.

We require that the owner of a role   be the rst argument of all the role state predicates in the rules that constitute it. This object shall also be the subscript of every memory predicate in   . These constraints are formally expressed in the typing and access control policy of   MSR [7], [6].

A protocol theory , written P , is a nite collection of roles:

$$P ::= \cdot \mid P; \ {}^{8A} \mid P; \ {}^{A}$$

It should be observed that we do not make any special provision for the intruder. The adversary is expressed as one or more roles in the same way as proper protocols. We have illustrated in [8], [6] how this is achieved for the standard Dolev-Yao intruder.

3   Simpli ed Needham-Schroeder Authentication Protocol

As our rst example using MSR as a speci cation language, we will formalize the Needham-Schroeder public-key authentication proto- col [19]. We familiarize the reader with MSR by rst considering the two-party nucleus of this protocol. We will tackle the full proto- col, which relies on a server to generate session keys in Section 4.

The server-less variant of the Needham-Schroeder public-key protocol [19] is a two-party crypto- protocol aimed at authenticating the initiator A to the responder B (but not necessarily vice versa).

1. A ! B : $\{\!\{ n_A \ A \}\!\}_{k_B}$
2. B ! A : $\{\!\{ n_A \ n_B \}\!\}_{k_A}$
3. A ! B : $\{\!\{ n_B \}\!\}_{k_B}$

It is expressed as the expected run on the right in the "usual notation" (where we have used our syntax for messages). In the rst line, the initiator A encrypts a message consisting of a nonce n_A and her own identity with the public key k_B of the responder B, and sends it (ideally to B). The second line describes the action that B undertakes upon receiving and interpreting this message: he creates a nonce n_B, combines it with A's nonce n_A, encrypts the outcome with A's public key k_A, and sends the resulting message out. Upon receiving this message in the third line, A accesses n_B and sends it back encrypted with k_B. The run is completed when B receives this message.

MSR and most modern security protocol speci cation languages focus on the sequence of actions that each principal involved in a protocol executes. We called such sequences roles. Strand spaces [16] are a simple and intuitive notation that emphasize this notion. The strand representation of this protocol is given by the following picture:

Initiator: $\{\!\{ n_A \ A \}\!\}_{k_B}$ +   Responder: + f $\{\!\{ n_A \ A \}\!\}_{k_B}$

+f $\{ n_A \ n_B \}_{k_A}$ $\{\!\{ n_A \ n_B \}\!\}_{k_A}$ +

$\{\!\{ n_B \}\!\}_{k_B}$ +   + f $\{ n_B \}_{k_B}$

Here incoming and outgoing single arrows respectively denote the reception and transmission of a message. The double arrows assign a temporal ordering on these actions.

We will now express each role in turn in the syntax of   MSR. For space reasons, we will typeset homogeneous constituents, namely the universal variable declarations and the predicate sequences in the antecedent and consequent, in columns within each rule; we will also rely on some minor abbreviation. We mark types that can be reconstructed from the other information present in a rule by denoting them in a   shaded font.

The initiator's actions are represented by the following two-rule role:

$$
\begin{array}{ll}
0 & \qquad\qquad\qquad\qquad\qquad\qquad\qquad\qquad\qquad\qquad 1\ _{8A}\\
\forall L : \text{principal} \times \text{principal}^{(B)} \times \text{pubK } B \times \text{nonce}. &
\end{array}
$$

$$
\left\{
\begin{array}{l}
\forall B : \text{principal}.\\
\forall k_B : \text{pubK } B.
\end{array}
\right.
\qquad \cdot \qquad \rightarrow \qquad \exists n_A : \text{nonce}. \;
\begin{array}{l}
N(\{\!|n_A\ A|\!\}_{k_B})\\
L(A,B,k_B,n_A)
\end{array}
$$

$$
\left\{
\begin{array}{l}
\forall \ldots\\
\forall k_A : \text{pubK } A.\\
@\forall k_A^0 : \text{privK } k_A.\\
\forall n_A, n_B : \text{nonce}.
\end{array}
\right.
\quad
\begin{array}{l}
N(\{\!|n_A\ n_B|\!\}_{k_A})\\
L(A,B,k_B,n_A)
\end{array}
\quad \rightarrow \quad
N(\{\!|n_B|\!\}_{k_B}) \qquad A
$$

Clearly, any principal can engage in this protocol as an initiator (or a responder). Our encoding is therefore structured as a generic role. Let   A be its postulated owner. The rst rule formalizes of the rst line of the \usual notation" description of this protocol from   A's point of view. It has an empty antecedent since initiation is unconditional in this protocol fragment. Its right-hand side uses an existential quanti er to mark the nonce n_A as fresh. The consequent contains the transmitted message and the role state predicate $L(A,B,k_B,n_A)$, necessary to enable the second rule of this protocol: it corresponds to the topmost double arrow in the strand speci cation on the left. The arguments of this predicate record variables used in the second rule.

The second rule encodes the last two lines of the "usual notation" and strand description. It is applicable only if the initiator has executed the rst rule (enforced by the presence of the role state predicate) and she receives a message of the appropriate form. Its consequent sends the last message of the protocol. The presence of both a message receptions and transmission in the same rule corresponds to the second double arrow in the strand speci cation of the initiator of this role.

Our notation provides a speci c type for each variable appearing in these rules. The equivalent \usual notation" speci cation relies instead on natural language and conventions to convey this same information, with clear potential for ambiguity. Observe that most declarations are grayed out, meaning that they can be reconstructed automatically: this simpli es the task of the author of the speci cation by enabling him or her to concentrate on the message flow rather than on typing details, and of course it limits the size of the speci cation. Algorithmic rules for this form of type reconstruction are the subject of a forthcoming paper.

The rationale behind the reconstructible types in this rule are as follows. The universal declarations for   B, k_B, and n_A and the type of the existential declaration for n_A in the rst rule can be deduced from the declaration of the role state predicate L. The declarations for k_A and k_A^0 can be omitted since k_A must be the public key of A, and k_A^0 be the corresponding private key. The possibility of reconstructing this information is intimately linked to the access control policy of MSR, formally de ned in [6]. The only universal declaration that cannot be reconstructed is " $8n_B$: nonce": n_B is clearly a universally quanti ed variable in this rule, but there is no hint that it should be a nonce. Let us now examine the declaration for L: the rst argument is always the rule owner, which is a principal. The third argument must be the public key of some principal B because of the way k_B is used. Therefore, we only need to indicate that B is bound in the second argument of L.

The responder is encoded as the generic role below, whose owner we have mnemonically called B. The rst rule of this role collapses the two topmost lines of the \usual notation" speci cation of this protocol fragment from the receiver's point of view. The second rule captures the reception and successful interpretation of the last message in the protocol by B: this step is often overlooked. This rule has no consequent.

$$
\begin{array}{l}
0 \\
9L : \text{principal}^{(B)} \times \text{principal} \times \text{pubK } B^{(k_B)} \times \text{privK } k_B \times \text{nonce}. \qquad 1_{8B}
\end{array}
$$

$$
\begin{array}{ll}
8k_B : \text{pubK } B. & \\
8k_B^0 : \text{privK } k_B. & \\
8A : \text{principal}. \quad N(\{n_A \ A\}_{k_B}) \qquad ! \quad 9n_B : \text{nonce}. & \dfrac{N(\{n_A \ n_B\}_{k_A})}{L(B,A,k_B,k_B^0,n_B)} \\
8n_A : \text{nonce}. & \\
8k_A : \text{pubK } A & \\
\end{array}
$$

$$
@8 \ldots \qquad \dfrac{N(\{n_B\}_{k_B})}{8n_B : \text{nonce}. \quad L(B,A,k_B,k_B^0,n_B)} \quad ! \qquad .
$$

Again, observe that most typing information has been grayed out since it can be reconstructed from the way variables are used and the few types left.

4   Full Needham-Schroeder Authentication Protocol

We will now specify the full version of the Needham-Schroeder public-key authentication protocol [19], which relies on a server S to generate the keys k_A and k_B used in the fragment discussed in the previous section. This protocol is written in the "usual notation" to the right of this text. The simpli ed version discussed in Section 3 corresponds to lines (3), (6) and (7) of this pro-

1. $A ! S: A B$
2. $S ! A: [k_B \ B]_{k_S}$
3. $A ! B: \{n_A \ A\}_{k_B}$
4. $B ! S: B A$
5. $S ! B: [k_A \ A]_{k_S}$
6. $B ! A: \{n_A \ n_B\}_{k_A}$
7. $A ! B: \{n_B\}_{k_B}$

tocol. In line (1), A asks the server for a key to communicate with B, which is obtained in the signed message on line (2). The responder issues and is granted a similar request in lines (4) and (5), respectively. The actions of the initiator are expressed in MSR by the following generic role, which consists of three rules that have to re in sequence, and consequently mentions two role state predicate declarations. The rst rule corresponds to line (1) in the \usual notation", the second to lines (2) and (3), and the third to lines (6) and (7).

$$
\begin{array}{l}
0 \qquad\qquad\qquad\qquad\qquad\qquad\qquad\qquad\qquad\qquad\qquad 1 \; 8\mathrm{A}\\
\quad 9\mathrm{L} : \mathrm{principal} \times \mathrm{principal}.\\
\quad 9\mathrm{L}^0 : \mathrm{principal} \times \mathrm{principal}^{(B)} \times \mathrm{pubK}\ B \times \mathrm{nonce}.\\
\end{array}
$$

$8\mathrm{B}$: principal. $\cdot$! $\qquad\qquad\qquad$ N(A B)
$\qquad\qquad\qquad\qquad\qquad\qquad\qquad\qquad\qquad$ L(A,B)

$8\ldots$
$8k_S$: sigK S. $\mathrm{N}([k_B\ B]_{k_S})$! $9n_A$: nonce. $\mathrm{N}(\{\!\{n_A\ A\}\!\}_{k_B})$
$8k_S^0$: verK k_S. $\mathrm{L}(A,B)$ $\qquad\qquad\qquad\qquad\qquad$ $\mathrm{L}^0(A,B,k_B,n_A)$
$8k_B$: pubK B.

$8\ldots$
$8k_A$: pubK A. $\mathrm{N}(\{\!\{n_A\ n_B\}\!\}_{k_A})$!
@$8k_A^0$: privK k_A. $\mathrm{L}^0(A,B,k_B,n_A)$ $\qquad\qquad\qquad\qquad$ $\mathrm{N}(\{\!\{n_B\}\!\}_{k_B})$ A
$8 n_A, n_B$: nonce.

Observe again that most declarations and types can be reconstructed. Notice in particular that, since in the second rule the arguments of   L form a pre x of the arguments of L^0, the entire declaration for   L can be synthesized from the type of L^0.

The responder's actions are expressed in the following generic role. The rst rule corresponds to lines (3) and (4) in the "usual notation", the second to lines (5) and (6), and the third to line (7). Observe again that most declarations can be automatically reconstructed.

$$
\begin{array}{l}
0 \qquad\qquad\qquad\qquad\qquad\qquad\qquad\qquad\qquad\qquad\qquad 1 \; 8\mathrm{B}\\
\quad 9\mathrm{L} : \mathrm{principal}^{(B)} \times \mathrm{principal} \times \mathrm{pubK}\ B^{(k_B)} \times \mathrm{privK}\ k_B \times \mathrm{nonce}.\\
\quad 9\mathrm{L}^0 : \mathrm{principal}^{(B)} \times \mathrm{principal} \times \mathrm{pubK}\ B^{(k_B)} \times \mathrm{privK}\ k_B \times \mathrm{nonce}.\\
\end{array}
$$

$8k_B$: pubK B.
$8k_B^0$: privK k_B . $\mathrm{N}(\{\!\{n_A\ A\}\!\}_{k_B})$! $\qquad\qquad$ N(B A)
$8\mathrm{A}$: principal. $\qquad\qquad\qquad\qquad\qquad\qquad\qquad\qquad$ L(B,A,k_B,k_B^0,n_A)
$8n_A$: nonce.

$8\ldots$
$8k_S$: sigK S. $\mathrm{N}([k_A\ A]_{k_S})$! $9n_B$: nonce. $\mathrm{N}(\{\!\{n_A\ n_B\}\!\}_{k_A})$
$8k_S^0$: verK k_S. $\mathrm{L}(B,A,k_B,k_B^0,n_A)$ $\qquad\qquad\qquad\qquad$ $\mathrm{L}^0(B,A,k_B,k_B^0,n_B)$
$8k_A$: pubK A.

$8\ldots$
@$8k_A^0$: privK k_A . $\mathrm{N}(\{\!\{n_B\}\!\}_{k_B})$! $\cdot$
$\qquad\qquad\qquad\qquad\quad$ $\mathrm{L}^0(B,A,k_B,k_B^0,n_B)$ $\qquad\qquad\qquad\qquad\qquad\qquad$ A
$8n_B$: nonce.

The last role in this protocol encompasses the actions of the server. Assuming that there is a single server,   S, they can conveniently be expressed by the following anchored role, which consists of a single rule. The "usual notation" speci cation of this protocol makes use of this role twice: in lines (1) and (2) to create   B 's keys for A, and then in lines (4) and (5) for the dual operation.

$$
\begin{array}{l}
\qquad\qquad\qquad\qquad\qquad\qquad\qquad\qquad\qquad\qquad\qquad\qquad\qquad\ s\\
8\mathrm{A}, \mathrm{B} : \mathrm{principal}.\quad \mathrm{N}(A B)\ !\quad 9k_B : \mathrm{pubK}\ B.\quad \mathrm{N}([k_B\ B]_{k_S})\\
8k_S : \mathrm{sigK\ S}. \qquad\qquad\qquad\qquad\ 9k_B^0 : \mathrm{privK}\ k_B .\\
\end{array}
$$

Upon receiving a message of the form   N(A B), the server constructs a public/private key pair for principal   B and noti es   A by sending the signed message N([k$_B$ B]$_{k_S}$). It should be observed how key generation is speci ed as existential quanti cation. The use of dependent types makes this process particularly elegant.

5   Neuman-Stubblebine Repeated Authentication Protocol

In this section, we devise an   MSR speci cation of the Neuman-Stubblebine repeated authentication protocol [20]. Similarly to Kerberos, this protocol consists of two phases. In a rst phases, a principal   A negotiates a "ticket" with a server in order to use services provided by another principal   B. In the second phases, A can reuse the ticket over and over to request this same service from   B until the ticket expires.

5.1   Initialization Subprotocol

The Neuman-Stubblebine protocol [20] is intended to enable a principal   A to repeatedly authenticate herself to another principal   B. Typically, B provides a service that A is interested in using repeatedly. Each time   A intends to use this service, she authenticates herself to   B by presenting a ticket he is expected to honor. The responder   B marks the ticket with a timestamp and will accept it within some expiration period.

The initialization phase of this protocol involves an interaction with a server S to obtain the ticket, as well as its

1. A ! B: A n$_A$
2. B ! S: B {A n$_A$ T$_B$ }$_{k_{BS}}$ n$_B$
3. S ! A: {B n$_A$ k$_{AB}$ T$_B$ }$_{k_{AS}}$ {A k$_{AB}$ T$_B$ }$_{k_{BS}}$ n$_B$
4. A ! B: {A k$_{AB}$ T$_B$ }$_{k_{BS}}$ {n$_B$ }$_{k_{AB}}$

rst use to request the service provided by   B. The expected trace in the "usual notation" is given to the right of this text. In the rst line,   A manifests her intention to use   B's service by sending him her identity and a nonce   n$_A$. In the second line, B forwards this information to the server   S together with his identity, a nonce of his own   n$_B$, and a timestamp   T$_B$. In the third line, the server constructs the ticket   {A k$_{AB}$ T$_B$ }$_{k_{BS}}$ by combining   A's name, a freshly generated key for communication between   A and B, and B's time stamp. It is encrypted with the key   k$_{BS}$ that B shares with S so that A cannot modify it. The server also informs   A of the extremes of the ticket in the message {B n$_A$ k$_{AB}$ T$_B$ }$_{k_{AS}}$ and forwards B's nonce to her. In the last line,   A identi es herself to B by sending him the ticket and his nonce   n$_B$ encrypted with the newly created   k$_{AB}$. Although part of no messages, this protocol assumes that B assigns a validity period to the ticket and will honor it until it expires. Therefore, upon receiving any message from   A, B will verify if the timestamp is still valid.

This initialization subprotocol is encoded in   MSR by means of three roles, one for A, one for B, and one for S. We start by giving a speci cation of   A's actions, reported in the following role:

$$\text{0}\quad 9L : \text{principal} \times \text{nonce}. \qquad\qquad\qquad\qquad\qquad\qquad\qquad\qquad 1\ {}_{8A}$$

$$! \quad 9n_A : \text{nonce}. \quad \begin{array}{l} N(A\,n_A) \\ L(A, n_A) \end{array}$$

$$\begin{array}{l} 8B : \text{principal}. \\ 8n_A, n_B : \text{nonce}. \\ 8k_{AB} : \text{shK } A\ B. \\ @8k_{AS} : \text{shK } A\ S. \\ 8X : \text{msg}. \end{array} \quad \begin{array}{l} N(\{B\,n_A\,k_{AB}\}_{k_{AS}}\,X\,n_B) \\ L(A, n_A) \end{array} \quad ! \quad \begin{array}{l} N(X\,\{n_B\}_{k_{AB}}) \\ \text{Ticket}_A(B, k_{AB}, X) \end{array}$$

The rst rule is a straightforward encoding of line (1) of the \usual notation" description of this subprotocol. The more interesting second rule corresponds to lines (3) and (4). Notice that   A is not entitled to observe the inner structure of the ticket. We express this fact by placing the variable   X in the second component of the received message. Expanding this object as $\{A\,k_{AB}\,T_B\}_{k_{BS}}$ to expose its structure would violate the access control policy [6]. In the consequent of this same rule, A sends the message on line (4) of the informal presentation to   B. She also needs to memorize some information to be able to reuse the ticket in the future, namely the ticket itself, the associated key k_{AB}, and B's identity. This is achieved by means of the memory predicate $\text{Ticket}_A(\_, \_, \_)$. The type of this predicate is " principal$^{(A)} \times$ principal$^{(B)} \times$ shK $A\,B \times$ msg" where the last argument corresponds to the ticket.

The responder's actions in this subprotocol are speci ed by the following role. Its two rules correspond to lines (1) and (2), and line (4) of the "usual notation" speci cation above.

$$\text{0}\quad 9L : \text{principal}^{(B)} \times \text{principal} \times \text{nonce} \times \text{shK } B\ S \times \text{nonce} \times \text{time}. \qquad\qquad 1\ {}_{8B}$$

$$\begin{array}{l} 8A : \text{principal}. \\ 8n_A : \text{nonce}. \\ 8k_{BS} : \text{shK } B\ S. \\ 8T_B : \text{time}. \end{array} \quad \begin{array}{l} N(A\,n_A) \\ \text{Clock}_B(T_B) \end{array} \quad ! \quad \begin{array}{l} 9n_B : \text{nonce}. \\ N(B\,\{A\,n_A\,T_B\}_{k_{BS}}\,n_B) \\ \text{Clock}_B(T_B) \\ L(B, A, n_A, k_{BS}, n_B, T_B) \end{array}$$

$$\begin{array}{l} 8\ldots \\ 8k_{AB} : \text{shK } A\ B \\ 8n_B : \text{nonce}. \\ @8T_B, T_{now} : \text{time}. \\ 8T_V, T_{exp} : \text{time}. \end{array} \quad \begin{array}{l} N(\{A\,k_{AB}\,T_B\}_{k_{BS}}\,\{n_B\}_{k_{AB}}) \\ L(B, A, n_A, k_{BS}, n_B, T_B) \\ \text{Valid}_B(A, T_B, T_V) \\ (T_{exp} = T_B + T_V) \end{array} \quad ! \quad \begin{array}{l} \text{Auth}_B(A, k_{AB}, T_B, T_{exp}) \\ \text{Valid}_B(A, T_B, T_V) \end{array}$$

Upon receiving the message $N(A\,n_A)$, the responder B reads the timestamp T_B o his local clock, which we model by means of the memory predicate $\text{Clock}_B(\_)$, of type " principal $\times$ time". A speci cation of how local clocks are updated is outside of the scope of this paper. A technique akin to the handling of time in [17] is particularly appealing for the elegant form of automated reasoning about temporal entities it supports.

In the second rule of this role, B receives the ticket $\{A\,k_{AB}\,T_B\}_{k_{BS}}$ and the response $\{n_B\}_{k_{AB}}$ to the challenge he issued in the rst rule by creating a nonce. He can clearly access the contents of the ticket and therefore verify this latter message. The timestamp must have the same value T_B memorized in the last argument of the role state predicate L in the rst rule. The responder now assigns an expiration date to the ticket: he consults the memory predicate $Valid_B\,(A, T_B, T_V)$ to decide on the length of time T_V it should be valid for (possibly on the basis of the initiator's identity A and the time of the day T_B it was requested), and then uses the arithmetic constraint ($T_{exp} = T_B + T_V$) to compute its expiration date T_{exp}. The components of the ticket together with its expiration date are stored in the memory predicate $Auth_B\,(\_,\_,\_,\_)$, of type "principal$^{(B)}$ × principal$^{(A)}$ × shK A B × time × time".

Finally, we have a single rule that formalizes the actions of the server. Upon receiving a request from B, the server generates the shared key k_{AB}, constructs the ticket and the noti cation message for A, and transmits this information.

$$
\begin{array}{ll}
0 & 1\;\text{s} \\[4pt]
\begin{array}{l}
\exists A, B : principal. \\
\exists k_{AS} : shK\ A\ S. \\
\exists k_{BS} : shK\ B\ S. \quad N(B\,\{A\,n_A\,T_B\}_{k_{BS}}\,n_B) \quad ! \\
@\exists n_A, n_B : nonce. \\
\exists T_B : time.
\end{array}
&
\begin{array}{l}
\exists k_{AB} : shK\ A\ B. \\[4pt]
N(\{B\,n_A\,k_{AB}\,T_B\}_{k_{AS}} \\
\quad \{A\,k_{AB}\,T_B\}_{k_{BS}} \qquad A \\
\quad n_B)
\end{array}
\end{array}
$$

5.2   Repeated Authentication Subprotocol

The second phase of the Neuman-Stubblebine protocol allows the initiator A to repeatedly use the ticket she has acquired in the rst phase to access the

1. A ! B: $n_A^0\,\{A\,k_{AB}\,T_B\}_{k_{BS}}$
2. B ! A: $n_B^0\,\{n_A^0\}_{k_{AB}}$
3. A ! B: $\{n_B^0\}_{k_{AB}}$

service provided by B, as long as the ticket has not expired. It is expressed in the "usual notation" by the three-step subprotocol displayed to to the right of this text. In the rst line, A generates a new nonce n_A^0 and sends it to B together with the ticket $\{A\,k_{AB}\,T_B\}_{k_{BS}}$ she has acquired in the initial phase of the protocol. Upon receiving this message, B checks that the ticket is still valid, creates a nonce of his own n_B^0, and transmits it to A in line (2) together with the encryption of n_A^0 with the key k_{AB} embedded in the ticket. In the last line, A sends B's nonce back after encrypting it with their shared key k_{AB}.

This subprotocol is formalized in MSR by means of the three roles below. The initiator's actions are expressed by the following generic role. In its rst rule, corresponding to line (1) of the informal speci cation, A accesses the ticket she has stored in the memory predicate Ticket_ during the initialization phase. The second rule corresponds to the remaining lines of the "usual notation" spec-i cation.

$$\forall L : principal^{(A)} \times principal^{(B)} \times shK\ A\ B \times nonce.$$

$$
\begin{array}{l}
\exists B : principal. \\
\exists X : msg. \qquad Ticket_A(B, k_{AB}, X) \quad \rightarrow \quad \exists n_A^0 : nonce. \\
\exists k_{AB} : shK\ A\ B.
\end{array}
\quad
\begin{array}{l}
N(n_A^0\ X) \\
Ticket_A(B, k_{AB}, X) \\
L(A, B, k_{AB}, n_A^0)
\end{array}
$$

$$
@ \dots \qquad N(n_B^0\ \{n_A^0\}_{k_{AB}})
$$
$$
\exists n_A^0, n_B^0 : nonce. \quad L(A, B, k_{AB}, n_A^0) \quad \rightarrow \quad N(\{n_B^0\}_{k_{AB}})
$$

The actions of the service provider B are given by the following two generic roles. The first rule of the first of them captures lines (1) and (2) of the informal specification, while the second rule formalizes the remaining line.

$$\forall L : principal^{(B)} \times principal^{(A)} \times shK\ A\ B \times nonce.$$

$$
\begin{array}{ll}
\exists n_A^0 : nonce. & \\
\exists k_{BS} : shK\ B\ S. & N(n_A^0\ \{A\ k_{AB}\ T_B\}_{k_{BS}}) \\
\exists A : principal. & Auth_B(A, k_{AB}, T_B, T_{exp}) \\
\exists k_{AB} : shK\ A\ B. & Clock_B(T_{now}) \\
\exists T_B, T_{exp}, T_{now} : time. & (T_{now} < T_{exp})
\end{array}
\rightarrow
\begin{array}{l}
\exists n_B^0 : nonce. \\
N(n_B^0\ \{n_A^0\}_{k_{AB}}) \\
Auth_B(A, k_{AB}, T_B, T_{exp}) \\
Clock_B(T_{now}) \\
L(B, A, k_{AB}, n_B^0)
\end{array}
$$

$$
@ \dots \qquad N(\{n_B^0\}_{k_{AB}})
$$
$$
\exists n_B^0 : nonce. \qquad L(B, A, k_{AB}, n_B^0) \quad \rightarrow \quad .
$$

Upon receiving each new request, B checks that the ticket has not expired yet. This is achieved by means of the constraint ($T_{now} < T_{exp}$) that verifies whether the current time T_{now} (read from his "$Clock\_$" memory predicate) is less than the ticket's expiration time T_{exp}.

Although there is no harm in keeping stale tickets, it is easy to write an MSR rule that removes expired tickets: whenever B notices that a ticket has expired (by means of the constraint ($T_{now} \geq T_{exp}$)), he simply retracts the corresponding "$Auth\_$" predicate.

$$
\begin{array}{ll}
\exists A : principal. & Auth_B(A, k_{AB}, T_B, T_{exp}) \\
@ \exists k_{AB} : shK\ A\ B. & Clock_B(T_{now}) \quad \rightarrow \quad Clock_B(T_{now}) \\
\exists T_B, T_{exp}, T_{now} : time. & (T_{now} \geq T_{exp})
\end{array}
$$

This concludes our MSR specification of the Neuman-Stubblebine repeated authentication protocol. The two phases that constitute it have been modeled by providing two sets of roles. The connection between them is given by a number of memory predicates used by both the client A and the service provider B. It should be noted that this protocol lies outside of the scope of the previous version of MSR [9], [10], which did not provide any secure means to share data across different roles.

6   Conclusions and Future Work

In this paper, we have presented the syntax of MSR, a strongly typed specification language for security protocol. The typing infrastructure, based on the theory of dependent types with subsorting, yields elegant and precise formalizations. The underlying methodology does not prescribe a fixed set of types to be used for every protocol, but rather allows de ning the objects (both types and term constructors) needed in each individual circumstance. This typing information is mostly used statically to discover simple but potentially harmful mistakes in a speci cation: for example, assuming appropriate declarations, type-checking would catch the unduly transmission of a long-term key in a network message. On the other hand, access control veri cation will point at attempts to use keys that do not belong to a principal. These two applications are presented in detail in [7] and [6], respectively. Static checks of this kind are particularly useful when modeling complex crypto-protocols. Previous versions of   MSR were mostly aimed at investigating decidability problems for crypto-protocols [9], [15] and at establishing the relative expressive power of di erent formalisms [11], [10]. The present work makes   MSR usable as a speci cation language for a large class of security protocols thanks to the introduction of a few key constructs and a flexible typing infrastructure. Memory predicates, in particular, allow a principal to share data and control among di erent role instances. This makes our formalism applicable to protocols structured as a collection of subprotocols. Constraints allow instead factoring recurrent operations on complex domain as external modules, keeping in this way protocol speci cations simple.

We have undertaken a formal study of various aspects of   MSR. Besides the general discussion and case studies presented in this paper, its type-checking rules and their properties are analyzed in [7], access control is the subject of [6], while [8] starts examining parallel executions and implements di erent formulations of the Dolev-Yao intruder. A number of problems are however still open and subject of current investigation. First, a number of issues need to be solved in order to make   MSR practical. In particular, a reliable type reconstruction algorithm is necessary to shelter users from the often tedious process of providing all the type declarations, and also to make formalizations reasonably sized. We are also extending our current collection of case studies to encompass not only the most common authentication protocols [12], but also complex schemes such as key management protocols for group multicast [4] and fair exchange protocols. Among other results, the formalization of these examples will allow us to experiment with numerous constructs and type layouts. We hope that this activity will enable us to extract useful speci cation techniques for the constructions needed in the formalization of a protocol. For example, we would like to be able to give a speci cation of hash functions from which the appropriate typing and access control rules can be automatically generated together with arguments that extend the validity of their various properties to these objects.

References

1. Abadi, M. and Gordon, A.: A calculus for cryptographic protocols: the spi calculus. Information and Computation, 148, 1, (1999) 1–70
2. Abadi, M. and Needham, R.: Prudent Engineering Practice for Cryptographic Protocols. Research Report 125, Digital Equipment Corp., System Research Center, (1994)

3. Aspinall, D. and Compagnoni, A.: Subtyping Dependent Types. In E. Clarke, editor, Proc. LICS'96. New Brunswick, NJ. IEEE Computer Society Press (1996) 86–97

4. Balenson, D., McGrew, D. and Sherman, A.: Key Management for Large Dynamic Groups: One-Way Function Trees and Amortized Initialization. Internet Draft (work in progres), draft-irtf-smug-groupkeymgmt-oft-00.txt, Internet Engineering Task Force (August 25, 2000)

5. Burrows, M., Abadi, M. and Needham, R.: A Logic of Authentication. Proceedings of the Royal Society, Series A, 426, 1871 (1989) 233–271

6. Cervesato, I.: MSR, Access Control, and the Most   Powerful Attacker. Submitted to LICS'01, Boston, MA, 2001.   http://www.cs.stanford.edu/~iliano

7. Cervesato, I.: A Speci cation Language for Crypto-Protocol based on Multiset Rewriting, Dependent Types and Subsorting.

8. Cervesato, I.: Typed Multiset Rewriting Speci cations of Security Protocols. Submitted to Proc. MFCSIT'00 , ENTCS. http://www.cs.stanford.edu/~iliano

9. Cervesato, I., Durgin, N., Lincoln, P., Mitchell, J. and Scedrov, A.: A Meta-Notation for Protocol Analysis. In Proc. CSFW'99. Mordano, Italy, IEEE/CS Press, (1999) 55–69

10. Cervesato, I., Durgin, N., Lincoln, P., Mitchell, J. and Scedrov, A.: Relating Strands and Multiset Rewriting for Security Protocol Analysis. In Proc. CSFW'00, (2000) 35–51

11. Cervesato, I., Durgin, N.A., Kanovich, M. and Scedrov, A.: Interpreting Strands in Linear Logic. In Proc. FMCS'00. Chigaco, IL, (2000)

12. Clark, J. and Jacob, J.: A Survey of Authentication Protocol Literature. Department of Computer Science, University of York. (1997) Web Draft Version 1.0 available from http://www.cs.york.ac.uk/~jac/

13. Denker, G. and Millen, J.K.: CAPSL Intermediate Language. In N. Heintze and E. Clarke (eds.): Proc. FMSP'99. Trento, Italy (1999)

14. Dolev, D. and Yao, A.C.: On the security of public-key protocols. IEEE Transactions on Information Theory. (1983) 2(29): 198–208

15. Durgin, N., Lincoln, P., Mitchell, J. and Scedrov, A.: Undecidability of bounded security protocols. In Heintze, N. and Clarke, E. (eds.): Proc. FMSP'99. Trento, Italy (1999)

16. F'abrega, F.J.T., Herzog, J.C. and Guttman, J.D.: Strand Spaces: Why is a Security Protocol Correct?. In Proc. SSP'98. (1998) 160–171 Oakland, CA, IEEE/CS Press

17. Kanovich, M.I., Okada, M. and Scedrov, A.: Specifying real-time nite-state systems in linear logic. In Proc. COTIC'98. Nice, France (1998) ENTCS 16(1)

18. Meadows, C.: The NRL protocol analyzer: an overview. J. Logic Programming, (1996) 26(2): 113–131

19. Needham, R.M. and Schroeder, M.D.: Using Encryption for Authentication in Large Networks of Computers. Communications of the ACM. (1978) 21(12): 993–999

20. Neuman, B.C. and Stubblebine, S.G.: A Note on the Use of Timestamps as Nonces. Operating Systems Review, (1993) 27(2) 10–14

21. Pfenning, F.: Re nement Types for Logical Frameworks. In Geuvers, H. (ed.): Proc. TYPES'93. Nijmegen, The Netherlands (1993) 285–299

22. Syverson, P.F.: A Di erent Look at Secure Distributed Computation. In Proc. CSFW-10, (1997) 109–115 IEEE Computer Society Press

TRBAC^N: A Temporal Authorization Model

Steve Barker

Cavendish School of Computer Science, University of Westminster
London, U.K, W1W 6UW
barkers@westminster.ac.uk

Abstract. We show how the family of temporal role-based access control
(TRBAC) models from [6], the TRBAC^O models, may be equivalently
represented in a considerably simpler and more efficiently implemented way.
We call the latter the TRBAC^N models. To specify TRBAC^N models, stratified
normal clause logic is sufficient. To compute with TRBAC^N models, any
procedural semantics that enables the perfect model of a stratified theory to be
generated may be used. Although TRBAC^N security models have a much
simpler representation than TRBAC^O models, we show that TRBAC^N and
TRBAC^O models are equivalent in terms of their expressive power.

1 Introduction

Temporal authorization models have been widely recognized as being important in
practice [19]. Temporal authorizations enable a security administrator (SA) to specify
that a user's permission to access a data item is to hold for a restricted interval of time,
and automatically expires as soon as the maximum time point in the interval is
reached. Temporal authorizations provide a SA with a fine level of control on
permission assignments, and can limit the damage that unauthorized users may wreak
if they do gain access to a system [18].

Although the emphasis thus far in the literature has been on temporal authorization
models that assume that a discretionary access control policy is to be used to protect
information [5, 8], a family of temporal role-based access control (TRBAC) models
has recently been described in [6]. Like temporal authorizations, RBAC has a number
of well-documented attractions [13] and has been shown to be an appropriate choice
of security policy for many types of organization to adopt [14].

The approach presented in [6] enables a wide range of temporal RBAC models to
be formally specified and proved to satisfy technical and organizational requirements
prior to implementation. These models are expressed using normal clause logic [15].
Henceforth, we refer to the TRBAC models from [6] as the TRBAC^O theories or
simply TRBAC^O (where O is used to specify the original TRBAC theories). In this
paper, we show how the TRBAC^O theories may be represented in a much simpler
way. Henceforth, we will refer to our re-interpretation of the TRBAC^O theories as the
TRBAC^N theories (where N is used to specify the new TRBAC theories).

TRBAC^N theories have the same attractions as TRBAC^O theories. They are
formally well defined and can be proved to satisfy essential security criteria prior to
implementation. What is more the formal specification of a TRBAC^N theory can be

V.I. Gorodetski et al. (Eds.): MMM-ACNS 2001, LNCS 2052, pp. 178–188, 2001.

translated into a variety of practical languages for implementing a chosen temporal authorization model. As such, the approach enables verifiably correct temporal authorization policies to be represented in practice. As we will show, although TRBAC^N theories have a much simpler representation than TRBAC^O theories the two forms of temporal RBAC model have precisely the same expressive power.

The rest of the paper is organized in the following way. In Section 2, we describe some basic notions from clause form logic and database security. In Section 3, we give a brief overview of TRBAC^O. In Section 4, we describe the core axioms that define a TRBAC^N theory. In Section 5, the issues relating to access control mechanisms for TRBAC^N are discussed. In Section 6, we show that TRBAC^O and TRBAC^N are equivalent. Finally, in Section 7, we draw some conclusions and make suggestions for further work.

2   Preliminaries

Any instance of a TRBAC^O or TRBAC^N theory may be expressed using a finite set of function-free normal clauses, a normal clause theory. A normal clause takes the form: H <- L1,L2,...,Lm. The head of the clause, H, is an atom and L1,L2,...,Lm is a conjunction of literals that constitutes the body of the clause. The conjunction of literals L1,L2,...,Lm must be true (proven) in order for H to be true (proven). A literal is an atomic formula or its negation; in this paper negation is negation as failure [12] and the negation of the atom A is denoted by not A. A clause with an empty body is an assertion or a fact. A clause with an empty head and a non-empty body is a denial. A definite clause is a normal clause that has no negative literals in its body; a definite clause theory is a finite set of definite clauses.

Since we consider function-free theories, the only terms that appear in TRBAC^O or TRBAC^N theories are constants and variables. Henceforth, we will denote constants using symbols that appear in the lower-case in literals. The variables that appear in literals will be denoted using upper-case symbols.

The security theories that we discuss in subsequent sections are based on the RBAC$_1$ model described in [17]. That is, we assume that an RBAC theory includes role and permission assignments (i.e. RBAC$_0$), and may additionally include specifications of role hierarchies (i.e. RBAC$_1$). We will denote TRBAC^O theories that are based on RBAC$_1$ by TRBAC$^O_1$ and TRBAC^N theories that are based on RBAC$_1$ by TRBAC$^N_1$. Whilst it is possible to include constraints in specifications of TRBAC^O and TRBAC^N theories, space restrictions prevent us considering this issue in this paper. Details of constraint representation and checking on TRBAC^O theories can be found in [6].

The TRBAC^O and TRBAC^N security theories that we describe below are based on the specification of RBAC as a normal clause theory from [6]. In this context, a user U is specified as being assigned to a role R by defining instances of a ura(U,R) predicate. For example, ura(bob,r1) is used to record the assignment of the user Bob to the role r1. Similarly, to record that a P permission on an object O is assigned to a role R, clause form definitions of an rpa(R,P,O) predicate are used.

An RBAC$_1$ role hierarchy is expressed by a set of clauses that define a senior-to relation as the reflexive-transitive closure of an irreflexive-intransitive d-s relation that defines the set of pairs of roles (r_i, r_j) such that r_i is directly senior to role r_j in an

$RBAC_1$ role hierarchy. The d-s and senior-to relations are related thus (where $\wedge$ is logical 'and', - is classical negation, and r_i, r_j and r_k are arbitrary roles):

$$\forall r_i \, r_j \, [\text{d-s}(r_i, r_j) \text{ iff senior-to}(r_i,r_j) \wedge -(r_i=r_j) \wedge$$
$$-\exists r_k \, [\text{senior-to}(r_i,r_k) \wedge \text{senior-to}(r_k,r_j) \wedge -(r_i=r_k) \wedge -(r_j=r_k)]]$$

In clause form, the senior-to relation is defined in terms of d-s thus (where '_' is an anonymous variable):

senior-to(R1,R1) <- d-s(R1,_)
senior-to(R1,R1) <- d-s(_,R1)
senior-to(R1,R2) <- d-s(R1,R2)
senior-to(R1,R2) <- d-s(R1,R3),senior-to(R3,R2)

The senior-to predicate is used in the definition of permitted that follows:

permitted(U,P,O) <- ura(U,R1),senior-to(R1,R2),rpa(R2,P,O)

The permitted clause expresses that a user U is authorized to perform the P operation on an object O if U is assigned to a role R1 that is senior to the role R2 in an $RBAC_1$ role hierarchy associated with the $RBAC_1$ theory, and R2 has been assigned the P permission on O.

We assume that a closed policy [10] is to be used when a $TRBAC^O$ or $TRBAC^N$ theory is used to protect a system. That is, a user's permission to access information must be specifically authorized in the security theory; users do not automatically have a permission on an object in the absence of a denial of this access (an open policy [10]). Only minor modifications to the approach we describe below are required to implement open or any number of hybrid (i.e. open/closed) policies.

Although the representation of arbitrary instances of $TRBAC^O$ and $TRBAC^N$ theories requires that full normal clause logic be used, the stratified [2] subset of normal clause logic is sufficient to represent the core set of axioms that define $TRBAC^O$ and $TRBAC^N$. Stratified theories have a unique (2-valued) perfect model [16]. An important corollary of $RBAC_1$ theories being categorical and having a total model is that these theories define a consistent and unambiguous set of authorizations.

3   Temporal Authorization in $TRBAC^O_1$

The $TRBAC^O_1$ security theories from [6] extend $RBAC_1$ by enabling a SA to specify that authorizations are to apply to role and permission assignments for restricted periods e.g. Bob is assigned to role r1 until 2001/02/01 and members of the role r1 are authorized to update the object o1 between 2000/12/01 and 2000/12/31. Here and henceforth we assume that times have a DAY granularity, that they are always expressed in YYYY/MM/DD form, and that two times in this form can be compared using the $\leq$ or $<$ comparison operators where $\leq$ is read as "earlier or the same time as" and $<$ is "earlier than". Axioms defining $\leq$ and $<$ are assumed to be included in $TRBAC^O_1$ and $TRBAC^N_1$.

In TRBAC$^O_1$, a security event description [6] is used to specify a user-role or permission-role assignment and the interval of time for which these assignments hold. For example, to record that, as of 2000/03/01, Bob is assigned to role r1 indefinitely into the future and that the role r1 is assigned the insert permission on the object o1 from 2000/12/01 until 2000/12/31, the following security event descriptions are used (where e1 and e2 are event identifiers, ';' is used to separate clauses, and max is the maximum permitted future time):

{happens(e1,2000/03/01)<-; act(e1,ura)<-;
                    subject(e1,bob)<-; role(e1,r1)<-; stop(e1,max)<-}

{happens(e2,2000/12/01)<-; act(e2,rpa)<-; role(e2,r1)<-;
            object(e2,o1)<-; permission(e2,insert)<-; stop(e2,2000/12/31)<-}

In [6], a ground instance of the goal clause permitted(access(U,P,O),T) is evaluated using SLDNF-resolution [15] with respect to an instance of a TRBAC$^O_1$ theory in order to decide whether a user U has the P permission on an object O at the time T at which access is sought by U. For permitted(access(U,P,O),T) to be proved from an instance of a TRBAC$^O_1$ theory it is necessary to prove that: U's assignment to a role R1 has been initiated by a security event E1; the P permission on O is assigned to a role R2 by an initiating event E2; R1 inherits permissions from R2 (defined via the definition of senior-to); and that neither U's assignment to the role R1 nor the assignment of the P permission on O to R2 has been revoked or has expired at the time T. The core axiom of TRBAC$^O_1$ that defines permitted is expressed thus:

(C1^O) permitted(access(U,P,O),T) <- happens(E1,T1),initiates(E1,ura(U,R1)),
                        happens(E2,T2),initiates(E2,rpa(R2,P,O)),
                        T1 ≤ T,T2 ≤ T,senior-to(R1,R2),
                        not ended(ura(U,R1),T,T1),
                        not stopped(E1,ura(U,R1),T),
                        not ended(rpa(R2,P,O),T,T2),
                        not stopped(E2,rpa(R2,P,O),T)

To perform a P operation on an object O in the system at time T, we assume that U has activated the roles that are necessary for U to be able to perform the P action on O. Additional conditions may be added to C1^O to incorporate the notions of role activation/deactivation and sessions (see [6]).

The following set of axioms defines the initiates, terminates, ended and stopped predicates in C1^O, and are included in every instance of a TRBAC$^O_1$ theory:

(C2^O) initiates(E,ura(U,R)) <- subject(E,U),role(E,R),act(E,ura)

(C3^O) initiates(E,rpa(R,P,O)) <- role(E,R),permission(E,P),object(E,O),act(E,rpa)

(C4^O) terminates(E,ura(U,R)) <- subject(E,U),role(E,R),act(E,revokeura)

(C5^O) terminates(E,rpa(R,P,O))<-
                    role(E,R),permission(E,P),object(E,O),act(E,revokerpa)

(C6^O) ended(ura(U,R1),T,T1) <-
                happens(E2,T2),terminates(E2,ura(U,R1)),T1 < T2,T2 $\leq$ T

(C7^O) ended(rpa(R2,P,O),T,T2) <-
                happens(E3,T3),terminates(E3,rpa(R2,P,O)),T2 < T3,T3 $\leq$ T

(C8^O) stopped(E1,ura(U,R1),T) <- stop(E1,T1),T1 $\leq$ T

(C9^O) stopped(E2,rpa(R2,P,O),T) <- stop(E2,T1),T1 $\leq$ T

The clauses in C^O={C1^O,C2^O,C3^O,C4^O,C5^O,C6^O,C7^O,C8^O,C9^O} define the consequences of performing a security action expressed via a security event description. Moreover, C^O comprises the core set of axioms that every TRBAC$^O_1$ theory must include if a history of role and permission assignments is maintained (see below). A set of application-specific security event descriptions that define an authorization history [6] is added to C^O to define an instance of a TRBAC$^O_1$ theory.

In TRBAC$^O_1$, a history of user-role and permission-role assignments and revocations is assumed to be maintained. As with assignments, revocations are described using security event descriptions. For example, to record that, as of 2000/12/10, Bob's membership of the role r1 is revoked, the following security event description is used:

{happens(e3,2000/12/10)<-; act(e3,revokeura)<-; subject(e3,bob)<-; role(e3,r1)<-}

Maintaining a history of assignments and revocations is useful for a number of reasons; for instance, it enables the initiation and termination of assignments to be defined in terms of terminates conditions and enables temporal constraints on an authorization history to be specified. A SA may choose not to record a history of assignments and revocations. In this case, revocation may be treated by physically deleting the security event descriptions that initiated the user-role or permission-role assignment. Moreover, security event descriptions that assign users or permissions to roles and include a stop(ei,t) assertion (where ei is an event identifier and t is the time at which the assignment expires) may be physically deleted from a TRBAC$^O_1$ theory when t is earlier than the current time. The axioms in TRBAC$^O_1$ are modified to reflect whatever deletes policy a SA chooses to implement.

4   Temporal Authorization in TRBAC$^N_1$

In contrast to TRBAC$^O_1$, the approach that is adopted in TRBAC$^N_1$ is to move time from the security event descriptions and to the ura and rpa specifications. To specify the interval of time during which a definition of a ura or rpa predicate is to hold, a conjunction of linear constraints on a time variable, T, is used. For example, to express that U is assigned to the role R from a point in time t_{Start} until some future time point t_{Stop}, the specification of ura takes the following form:

(C2^N) ura(u,r,T) <- $t_{Start} \leq$ T,T $\leq t_{Stop}$

This clause specifies that a user with the identifier u will be assigned to a role r at time T if the $t_{Start} \leq T$ and $T \leq t_{Stop}$ conditions hold i.e. if T is in the interval $[t_{Start}, t_{Stop}]$. In this case, and henceforth, we use CK^N (where K e {1,2,3}) to denote that CK^N is the Kth core axiom of TRBAC$^N_1$. The u and r terms that appear in the head of a ura predicate are constants, and T is always a variable.

To express that the P permission on object O is assigned to the role R from a time t_{Start} until some future time point t_{Stop}, the definition of rpa clauses will be of the following form:

(C3^N) rpa(R^t,P^t,O^t,T) <- $t_{Start} \leq T, T \leq t_{Stop}$

Whilst constants always appear in the first and second arguments of a definition of ura, some or all of the R^t, P^t, and O^t terms may be either constants or variables in a definition of rpa. We use R^t, P^t, and O^t to denote that each of these terms can be either a variable or a constant; T is always a variable.

Example
To represent, in an instance of TRBAC$^N_1$, that, as of 2000/03/01, Bob is assigned to the role r1 and that the role r1 is assigned all permissions on the object o1 from 2000/12/01 until 2000/12/31, the following pair of clauses may be used:

ura(bob,r1,T) <- 2000/03/01 $\leq$ T
rpa(r1,P,o1,T) <- 2000/12/01 $\leq$ T, T $\leq$ 2000/12/31

As in TRBAC$^O_1$, it is possible to express proactive ura (or rpa) specifications in an instance of a TRBAC$^N_1$ theory. That is, it is possible to append, at a time point T1, a definition of ura (or rpa) with a t_{Start} time that is later than T1. In this case, the definition of ura (or rpa) will become effective when $T=t_{Start}$ and T is the current time.

A major attraction of the approach adopted in TRBAC$^N_1$ is that it precludes the need for the set of security event descriptions that are required in TRBAC$^O_1$. As such, there is no need to use the core initiates and terminates clauses that are included in TRBAC$^O_1$. This implies that the core permitted axiom in TRBAC$^O_1$ can be considerably simplified. More specifically, the core axiom C1^O from TRBAC$^O_1$ can be simplified to the following core permitted axiom in TRBAC$^N_1$:

(C1^N) permitted(access(U,P,O),T) <- ura(U,R1,T),senior-to(R1,R2),rpa(R2,P,O,T)

The reading of this axiom is that: U has the P permission on object O at time T if U is assigned to the role R1 at time T, R1 inherits permissions from the role R2, in the role hierarchy defined by the senior-to relation, and R2 is assigned the P permission on O at time T.

In TRBAC$^N_1$, the revocation of a user-role or permission-role assignment is treated by physically deleting appropriate definitions of ura(U,R,T) or rpa(R,P,O,T) from an instance of a TRBAC$^N_1$ theory; this corresponds to the ending of user-role or permission-role assignments in TRBAC$^O_1$. Definitions of ura(U,R,T) and rpa(R,P,O,T) are also physically deleted from an instance of a TRBAC$^N_1$ theory whenever there is a $T \leq t_{Stop}$ condition in the body of a clause that defines ura(U,R,T) or rpa(R,P,O,T) and t_{Stop} is earlier than the current time. In this case the user-role or permission-role assignment has expired (or has stopped in the language of TRBAC$^O_1$).

If a history of user-role or permission-role assignments needs to be then assertions of the following form are included in a $TRBAC^N_1$ theory: terminated(A,T). In terminated(A,T), T is the time of revocation of the assignment A where A is a user-role (ura) or permission-role (rpa) assignment. For example, if Bob's assignment to the role r1 is revoked on 2000/12/10 then the clause defining ura(bob,r1,T) is physically deleted from the $TRBAC^N_1$ theory and the assertion terminated(ura(bob,r1),2000/12/10) is added. Similarly, the assertion terminated(rpa(r1,read,o1),2000/11/10) specifies that the read permission on object o1, that was assigned to the role r1, is removed on 2000/11/10. In the case where a ura(U,R,T) or rpa(R,P,O,T) definition is removed from an instance of $TRBAC^N_1$ when $T > t_{Stop}$ holds and T is the current time, a terminated(A,t_{Stop}) assertion is included in the $TRBAC^N_1$ theory.

As in the case of $TRBAC^O_1$, role activation and deactivation may be easily accommodated in $TRBAC^N_1$ by dynamically asserting ground instances of an active-in(U,R,T) predicate if a user U is allowed to activate a role R and does so at time T; the instance of active-in(U,R,T) is dynamically retracted whenever U deactivates the role R. Since role activation is not part of the formal specification of $TRBAC^N_1$ we do not consider it in any further detail in this paper.

5   Access Control and TRBAC $^N_1$ Theories

In $TRBAC^O_1$, Clark's (2-valued) completion [12] and SLDNF-resolution are, respectively, the principal declarative and operational semantics that are considered for evaluating access requests with respect to a $TRBAC^O_1$ theory. The soundness and completeness of SLDNF-resolution for the evaluation of permitted queries on $TRBAC^O_1$ theories is shown in [6] and left termination [4] of SLDNF-resolution for computing with $TRBAC^O_1$ theories follows from the fact that $TRBAC^O_1$ theories without the recursive senior-to relation are acyclic [3], and with senior-to are acceptable [4]. The left termination of SLDNF-resolution for computing with $TRBAC^O_1$ theories is implied by their acceptability [4]. The presence of the built-in comparison operators $<$ and $\leq$ in a $TRBAC^O_1$ theory does not affect the general termination results for SLDNF-resolution applicable to this class of theory.

As noted in [6], for $TRBAC^O_1$ theories the soundness and completeness results for SLDNF-resolution (with respect to Clark's completion) may be alternatively expressed in terms of SLG-resolution [11] and the well-founded semantics [21]. Since SLDNF-resolution and Clark's completion have well-known shortcomings and the well-founded and perfect model semantics coincide for $TRBAC^O_1$ and $TRBAC^N_1$ (the core axioms of $TRBAC^O_1$ and $TRBAC^N_1$ theories are stratified), in the rest of the discussion we will use SLG-resolution for the operational semantics and the well-founded semantics for the declarative semantics for $TRBAC^O_1$ and $TRBAC^N_1$ theories.

In the case of $TRBAC^N_1$, whenever U requests P access on O at time T, a permitted(access(U,P,O),T) <- permitted(access(U,P,O),T) goal clause is evaluated using SLG-resolution on the instance, Y, of the $TRBAC^N_1$ theory that defines the security that applies to a system that Y protects. In an instance of a permitted(access(U,P,O),T) <- permitted(access(U,P,O),T) goal clause on Y, the U variable is instantiated with the identifier of the user that makes the request to perform a P operation on O at the time T (a user identifier is assumed to be required for a user to be able to connect to the system). The P variable is instantiated with a constant from the set of operations that may be performed on objects within the system to be protected, and the T variable is instantiated with the time at which U requests P access on O (T is the current time taken from the system clock). The form that O takes depends on the type of object that needs to be protected. To simplify the ensuing discussion we will assume that each object in the system has a unique identifier and that a permitted(access(U,P,O),T) <- permitted(access(U,P,O),T) goal clause will always be ground when it is evaluated. When a clause is ground we will add the superscript G to the clause; square brackets will be used to denote the scope of the grounding e.g. [permitted(access(U,P,O),T) <- permitted(access(U,P,O),T)] ^{G}. Henceforth, we will also denote an arbitrary ground instance of a predicate Q by Q^G and a ground instance of a conjunction of predicates $Q_1 \wedge Q_2 \wedge \ldots \wedge Q_n$ by $[Q_1 \wedge Q_2 \wedge \ldots \wedge Q_n]^G$.

Whenever a user U requests P access on object O at time T, an SLG-forest [11] is generated for the evaluation of the goal clause permitted(access(U,P,O),T) <- permitted(access(U,P,O),T)^G on Y. From the soundness of SLG-resolution [11] with respect to the well-founded semantics we have that: if permitted(access(U,P,O),T)^G<- is an answer clause by SLG-resolution with respect to the TRBAC$^N_1$ theory Y then WFM(Y) |= permitted(access(U,P,O),T)^G (where WFM(Y) is used to denote the well-founded model of Y). Since SLG-resolution is (search space) complete for non-floundering query evaluation on normal clause theories [14] it follows that it is (search space) complete for the evaluation of non-floundering permitted(access(U,P,O),T)^G queries with respect to an arbitrary TRBAC$^N_1$ theory, Y. Thus, if WFM(Y) |=permitted(access(U,P,O),T)^G then permitted(access(U,P,O),T)^G<- is an answer clause by SLG-resolution with respect to Y.

In addition to its attractive soundness and completeness results, SLG-resolution is guaranteed to terminate for the evaluation of permitted(access(U,P,O),T)^G queries on TRBAC$^N_1$ theories since these theories satisfy the bounded-term-size property [20]. Moreover, for theories that satisfy this property, SLG-resolution has polynomial time data complexity [22] for computing the well-founded semantics.

Since the TRBAC$^N_1$ theories we have thus far described are always definite it follows that SLG-resolution is guaranteed not to flounder [15] when a permitted query is evaluated on TRBAC$^N_1$. What is more, the termination results for SLG-resolution imply that for the evaluation of any permitted query on a TRBAC$^N_1$ theory, the search forest will be finite and, as such, all answer clauses can be found in finite time. It follows then that the search space and flounder-free conditions in the completeness result for SLG-resolution can be dropped in the case of evaluating permitted queries on TRBAC$^N_1$ theories. A consequence of this observation is that the following strong completeness result holds for SLG-resolution used for evaluating access requests on an arbitrary TRBAC$^N_1$ theory, Y: WFM(Y) |= permitted(access(U,P,O),T)^G iff permitted(access(U,P,O),T)^G<- is an answer clause by SLG-resolution. Since TRBAC$^O_1$ is an allowed theory [15], it follows from this and the results in [6] that the strong completeness result also holds for SLG-resolution used to evaluate permitted(access(U,P,O),T)^G queries with respect to an arbitrary TRBAC$^O_1$ theory S viz.: WFM(S) |= permitted(access(U,P,O),T)^G iff permitted(access(U,P,O),T)^G<- is an answer clause by SLG-resolution with respect to S.

6   The Equivalence of TRBAC$^O_1$ and TRBAC$^N_1$

The strong completeness result for SLG-resolution used to evaluate a permitted query on a TRBAC$^O_1$ or TRBAC$^N_1$ theory enables us to show that TRBAC$^O_1$ and TRBAC$^N_1$ are equivalent. We interpret equivalence as meaning that the set of ground instances of permitted(access(U,P,O),T) that are logical consequences of a TRBAC$^O_1$ theory S is identical to the set of ground instances of permitted(access(U,P,O),T)^G that are logical consequences of the representation of S as a TRBAC$^N_1$ theory Y (equivalently, WFM(S)=WFM(Y)). We assume that the axioms defining < and ≤ are identical for S and Y. Hence, if d_1 and d_2 are any pair of dates then $d_1 < d_2$ e WFM(S) iff $d_1 < d_2$ e WFM(Y) and $d_1 \leq d_2$ e WFM(S) iff $d_1 \leq d_2$ e WFM(Y).

By inspection of the definitions of permitted(access(U,P,O),T) in TRBAC$^O_1$ and TRBAC$^N_1$ it is possible to see that happens(E1,T1) ∧ initiates(E1,ura(U,R1)) ∧ T1 ≤ T ∧ - ended(ura(U,R1),T,T1) ∧ - stopped(E1,ura(U,R1),T) in TRBAC$^O_1$ (where stopped(E1,ura(U,R1),T) iff stop(E1,T2) ∧ T2 ≤ T and - A denotes that the atom A is false in the well-founded model of a TRBAC$^O_1$ or TRBAC$^N_1$ theory) is represented in TRBAC$^N_1$ by the clause: ura(U,R,T) <- T1 ≤ T, T ≤ T2 (where T1=t_{Start}, T2=t_{stop} and T is the current time). Similarly, happens(E2,T2) ∧ initiates(E1,rpa(R2,P,O)) ∧ T2 ≤ T ∧ - ended(rpa(R2,P,O),T,T2) ∧ - stopped(R2,rpa(R2,P,O),T) in TRBAC$^O_1$ (where stopped(E2,rpa(R2,P,O),T) iff stop(E1,T2) ∧ T2 ≤ T) is represented in TRBAC$^N_1$ by the clause: rpa(R,P,O,T) <- T1 ≤ T, T ≤ T2 (where T1=t_{Start}, T2=t_{stop} and T is the current time). More fully, the equivalence of TRBAC$^O_1$ and TRBAC$^N_1$ is demonstrated in the proof of the following result:

Theorem : TRBAC$^O_1$ and TRBAC$^N_1$ are equivalent.

Proof: We show that, for an arbitrary instance of permitted(access(U,P,O),T)^G, WFM (TRBAC$^O_1$) |= permitted(access(U,P,O),T)^G iff WFM(TRBAC$^N_1$) |= permitted(access(U,P,O),T)^G. The proof is in two main parts. We show first that WFM(TRBAC$^O_1$) |= [happens(E1,T1) ∧ initiates(E1,ura(U,R1)) ∧ T1 ≤ T ∧ - stopped(E1,ura(U,R1),T) ∧ - ended(ura(U,R1),T,T1)]^G => WFM(TRBAC$^N_1$) |= ura(U,R,T)^G (where => is material implication). Secondly, we show that WFM(TRBAC$^O_1$) |= [happens(E2,T2) ∧ initiates(E2,rpa(R2,P,O)) ∧ T2 ≤ T ∧ - stopped(E2,rpa(U,R1),T) ∧ - ended(rpa(U,R1),T,T1)]^G => WFM(TRBAC$^N_1$) |= rpa(R,P,O,T)^G. This enables us to show that: WFM(TRBAC$^O_1$) |= permitted(access(U,P,O),T)^G => WFM(TRBAC$^N_1$) |= permitted(access(U,P,O),T)^G. Since the converse also holds it follows that TRBAC$^O_1$ and TRBAC$^N_1$ are equivalent.

If WFM(TRBAC$^O_1$) |= [happens(E1,T1) ∧ initiates(E1,ura(U,R1)) ∧ T1 ≤ T]^G then E1 initiates (i.e. makes true) ura(U,R1)^G at time T1 and if T1 ≤ T then ura(U,R1)^G holds at the current time T provided that WFM(TRBAC$^O_1$) |= [- stopped(E2,ura(U,R1),T) ∧ - ended(ura(U,R1),T,T1)]^G. If WFM(TRBAC$^O_1$) |= - ended(E1,ura(U,R1),T,T1)^G then U's membership of R1 has not been terminated by an act of revocation prior to T. Hence if TRBAC$^O_1$ includes a definition of ura(U,R1,T) at a time T3 such that T3 $\leq$ T then this definition is in TRBAC$^N_1$ at T. If WFM(TRBAC$^O_1$) $\models$ stopped(E1,ura(U,R1),T)^G then from the axiom C8^O, there exists a time T2 such that [stop(E1,T2) $\wedge$ T2 $\leq$ T]^G holds in TRBAC$^O_1$. Thus, if WFM(TRBAC$^O_1$) $\models$ - stopped(E1,ura(U,R1),T)^G then T2 $\geq$ T or equivalently T $\leq$ T2. Hence, since [T1 $\leq$ T $\wedge$ T $<$ T2] and T1=t_{Start} and T2=t_{Stop} it follows that WFM(TRBAC$^N_1$) $\models$ ura(U,R,T)^G from C2^N.

If WFM(TRBAC$^O_1$) $\models$ [happens(E2,T2) $\wedge$ initiates(E2,rpa(R2,P,O)) $\wedge$ T2 $\leq$ T]^G then E2 initiates (i.e. makes true) rpa(R2,P,O)^G at time T2 and if T2 $\leq$ T then rpa(R2,P,O)^G holds at the current time T provided that WFM(TRBAC$^O_1$) $\models$ [- stopped(E2,rpa(R2,P,O),T) $\wedge$ - ended(rpa(R2,P,O),T,T2)]^G. If WFM(TRBAC$^O_1$) $\models$ - ended(E2,rpa(R2,P,O),T,T2)^G then the assignment of the P permission on O to R2 has not been terminated by an act of revocation prior to T. Hence if TRBAC$^N_1$ includes a definition of rpa(R2,P,O,T) at a time T4 such that T4 $\leq$ T then this definition is included in TRBAC$^N_1$ at T. If WFM(TRBAC$^O_1$) $\models$ stopped(E1,rpa(R2,P,O),T)^G then from the axiom C9^O, there exists a time T5 such that WFM(TRBAC$^O_1$) $\models$ [stop(E2,T5) $\wedge$ T5 $\leq$ T]^G. Thus, if WFM(TRBAC$^O_1$) $\models$ - stopped(E2,rpa(R2,P,O),T)^G then T5 $\geq$ T or equivalently T $\leq$ T5. Hence, since [T2 $\leq$ T $\wedge$ T $\leq$ T5] and T2=t_{Start} and T5=t_{Stop} it follows that WFM(TRBAC$^N_1$) $\models$ rpa(R2,P,O,T)^G from C3^N.

By combining results, we have that if WFM(TRBAC$^O_1$) $\models$ [happens(E1,T1) $\wedge$ initiates(E1,ura(U,R1)) $\wedge$ T1 $\leq$ T $\wedge$ - stopped(E1,ura(U,R1),T) $\wedge$ - ended(ura(U,R1),T,T1) $\wedge$ happens(E2,T2) $\wedge$ initiates(E2,rpa(R2,P,O)) $\wedge$ T2 $\leq$ T - stopped(E2,rpa(R2,P,O),T) $\wedge$ - ended(rpa(R2,P,O),T,T2)]^G then WFM(TRBAC$^N_1$) $\models$ ura(U,R1,T)^G and WFM(TRBAC$^N_1$) $\models$ rpa(R2,P,O,T)^G. Hence, since senior-to(R1,R2), $<$ and $\leq$ are defined identically in TRBAC$^O_1$ and TRBAC$^N_1$, it follows that: WFM(TRBAC$^O_1$) $\models$ permitted(access(U,P,O),T)^G => WFM(TRBAC$^N_1$) $\models$ permitted(access(U,P,O),T)^G.

That WFM(TRBAC$^N_1$) $\models$ permitted(access(U,P,O),T)^G => WFM(TRBAC$^O_1$) $\models$ permitted(access(U,P,O),T)^G holds may be proved using a similar argument to that given above. Hence, TRBAC$^O_1$ and TRBAC$^N_1$ are equivalent.

7   Conclusions and Further Work

We have shown how the family of TRBAC$^O_1$ theories from [6] may be represented in an equivalent but much simpler form, as TRBAC$^N_1$ theories. By writing application-specific instances of the core axioms C1^N, C2^N and C3^N, any number TRBAC$^N_1$ theories may be formulated. Moreover, any operational semantics that can compute the perfect model of a stratified theory may be used to evaluate access requests with respect to TRBAC$^N_1$ theories (e.g. SQL2 together with the type of WITH RECURSIVE statements proposed for inclusion in SQL3). It follows that the specification of TRBAC$^N_1$ can be translated into a variety of practical programming languages to implement any number of TRBAC authorization models.

Due to space limitations, we have not considered extending TRBAC$^N_1$ theories to include definitions of ura and rpa predicates with arbitrary literals in their body and we have not been able to discuss the notion of constraints on TRBAC$^N_1$ theories.

Extended forms of TRBAC$^N_1$ theories will be described in future work. We will also show how other types of temporal authorization model that have been described in the literature can be reinterpreted from a TRBAC^N perspective, how the types of periodicity constraint that are described in [9] may be incorporated into TRBAC^N theories, and how the ideas presented here may be used in our previous work on the security of deductive databases [7].

References

1. Abiteboul, S., Hull, R., and Vianu, V.: Foundations of Databases. Addison-Wesley (1995)
2. Apt, K., Blair, H., and Walker, A.: Towards a Theory of Declarative Knowledge. In Minker, J. (ed.): Foundations of Deductive Databases and Logic Programming. Morgan-Kaufmann Publishers (1988)
3. Apt, K., and Bezem, M.: Acyclic Programs. New Generation Computing (1990)
4. Apt., K., and Pedreschi., D.: Reasoning about Termination of Pure Prolog Programs. Information and Computation, 106 (1993)
5. Barker, S.: Temporal Authorization in the Simplified Event Calculus. In Atluri V., and Hale, J., Hale J. (eds): Research Advances in Database and Information Systems Security. Kluwer Academic Publishers (2000)
6. Barker, S.: Data Protection by Logic Programming, 1st International Conference on Computational Logic, LNAI 1861, Springer-Verlag (2000)
7. Barker, S.: Secure Deductive Databases. PADL'01 (2001)
8. Bertino, E., Bettini, C., Ferrari, E., and Samarati, P.: A Temporal Access Control Mechanism for Database Systems. IEEE TKDE, 8(1) (1996)
9. Bertino, E., Bettini, C., Ferrari, E., and Samarati, P.: An Access Control Model Supporting Periodicity Constraints and Temporal Reasoning. TODS, 23(3) (1998)
10. Castano, S., Fugini, M., Martella, G., and Samarati, P.: Database Security, Addison Wesley. (1995)
11. Chen, W., Swift, T., and Warren, D.: Efficient Top-Down Computation of Queries Under the Well-Founded Semantics. JLP, 24(3) (1995)
12. Clark, K.: Negation as Failure. In H. Gallaire and J. Minker (eds): Logic and Databases. Plenum (1978)
13. Ferraiolo, D., Cugini, J., and Kuhn, R.: Role-Based Access Control: Features and Motivations. Proc. 11th Annual Computer Security Applications Conf., (1995)
14. Ferraiolo, D., Gilbert, D., and Lynch, N.: An Examination of Federal and Commercial Access Control Policy Needs. Proc. NIST-NCSC National Security Conf.(1993)
15. Lloyd, J.: Foundations of Logic Programming, 2nd Ed., Springer (1987)
16. Przymusinski, T.: Perfect Model Semantics. Proc. 5th ICLP, (1988)
17. Sandhu, R., Coyne, E., Feinstein, H., and Youman, C.: Role-Based Access Control Models. IEEE Computer, (1996)
18. Sandhu, R., Coyne, E., Feinstein, H., and Youman, C.: Role-Based Access Control: A Multi-Dimensional View. Proc. 10th Annual Computer Security Applications Conf. (1994)
19. Thomas, R., and Sandhu, R.: Discretionary Access Control in Object-Oriented Databases: Issues and Research Directions, Proc. 16th National Computer Security Conf. (1993)
20. Van Gelder, A.: Negation as Failure Using Tight Derivations for General Logic Programs. In Minker, J. (ed.): Foundations of Deductive Databases and Logic Programming. Morgan-Kaufmann Publishers (1988)
21. Van Gelder, A., Ross, K., and Schlipf, J.: The Well-Founded Semantics for General Logic Programs. J. ACM, 38(3) (1991)
22. Vardi, M.: The Complexity of Query Languages. ACM Symp. on the Theory of Computing (May, 1982)

The Set and Function Approach to Modeling Authorization in Distributed Systems

Tatyana Ryutov and Clifford Neuman

Information Sciences Institute University of Southern California
4676 Admiralty Way suite 1001
Marina del Rey, CA 90292
{tryutov, bcn }@isi.edu
(310)822-1511 (voice) (310)823-6714 (fax)

Abstract.   We present a new model that provides clear and precise semantics for authorization. The semantics is independent from underling
security mechanisms and is separate from implementation. The model
is capable of representing existing access control mechanisms. Our approach is based on set and function formalism. We focus our attention
on identifying issues and use our model as a general basis to investigate
the issues.

1 Introduction

The Internet has rapidly evolved to a platform that supports business and services such as e-commerce, electronic publishing, and health care. Security com-
promises now have real world consequences, resulting in release of sensitive or
protected information and monetary loss. Attacks on medically critical computing capabilities might even result in loss of human life. The ability to define and
enforce fine-grained security policies for systems and services is important in
such systems. The ability to understand such security policies is critical if they
are to be correctly written or implemented. Unfortunately, as the complexity
of the systems grow these policies are becoming harder to correctly define and
more difficult to enforce.

To cope with the growing complex ity of policy specification it is useful to
design a conceptual model that gives a structured way to think about policies.
A model enables one to better understand the domain of study, visualize the
main elements and their behavior at some chosen level of detail and use a short
hand notation for precise description and decreased ambiguity. Furthermore, the
conceptual integrity of a system derives from a coherent high-level view of the
system organization and functionality. Thus, one of the main objectives of this
work is to construct a conceptual model for policy representation and evaluation.
For doing so, we use a methodology based on concepts of sets and functions.

In our paper we are only interested in the class of authorization policies versus a wider range of policies, such as distributed system management policies.
The goal of authorization polices is to govern access to objects. Supporting such

V.I. Gorodetski et al. (Eds.): MMM-ACNS 2001, LNCS 1452, pp. 189–207, 2001.

190   T. Ryutov and C. Neuman

policies takes the form of monitoring and restricting the user activity within the distributed system (access control), making authorization decisions (authorization) and performing necessary actions to modify the behavior of the system (policy enforcement).

An authorization policy specifies conditions, which must be satisfied before, during or after the access right is exercised. For example, it may be desirable to enforce the following policy: "A process can be run on the host   A if the request originates from a domain   B and the process does not use more then 20% of the CPU time. An audit record about the started process must be generated".

This policy specifies several conditions:

1. location of the requester
   This condition must be satisfied before the access right "process run" is granted.
2. system load
   This condition must hold while the process is running.
3. audit record generation
   This condition must be met after the process is started.

Our model captures this intuitive notion of authorization policy and provides a formalism for the policy representation and evaluation.

There has been extensive research in authorization and a number of formal models have been developed.

Some of these contributions focus on addressing authorization requirements for specific policy domains, eg, database systems[3] , collaborative environment [17] or separation of duty [2]. Others are concerned with a particular access control mechanism, such as an ACL [1].

What is still missing is a unified view of authorization in a distributed multi-policy environment. Such a environment is composed of connected independent computer systems managed by separate administrative authorities. In a multi-policy environment the policy integration should incorporate diverse authorization models, which can coexist in a distributed system. Administrators of each domain might ex press security policies by means of d erent formalism

Generalizing the way that applications define their authorization requirements provides the means for integration of local and distributed security policies and translation of security policies across multiple authorization models.

Our paper describes an authorization model designed to meet these needs. In particular, our model allows us to represent existing access control models (e.g., ACL and capability) in a uniform and consistent manner.

The model simplifies the specification of complex   authorization policies and provides a generic policy evaluation environment. Furthermore, the model provides a general basis for identifying and resolving issues, not well-understood before, such as side e ects of the policy evaluation on the system state and related policies.

By separating generic from domain specific elements, we ensure that the model is extensible to arbitrary (authorization policy) domains.

We keep our model simple and practical to serve as an aid to implementation. We have found that the model suggested ideas for implementations, for example that condition implementation should be based on three phases.

Our final goal is to implement a subset of our conceptual model and provide a programmable framework for d erent kinds of policies. The framework maps real-world policy entities such as users, resources, and organizational policies, to the representation of these entities in the programming environment. The discussion of the initial implementation can be found in [14].

2   Related work

In this section we review prior research in representation and evaluation of authorization. Formal semantics for policy representation and evaluation has been used by other researches, in particular Woo and Lam [15].

Their work addresses general concerns as ours, in particular, positive and negative authorizations and providing computable semantics. In our model, authorization is given a precise semantics independent of underlying policy requirements. This distinguishes our work from [15] where a formal notion of an authorization policy has d erent semantics for each set of authorization require-ments.

The Policy Maker system described in the papers by Blaze, et al. [4], [5] focuses on construction of a practical algorithm for determining trust decisions. Policies and credentials encode a set of trust relationships among the issuing sources.

In Policy Maker's terminology, "proof of compliance question" asks if the request q, supported by a set of credentials complies with a policy   p. This is equivalent to the authorization question that we consider in our work: "is request q authorized by the policy   p (in our model credentials are contained in the request) ". Their approach, however, is d erent from ours.

In our approach, the information passed to the authorization engine with the authorization request is used to evaluate conditions in the relevant policy statements. Each condition is evaluated just one time. The order of condition evaluation is important.

In Policy Maker, the credentials and policy (called assertions) are used collectively to compute a proof of compliance. The assertions can be run in arbitrary order (and possibly many times) and produce intermediate results, that then can be fed into other assertions. Policies, representable in the Policy Maker, are restricted to the set of policies which do not produce side-effects, resulting in change of the system state. The Policy Maker can be integrated in our model as a component for evaluation of the   trust constraints   conditions.

Detailed formal language specification based on set and function formalisms is given in the paper by Sandhu [2] for specific constraints of separation of duty in rde based environment. The language semantics is defined by a restricted form of the first order logic. The formal language provides a useful model to study properties of conflict of interests, in particular separation of duty

The paper by Abadi, et al. [1] presents a logical language for access control lists. They study the notions of delegation, roles and groups using their logical language and rules for making access control decisions.

The exploratory work by Moffet and Sloman [11] is aimed to understanding policy semantics. The two aspects of a policy are considered: motivation and actual ability to carry out actions.

3 Basic Conceptual Model

The conceptual model presents the high level organizing principles of the authorization model and defines the strategy chosen to realize the model.

3.1 Policy Elements

In this section we explore the notion of a policy and abstract it into a conceptual model. This section prepares us for going to the more detailed specification given in the next section. We start the design of the conceptual model with specification of the components that are to be modeled. At a conceptual level a policy is a compound entity, which regulates access to objects.

The notion of an object is central to the policy definition. An object is a target of requests and it has to be protected. An object can be a physical resource such as a host or a communication channel, as well as an abstract, higher level entity, e.g., a bank account.

An access right is a particular type of access to a protected object, e.g., read or write. The notion of a negative access right is useful to specify many practical policies. Sometimes it is easier to allow access to all and explicitly disallow access for those who should not have access.

A condition describes the context in which each access right is granted. A condition must be satisfied in order to allow an operation to be performed on a target object [1]. Here are several of the more useful conditions [12].

- access identity
  Specifies an authenticated access identity (subject) on whose behalf request to access an object has been issued.
- time
  Time periods for which access is granted.
- location
  Location of the principal. Authorization is granted to the principals residing on specific hosts, domains, or networks.
- payment
  Specifies a currency and an amount that must be paid prior to accessing an object.

---

[1] However, if the access right is negative, the access is denied if all conditions are met.

- quota
  Specifies a currency and a limit. It limits the quantity of a resource that can
  be consumed or obtained.
- audit
  Enables automatic generation of an application level audit data in response
  to access requests.
- notification
  Enables automatic generation of notification messages in response to access
  requests. Specifies the notification method and a receiver.
- trust constraints
  Specifies restrictions placed on security credentials. Allows one to validate
  the legitimacy of the received certificate chain and the authenticity of the
  specified keys.
- attributes of subjects
  Defines a set of attributes that must be possessed by subjects in order to get
  access to the object, e.g., user age.

Traditional security thinking has been oriented toward authentication as a
prerequisite for authorization. Usually authorization applies after authenticated
requester identity has been established.

In our model policies are treated as the first class citizens. Authentication,
audit and accounting mechanisms are activated by explicit policy requirements,
expressed through conditions. If a policy does not require authenticated user
identity, authentication steps can be ignored or deferred until the policy explicitly
requests it. An example of a policy, which is not concerned with the identity is
"anyone can read file   A if $10 is paid".

Note that in the implementation, some of these conditions might have side
effects. For ex ample, evaluation of   payment   and quota   conditions reduces a
balance somewhere. Evaluation of   notification   condition results in sending a
message, which is useful in audit.

Unfortunately side effects might complicate the model. Ignoring the side
effects might cause problems when the side effects create a feedback loop, for
ex ample, when an audit record triggers a network threat detection which affects
the evaluation of subsequent policies, or where payment affects quotas which
affects the ability to perform other operations (one one runs out of money) .

Balancing the complexity this adds with the simplicity of the model is still
an open issue, which requires further investigation. Initial ideas on handling the
side effects are given in Section 4.2

3.2   Basic Definitions and Assumptions

We present our conceptual model based on set and function formalism, algebra
of sets and first order logic. The conceptual model specification is guided by
conventional authorization notions and expected authorization requests.

An elementary policy statement consists of an object component, a positive or
negative access right component and zero or more condition components. Thus,

to represent the components, we define sets of elements called objects   O, positive
rights R, negative rights $\overline{R}$ and conditions C. All existing policy statements are
contained in the set   P. In addition, we define a set of authorization requests   Q.

All the sets, except for C^2 , are finite dynamic and unordered. The dynamic
property means that sets are not fixed and new elements can be added and existing
elements can be deleted. The finite property assumption requires that at any
particular time, the sets are finite. Negation is applied only to the elements of
the set R to model negative rights. We do not define negative conditions. The
empty set is denoted by   .

O is finite dynamic non-empty unordered set of object elements:

$$O = \{o_1, o_2, \ldots, o_n\} . \tag{1}$$

R is finite dynamic non-empty unordered set of access right elements:

$$R = \{r_1, r_2, \ldots, r_n\} . \tag{2}$$

$\overline{R}$ is finite dynamic non-empty unordered set of negative access right elements. Set $\overline{R}$ is constructed from the set R by applying negation to each element
of the set R.

$$\overline{R} = \{\neg r_1, \neg r_2, \ldots, \neg r_n\} . \tag{3}$$

Note that R $\overline{R}$ =   .

C is dynamic unordered set of condition elements with a special condition
element c , which represents an empty condition:

$$C = \{c , c_1, c_2, \ldots, c_n\} . \tag{4}$$

P is finite dynamic unordered set of compound policy elements:

$$P = \{p_1, p_2, \ldots, p_n\} . \tag{5}$$

Each element p of the set P represents a set of three elements:

$$p = \{o, r, c\} , o \quad O, \ r \quad R \quad \overline{R}, c \quad C. \tag{6}$$

Note that a condition element can be   c . When c = c the rights are granted or
denied unconditionally. An example of a practical policy with an empty condition
is: "file A can be read by anyone".

Q is finite dynamic partially ordered set [3] of compound authorization request
elements:

$$Q = \{q_1, q_2, \ldots, q_n\} . \tag{7}$$

Each element q of the set Q represents a set of three elements:

$$q = \{o, r, c\} , o \quad O, \ r \quad R, c \quad C. \tag{8}$$

---

[2] The conditions can be represented by different entities, including numbers (see
Section 4.2), so we can not state finiteness property

[3] The reasoning behind the requirement of the partial ordering of the set   Q is discussed
in Section 4.2.

The elements correspond to the target object (o), requested access right (r) and a condition constant (c). The condition constant c represents information which is needed to the requirents specified in the condtion of the releat policy statement. In practice, this information can be represented by a set of credentials, e.g., authenticated user identity, For example, a policy statement "Ayone can read file   A from 8amtill 6 pm specifies a   time condition. The request "read (r) file A (o) at 5pm (c) " specifies arrent time and is needed to the time condition in the policy statement.

To make our model practical, special provisions should be made for dealing with the following situations:

- incomplete data, not known at the authorization time. During network fragmentation some data may be inaccessible.
- policy requires a certain event to happen in the future. Statements about the future do not have truth values until the event described takes place.
- the function used to evaluate conditions does not terminate for the arguments supplied. Incorrect implementation, bad parameters.

In order to properly deal with these situations we will adopt a three-valued logic [9], [13].

Three-valued logic is classical boolean (true/false) logic extended with a third truth value - undefined

We define an auxiliary set   B, consisting of the three constants: true, represented by T, false, represented by F and U, meaning uncertainty.

$$B = \{T, F, U\} . \tag{9}$$

Table 1 shows the truth tables, when at least one argument is equal to   U.

P	Q	P & Q	P ∨ Q
T	U	U	T
U	T	U	T
F	U	F	U
U	F	F	U
U	U	U	U

Table 1.

In addition, $\neg U = U$. Next we define functions to express an authorization process.

The by_object function takes a set of policy elements   P and request q, which contains particular object   o as an argument and returns a subset   P ⊆ P where this object appears.

$$P = by\_object(P, q),$$

o O, o q, q = {o, r, c}, q Q, P   P : p   P : p = {ô, r, c} .   (10)

The by_right function takes a set of policy elements P and request q, which contains particular access right r as an argument and returns a subset $P' \subseteq P$ where this right appears.

$$P' = \text{by\_right}(P, r), \quad r \in R,$$

$$\forall P' \subseteq P : p \in P' : p = \{o, r, c\} \text{ or } p = \{o, \neg r, c\}. \tag{11}$$

The eval_cond is a condition evaluation function.

$$b = \text{eval\_cond}(c, c), \quad c \in C, \ c \subseteq C, \ b \in B. \tag{12}$$

The function M defines positive or negative modality of the policy element.
If the access right, contained in the policy element is positive or negative, the modality is positive or negative, respectively.

$$M(p_i, q) = \begin{array}{l} \text{eval\_cond}(c, c), \quad r \in R \\ \neg \text{eval\_cond}(c, c), \quad r \in \overline{R}, \end{array}$$

$$c \in p_i, \ c \subseteq q, \ r \in p_i, \ p_i \in P, \ q \in Q. \tag{13}$$

The M function has to be applied to all elements $P' \subseteq P$. The evaluated modality of each policy element will be taken with or without the negation $\neg$ according to its right. After all the modalities are evaluated, we will take their disjunction. These operations are performed by the eval_conditions function.

$$b = \text{eval\_conditiods}(P', q) = M(p_1, q) \lor M(p_2, q) \lor \ldots \lor M(p_n, q),$$

$$p_i \in P', \ i = \overline{1, n}, \ n \text{ is the cardinality of } P',$$

$$P' \subseteq P, \ q \in Q, \ b \in B. \tag{14}$$

The resulting value b obeys to the $\lor$ operation for three-valued logic. That is, eval_conditions returns T if at least one modality gave the result T, F if all results were F, and U otherwise (i.e., at least one result was U, possible some F but none T).

The authorization is a composite function:

$$b = \text{authorization}(P, q) =$$

$$= \text{eval\_conditions}(P', q) \circ \text{by\_right}(P', q) \circ \text{by\_object}(P, q) =$$

$$= \text{eval\_conditions}(P', q) \circ \text{by\_object}(P', q) \circ \text{by\_right}(P, q). \tag{15}$$

The authorization function takes the set of policies P and an authorization request q as arguments. It returns F, T or U meaning authorized, not authorized or uncertain. Three-valued logic at the conceptual level has to be mapped to the two-valued logic at the implementation level. In the end, the access must be either granted or denied.

3.3   Time Dependency

Time dependency appears in our conceptual model implicitly. At each instant only the set of policies which exists at authorization time is considered. All future or past policies are irrelevant. Note that this does not mean that the current policy does not depend on the past or future events. Some policies must take into account the system execution history or the fact that particular event must have happened for some operation to take place. An example of practical policy taking into account occurrence of some event is "If one reads file A, then one can not send" [16]. Some policies may need to know precise time of the event occurrence, for example for audit purposes. This may require a time-stamping of certain occurrences and keeping record of them.

3.4   Changes in the Set Membership

Ex ercising access rights can result in creating new objects and defining new policies. In the conceptual schema this is represented as adding an element to the corresponding set. As we discussed in the previous section, changes in membership of the sets R and $\overline{R}$ depend entirely on the set   O.

The deletion of an element from the sets   O, R or $\overline{R}$ entails deletion of each element from P in which the deleted element appears. To simplify our model we require that rights can be applied only to the elements of set   O. If we allow rights to be applied to the elements of   P , we will have to consider a policy management model.

3.5   Policy Representation Issues

We do not allow use of the disjunction in representation of elements of the set P . The disjunctive form policies such as "Tom or Joe can read file A", "Tom can read either file   A or B " and "Tom can either read or write file   A" is modeled by using separate policy statements.

$$O = \{A, B\}, R = \{\text{read}, \text{write}\}, C = \{c, Tom, Joe\},$$

$$P = \{\{A, \text{read}, Tom\}, \{A, \text{read}, Joe\}, \{B, \text{read}, Tom\}, \{A, \text{write}, Tom\}\}.$$

However, disjunction of policy elements can be used in practice for optimization reasons. For example, in the implementation of an ACL we can combine several access rights which correspond to a particular access identity condition.

Let us consider the exclusive   OR policy representation "Tom can read files A or B, but not both". This policy is a variant of the Chinese wall policy [6], required in the operation of many financial services. The policy guards against the conflict of interest. A consultant can freely chose a company in order to offer an advice. However, once the company has been chosen, the consultant is mandatory denied access to the information about all other companies. This policy can

be implemented using an additional condition, let us call it   trigger _history . This
condition activates the history of execution.

$$P = \{\{A, \text{read}, \text{Tom}, \text{trigger}\ \_\text{history}\}, \{B, \text{read}, \text{Tom}, \text{trigger}\ \_\text{history}\}\}.$$

If Tom decides to read file   A first, the history is checked and since initially it
is empty, the right is granted and the information about it is stored. If he tries
to read file   B after that, the request will be denied. A history information is
maintained by the system. The history can be centralized or distributed. An
example of implementation of the condition is briefly described in [18]. More
detailed discussion of implementation of the history-dependent access control
policies is given in [10].

In conventional access control models, a subject has been a separate notion.
A subject is an entity on whose behalf a request to access an object has been issued. Traditionally, policy conceptualization is based on three basic entity types:
objects, access rights and subjects. Some of the possible logical groupings of these
entities, such as ACL and capability, have become practical implementations of
the Lampson matrix [8].

In the ACL based systems, policies are grouped by objects. A typical ACL is
associated with an object (or a group of objects) to be protected and enumerates
the list of authorized subjects and their rights to access the object.

In the capability-based systems, policies are grouped by subjects. A capability
lists sets of objects accessible by the subject along with the types of access rights.

These logical grouping can be represented in our model.

ACL   An ACL consists of a set of ACL entries. An ACL entry is analogous to
a policy element p, where all conditions are   access identity   .
Consider a policy: "Tom and Bob can read and write file   A". We can translate
this policy into our policy model as:
"Tom (condition c_1) and Bob (condition c_2) can read (positive right r_1) and
write (positive right r_2) file A (object o_1) ". We need four policy elements to
represent this policy:

$$p_1 = \{o_1, r_1, c_1\},$$
$$p_2 = \{o_1, r_2, c_1\},$$
$$p_3 = \{o_1, r_1, c_2\},$$
$$p_4 = \{o_1, r_2, c_2\}.$$

This way of specification and storage of the policy is tedious and inefficient.

To represent an ACL, we adopt three modifications to the representation of
a policy element p specified in (6):

1. An ACL is associated with each object, so the object is implicit and is
   omitted from the policy elements.
2 Conditions are listed first, then access rights. This order is closer to the
   traditional ACL specification
3. We allow disjunction of either positive or negative access rights.

Now we need only two ACL entries to represent the policy:

$$p_1 = \{c_1, r_1 \quad r_2\},$$
$$p_2 = \{c_2, r_1 \quad r_2\}.$$

Furthermore, if we allow conditions to be aggregated into a single entry when the same set of access rights applies to all of them, we need only one policy statement to represent the policy: $p_1 = \{c_1 \quad c_2, r_1 \quad r_2\}$.

by_object function returns all policy statements associated with the given object. The returned set of policies   P   conceptually represents an ACL associated with the object   o.

Capability   To demonstrate how capabilities can be represented, we define function by_condition, which takes the set of policies   P   and particular condition   c   as arguments and returns a subset   P , where this condition appears. Intuitively, this function returns all policy statements associated with the given condition.

$$P = by\_condition (P, c), c \quad C, P \quad P : p \quad P : p = \{o, r, c\}.$$

Note that if the condition constant   c specifies particular access identity (subject), then the returned set of policies   P   conceptually represents a capability possessed by the subject identified by the condition   c. Next the set   P   can be passed to the authorization   function along with an authorization request for further evaluation.

Representation of a capability is quite similar to that of an ACL. A capability is associated with each subject, so the subject is implicit and is omitted from the policy element. Thus, each policy statement contains only elements, which represent objects and access rights.

More detailed discussion of the implementation of ACL and capability can be found in [14].

4   Extended Conceptual Model

The extended conceptual model expands upon basic conceptual model entities and interactions. The notion of a policy hierarchy is introduced. The design work at this level addresses condition side-effects issues.

41   Refinements

In this section we describe further refinements of our basic entities. A policy statement may specify several conditions of different types, for example: "Tom can read file   A only between 9 am and 6 pm. This policy defines two conditions: access identity   and time . In (6) we have considered only one condition in the policy statement. All existing conditions were aggregated into one set (4). Now we ex tend the notion of a condition to be distinguished not only by an identifier but also by a type. Each condition element has just one type. We assume that at each instant S condition types exist. We represent these d erat condition types by S disjoint sets:

$$C = \quad \underset{k=\overline{1,S}}{C^k}, C^i \quad \underset{i,j=\overline{1,S}\, i=j}{\qquad} C^j = \qquad . \tag{16}$$

Now we define a totally ordered [4] set C. Each element of this set is constructed from one element of the S disjunctive sets. Intuitively this means that each element of C consists of S condition denots of d erat type, some of the elements can be c.

$$C = \{c^1, c^2, ..., c^S\}, c^i \quad C^i, i = \overline{1, S}. \tag{17}$$

We define S condition evaluation functions for each condition type. In our policy example we define two function for checking access identity and current time

$$b = eval_i(c^i, c^i), c^i \quad C^i, c^i \quad C^i, b \quad B, eval_i(c, c) = T, i = \overline{1, S}. \tag{18}$$

From (4.2) and (15) we observe that if at least one of the policy statements evaluates to T, the authorization will be granted. This behavior may not be always desirable. For example, we would want a policy assigned by the system administrator to take precedence over the one assigned by an individual user. This requires the means of specifying a hierarchical relationship among policy statements.

The hierarchy of policies is modeled by assigning priorities. We do not attempt to give a full theoretical development of the method of assigning priorities here. The essential requirements is that one should be able to decompose the whole policy into totally ordered policy statements. To express policy priorities, we define set W. W is a finite totally ordered set of denots that can be compared (e.g., integers).

$$W = \{w_1, w_2, ..., w_n\}, \quad w_i < w_j, \, i,j = \overline{1,L}, \, i < j, \, L \text{ isthe cardinality of } W.$$

We define denot q, given in (8) in the following way:

$$q = \{o, r, c^1, c^2, ..., c^S\}, o \quad O, r \quad R, c^i \quad C^i, i = \overline{1, S}, q \quad Q. \tag{19}$$

We extend (6) in two ways: 1) each element p has an additional component w, which denotes priority of this element. 2) condition component is represented by a set of condition constats of d erat types.

$$p = \quad o, r, C, w, o \quad O, r \quad R \quad \overline{R}, C \quad C, w \quad W, p \quad P. \tag{20}$$

Figure 1 illustrates representation of a policy element p.

---

[4] The reasoning behind the requirement of the total ordering of the set C is discussed in Section 4.2.

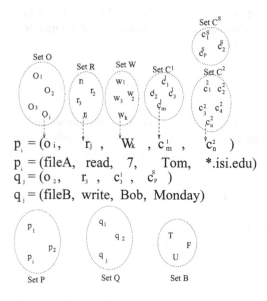

$$p_i = (o_i, \quad r_j, \quad W_k, \quad c_m^1, \quad c_n^2)$$
$$p_i = (\text{fileA}, \quad \text{read}, \quad 7, \quad \text{Tom}, \quad *.\text{isi.edu})$$
$$q_j = (o_2, \quad r_3, \quad c_3^1, \quad c_p^s)$$
$$q_j = (\text{fileB}, \quad \text{write}, \quad \text{Bob}, \quad \text{Monday})$$

Figure 1.

The by_priority function takes a set of policies P as an argument and returns a subset P with the maximum priority. The ordering in the set W determines the policy statement which is enforced if several policy statements are simultaneously satisfied. Note that if the set P contains more then one element, the elements have equal priorities. In this case, if any of the policy statements is satisfied, authorization is granted.

$$P = \text{by\_priority}(P),$$

$$P \quad P : p \quad P, p = \{o, r, C, w\}, w = \max(\quad w : p = \{o, r, C, w\} \quad P).$$

$$\tag{21}$$

We define eval_cond function given in (12) in the following way:

$$\text{eval\_cond} = \text{eval}_1(c^1, c^1) \& \text{eval}_2(c^2, c^2) \& ... \& \text{eval}_s(c^s, c^s),$$

$$c^i \quad C^i, c^i \quad C^i, i = \overline{1, S},$$

$$b = \text{eval\_cond}(p), p \quad P, b \quad B. \tag{22}$$

The eval_cond function is a short hand notation for representation of conjunction of the results, obtained by applying eval_i to corresponding condition constants from the policy element p. All conditions must be met simultaneously in order to satisfy the authorization request.

The resulting value b obeys to the & operation for three-valued logic. That is, eval_cond returns T if all elements gave the result T, F if at least one result was F, and U otherwise (i.e. at least one result was U, possible some T but none F.

We define   authorization   function given in (15) in the following way:

$$b = \text{authorization} \ (P, q) =$$

$$= \text{eval\_conditions} \ (P \quad , q) \quad \text{by\_priority} \ (P \ ) \quad \text{by\_right} \ (P \ , q) \quad \text{by\_object}(P, q) =$$

$$= \text{eval\_conditions} \ (P \quad , q) \quad \text{by\_priority} \ (P \ ) \quad \text{by\_object}(P \ , q) \quad \text{by\_right} \ (P, q),$$

$$P \quad P \quad P \quad P, q \quad Q, b \quad B \ . \tag{23}$$

Figure 2 illustrates the   authorization   function.

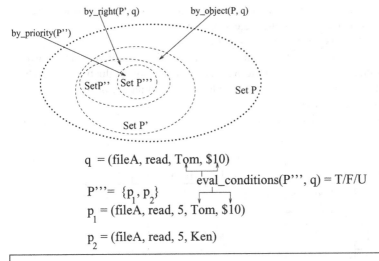

$$q = (\text{fileA, read, Tom, \$10})$$

$$\text{eval\_conditions}(P''', q) = T/F/U$$

$$P''' = \{p_1, p_2\}$$

$$p_1 = (\text{fileA, read, 5, Tom, \$10})$$

$$p_2 = (\text{fileA, read, 5, Ken})$$

authorization(P, q) = eval_conditions(P''', q) o by_priority(P'') o by_right(P', q) o by_object(P, q) =
= eval_conditions(P''', q) o by_priority(P'') o by_object(P', q) o by_right(P, q) = T/F/U

Figure 2.

42   Discussion of Condition Side-Effects

The total order property of the set   C defined in (16) requires that policy elements that differ only by the order of condition elements are considered to be distinct. This property is important to deal with possible side effects caused by the condition evaluation. Consider a policy "Tom can read file   A only if notification is set (notification   condition) and system threat condition is low (threat_level_low condition)". Assume that current system threat level is low. Assume that the notification about Tom reading file A triggers high system threat level. There are two ways to represent the policy in our model:

$$p_1 = \{A, \text{read}, \text{Tom}, \text{threat}\_\text{level\_low}, \text{notification} \ \},$$

$$p_2 = \{A, \text{read}, \text{Tom}, \text{notification}, \text{threat}\_\text{level\_low}\} \ .$$

The evaluation of p_1 results in access grant, however evaluation of p_2 results in denial.

In this section we will discuss determining the correct order of the condition elements in the policy statement p defined in (2).

System State Representation   To discuss side effects produced by evaluation of some conditions, we introduce time into our model explicitly. Time is discrete and is represented by a totally ordered set of natural numbers. Each number corresponds to a discrete time interval. A time interval is related to a condition evaluation process.

To simplify our presentation, we assume that dependent authorization requests do not overlap. The effects of the dependent requests are resolved by serialization, in which the requests are ordered by the cause-effect ordering.

Similarly, we assume that conditions are evaluated consecutively. These two assumptions enable us to concentrate on a single condition evaluation per each time interval and, therefore, avoid the problem of coordination of multiple condition evaluation processes.

Figure 3 illustrates our representation.

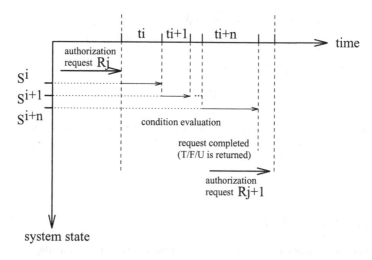

Figure 3.

A time interval begins when a condition evaluation starts and it ends when the condition evaluation is completed with the resulting $T/F/U$. This means that the duration of the time intervals can vary.

The general idea underlying our approach is that the system state can be formalized by a sequence of system states $S^1, S^2, ..., S^k$. Each system state S^i is labeled by the time interval i.

By a system state we mean not only information describing a particular computer system such as system load, network bandwidth consumption, number of

204   T. Ryutov and C. Neuman

available processors, but also all information about the real world which is representable in a computer system, for example: bank account balance, temperature, user identity.

Here any system state S^i is all the information that has been deduced up to the time interval i. The information is represented by a set of system variables. The information is partial, since some system variables can be undefined at some time intervals.

At each time interval i there is a transition $S^i \rightarrow S^{i+1}$ from the current system state S^i to the new system state S^{i+1}. Each transition is characterized by updating the values of some system variables. The variables can change not only as the result of condition evaluation but also because of other events, e.g., system load is altered. All side effects of condition evaluation are recorded in the corresponding system variables.

Classification of Conditions   In this section we present a taxonomy of conditions. We say that a condition writes system state, if the condition evaluation function changes values of some system variables.

The fact that evaluation of condition c changes value of the system variable j is represented by the notation $c(S^i) \rightarrow S^{i+1}_j$.

We say that a condition reads system state if the condition evaluation function requires reading of particular system variables.

The fact that evaluation of condition c requires the value of the system variable j is represented by the notation $c(S^i_j) \rightarrow S^{i+1}$.

We say that a condition c depends on condition c, if condition c requires reading of some system variables, which are written by the condition c.

The fact that condition c requires the value of the system variable j, which is written by the condition c is represented by the notation: $c(S^i_j) \leftarrow c(S^{i+1}_j)$.

Conditions are classified by the read/write system state property:

- read conditions read system state but do not write system state, for example time, location and system load.
- write conditions write system state and may read system state, for example, payment. Payment requires checking for the presence of required amount (read system variable k) and reducing the balance by the requested amount (write system variable k). This is represented as: $c(S^i_k) \rightarrow S^{i+1}_k$.

Note that write conditions must be evaluated before the read conditions that are dependent on them.

Designing the condition ordering algorithm that satisfies the ordering requirements falls into the realm of scheduling of processes with precedence constraints and is outside of the scope of this paper.

Condition Representation and Evaluation   Read conditions such as access identity and location appearing in the authorization request, specify a set of constants which must be matched against a corresponding set of constants found in the policy elements. These conditions are represented by a set C^1. This set is constructed from a set of all condition constants passed in authorization requests q defined in(19) , a set of all condition constants contained in policy elements   p defined in(20) and a set of operations   M . Condition evaluation function for this type of conditions returns   T if applying operation   m, (m   M) to the condition constants evaluates to   T , otherwise it returns   F .

For example, a set of operations   M may contain (   ). If m = , condition evaluation function returns   T if c^1 c^1,(c^1 C^1,c^1 C^1), otherwise it returns F .

Some conditions, such as   system load   , can be represented numerically. These conditions are evaluated by comparing numbers (natural, integer or real). Therefore, we can define the set of operations as M = { = , = , < , > , , } .

Write conditions, such as   notification   and audit   specify the name of a system variable, whose value must be changed, and the new value. Condition evaluation function for these conditions returns   T if the updating of the system variable succeeded [5], and F otherwise.

Unfortunately not all conditions can be represented in this way. In practice, conditions can be application-specific and complex . The problem is how an informal specification of the condition can be transformed into a precise formal mathematical structure, within which we can actually prove things about the properties, such as computability and polynomial-time decidability.

5   Conclusions and Future Work

In this paper we presented a conceptual model for authorization in distributed systems. We introduced precise semantics for policy representation and evaluation. The semantics is defined independently from underlying security mechanisms and is separate from implementation The flex ibility of the model makes it possible to represent existing access control mechanisms.

We believe that the model provides an effective way to understand and employ authorization policies in distributed systems.

We have begun to investigate the side-effects of the condition evaluation. Through the use of the side effects, in our current work we consider integrating intrusion and misuse detection systems with applications using our model.

We hope that this model will lead to other insights about authorization policies. We are looking for possible ways to restrict condition expressiveness to guarantee policy computability and polynomial-time decidability.

---

[5]   Updating the system variable can fail due to various reasons, for example we might be unable to append audit information to the audit log because the disc space has been exceeded.

6   Acknowledgments

This research was sponsored by the Defense Advanced Research Projects Agency (DARPA) and Air Force Research Laboratory, Air Force Materiel Command, USAF, under agreement number F30602-00-0595, Security Infrastructure for Large Distributed Systems (SILDS) Project under agreement number DABT63-94-C-0034 and by Xerox Corporation, XAUTH Project under agreement number HE1254-97. The U.S. Government is authorized to reproduce and distribute repreints for Governmental purposes notwithstanding any copyright annotation thereon.

The views and conclusions contained in this document are those of the authors and should not be interpreted as necessarily representing the o cial plicies or endorsements, either expressed or implied, of the Defense Advanced Research Projects Agency (DARPA), the Air Force Research Laboratory, the U.S. Government, or Xerox Corporation.

References

1. Abadi, M., Burrows, M., Lampson, B. and Plotkin, G.:   A calculus for Access Control in Distributed Systems.   ACM Transactions on Programming Languages and Systems, Vol. 15, No 4   (September 1993) 706–734
2. Gail-Joon Ahn and Sandhu, R.: The RSL99 Language for Role-Based Separation of Duty Constraints.   ACM Workshop on Role-Based Access Control   (1999) 43–54
3. Bertino, E. and Jajodia, S.:   Supporting Multiple Access Control Policies in Database Systems.   Procee dings of the 1996 IEEE Symposium on Security and Privacy (1996)
4. Blaze, M., Feigenbaum, J. and Lacy, J.: Decentralized Trust Management.   Proceedings IEEE Symposium on Security and Privacy   , IEEE Computer Press, Los Angeles (1996) 164–173
5. Blaze, M., Feigenbaum, J., Strauss, M.: Compliance Checking in the Policy Maker Trust Management System.   In Procee dings of the Financial Cryptography '98, Lecture Notes in Computer Science   , Vol. 1465 254–274
6. Brewer, D.F.C. and Nash, M.J.: The Chinese Wall Security Policy.   Procee dings of the 1989 IEEE Symposium on Security and Privacy   , pages (1989) 206–214
7. Jajodia, S., Samarati, P. and Subrahmanian, V.S.:   A logical Language for Expressing Authorizations.   Procee dings of the 1997 IEEE Symposium on Security and Privacy   (1997)
8. Lampson, B.:   Protection.   ACM Operation System review 8(1)   (January 1974) 18–24
9. Lukasiewicz, J.: On Three-Valued Logic. 1920.   Rch Filozficzny   1920, 5, pp.170-1. Englishtranslation in Borkowski, L. (ed.) Jan Lukasiewicz: Selected Works. Amsterdam: North Holland (1970)
10. Massimo, A., Cazzola, W., Fernandez, E.B.: A History-Dependent Access Control Mechanism Using Reflection   Procee dings of 5thECOOP Worksbp on Mobile Object Systems (EWMOS'99)   , (June 1999)
11 M et, J . D and Sonn MS :   The representation of Plicies as System obj ects Procee dings of the ACM Conference on Organizational Computing Systems   , Atlanta, GA (November 1991) 171–184

12. Neuman, B.C.: Proxy-based authorization and accounting for distributed systems. Procee dings of the 13th International Conference on Distributed Computing Systems, Pittsburgh   (May 1993)
13. Prior, A.N.: Three-Valued Logic and Future Contingents.   Philosophical Quarterly   . Vol. 3 (1953) 17–26
14. Ryutov, T.V. and Neuman, B.C.: Representation and Evaluation of Security policies for Distributed system Services.   In Procee dings of the DARPA Information Survivability Conference and Exposition   . Hilton Head, South Carolina (January 2000)
15. Woo, T.Y.C. and Lam, S.S.: Authorization in distributed systems: a new approach. Journal of Computer Security, 2   (1993) 107–136
16. Schneider, F.B.: Enforceable security policies.   Technical report TR98 1664   , Cornell University (January 1998)
17. Shen, W. and Dewan, P.: Access Control for Collaborative Environments.   Proceedings of CSCW   (November, 1992) 51–58
18. Simon, R.T. and Zurko, M.E.: Separation of Duty in Role-Based Environments Computer Security Foundations Workshop   (June 1997)

Fenix Secure Operating System: Principles, Models, and Architecture

Dmitry P. Zegzhda, Pavel G. Stepanov, and Alexey D. Otavin

Information Security Centre of Saint-Petersburg Technical University,
195273, Saint-Petersburg, K-273, P/B 290
Dmitry@ssl.stu.neva.ru

Abstract. The paper introduces design principles of Secure Operating System Fenix developed in Information Security Centre of Saint-Petersburg Technical University. Fenix is a special purpose secure operating system supposed to be a basis for secure information processing. Fenix is fully compliant with Russian national information security requirements and standards. Security was the main goal of this project, other aspects of operating system were subject to it. Security functions enforcement was the main factor in the operating system design. Microkernel architecture, client-server technology and object-oriented approach form a core of the Fenix operating system design.

Introduction

Awareness of the need for security in computing systems is growing rapidly as critical services are becoming increasingly dependent on interconnected computing systems but current efforts to provide security are unlikely to succeed. The need for secure operating systems is growing in today's computing environment due to substantial increases in connectivity and data sharing. In reality, operating system security mechanisms play a critical role in supporting security but the computer industry has not accepted the critical role of the operating system to security, as evidenced by the inadequacies of the basic protection mechanisms provided by current mainstream operating systems. The necessity of operating system security to overall system security is undeniable; the underlying operating system is responsible for protecting applications against tampering, bypassing, and others attacks. If it fails to meet this responsibility, system-wide vulnerabilities will result. No single technical security solution can provide total system security. Unsolved technical problems, implementation errors and flawed environmental assumptions will result in residual vulnerabilities. A proper balance of security mechanisms must be achieved. Each security mechanism provides specific security functions and should be designed to only provide those functions. It should rely on other mechanisms for support and for required security services. In a secure operating system, the entire set of mechanisms complement each other so that they collectively provide a complete security package. Systems that fail to achieve this balance will be vulnerable.

This paper describes the main principles and architecture of the Fenix secure operating system developed in Information Security Centre of Saint-Petersburg Technical

V.I. Gorodetski et al. (Eds.): MMM-ACNS 2001, LNCS 2052, pp. 207–218, 2001.

University. The Fenix operating system development was guided by the followed secure systems design principles:

1. Integrality principle. Security mechanisms form a complete package built into information processing system at the basic services level. OS security mechanisms (access control, identification/authentication, audit and others) maintain hold of all kind of interactions. In microkernel OS the control of message exchange provides total control of information flows as messages is the only way to interact.

2. Invariance principle. Operating system security mechanisms are security policy invariant as intercommunication of OS components and applications are based on client-server model. Client-server architecture allows combining global access control for common operations and additional object-specific control implemented by object managing services. Furthermore, usage of object-oriented technologies allows to extent security mechanisms by adding new method to a server. Standard object access protocols allows easy integration with other secure and non-secure systems.

3. Unification principle. Secure OS has the only implementation of security mechanisms for all kind of operations and all type of objects as security functions are unified by object-oriented programming technologies and common interfaces of access control. Unification of message format, operation types and resource properties allows implementing unified access control mechanisms. The design of all servers follows the same pattern, servers perform unified object access operations (creation, reading, writing, deletion, etc.).

4. Adequacy principle. Investigations demonstrate that one of the main source of vulnerabilities is no generic approach to access control. Common OS have privileged mechanisms that delegate authorities to non-privileged users and ignore standard control mechanisms. Any error in privileged program results in vulnerability. Secure OS architecture design provides unified access control model, privileged mechanisms and access control exceptions are excluded.

5. Correctness principle. The correctness of operating system security implementation depends on reliability of program code performing crucial security functions. Minimization of the security functions code is the best solution of this problem. Simple and compact access control mechanisms let to apply formal methods for program code verification and analysis.

According to these principles the Fenix secure OS architecture design is based on microkernel, client-server interaction model, security policy invariant access control mechanisms and common Object Access Unified Interface.

Related Works – Trusted Mach and DTOS

The security architecture of the Fenix system is derived from the architecture of the Trusted Mach and DTOS systems, which had similar goals. However, the Trusted Mach [1] and DTOS [3] security mechanisms were not sufficient to support dynamic security policies. Mach-like systems has no strict architecture separation of security and data processing functions.

Trusted Mach is running on top of MK++ kernel [2]. As is typical for microkernels, MK++ provides access control mechanisms while relying upon external servers to define the policy being enforced by these mechanisms. The basic division of duties is between the Root Name Server which maintains system namespace and makes security decisions and all of the various trusted servers (item managers) which enforce those decisions. All objects of a particular type are managed by the same item manager. The TMach design is extensible to allow the addition of new item managers and indirection mechanism for associating a particular client's access rights to server objects (fig. 1).

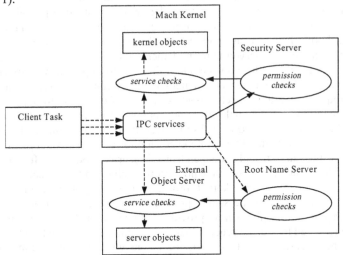

Fig. 1. Mach-based OS Access Control Mechanisms

To get access to any security object a client should open an object in the name space and provide a set of required access rights. Root Name Server (RNS) performs access rights check and requests the item manager to create an agent with specified access rights for the object. The item manager is expected to enforce further security decisions itself is upon object invocation. It is supposed that all trusted item managers uses code from the Class Library which performs this policy enforcement. But there is nothing to stop servers from providing specific access control rules or from avoiding to add it's objects to the root name space. Also TMach is expected to support untrusted item managers who still wish to make use of the services of the RNS but may fails to enforce security permissions.

If previously granted access rights to an object need to be invalidated, the RNS clears all internal agents associated with the object, item manager also is expected to revoke all of its agents. There is no way to recalculate granted access rights. Because of Mach's asynchronous messaging through a kernel buffered message queue, the sending process may have even been destroyed at the time that the message is being received and processed by item manager. It could be quite complex to identify that the permissions of the sender have changed. Those access control mechanisms are applied for any interaction whether client and server are trusted or not. So TMach uses the common DAC mechanism to mediate the rights of trusted and untrusted servers to perform privileged system operation.

The DTOS kernel was derived from the Mach 3 kernel, a predecessor of the MK++ kernel. A key feature of DTOS architecture is the amount of explicit security functionality within the kernel itself. The DTOS kernel provides explicit labelling of all kernel objects and controls all accesses according to security decisions made based upon those labels but it remains independent of any particular security policy, security decisions are made by an external security server which is queried by the kernel [4]. The DTOS kernel also provides no direct support for informing the security server when a previously granted permission is no longer being used.

In comparison with Mach-like systems the Fenix architecture provides more strict client-server model and separation of access control and data processing mechanisms. The Fenix kernel functions are minimised, all modules that manages objects (like of the Mach kernel device servers) are implemented as external trusted process. Kernel objects are available to their owners only, there is no opportunity to extend security policy upon the kernel objects.Each item manager(functional server) in Fenix OS services requests to a logical set of objects of various types, not to objects of the particular type or class. Functional servers are supposed to apply the unified access interfaces to their objects instead of enforcing security decisions. Functional servers perform all requests permitted by the reference monitor without any supplementary security checks, they are strongly independent of security model. All functional servers are trusted, untrusted processes have no "receive rights" and never act as servers.

Unlike MK++ the Fenix kernel provides message source authentication, any trusted server can attain valid sender process ID not sender security ID, available to the reference monitor only. Kernel provides authentication of recipient also, message would be received by trusted process that owns the target port identified by sender.

Like other secure OS the Fenix kernel interacts with the reference monitor, but the goal of these interactions is to enforce decisions made by the reference monitor about requests to the objects of trusted servers. Reference monitor checks every message sent by untrusted process just before it is delivered to receiver, messages from trusted processes are not checked. If reference monitor had failed, the kernel provides that all untrusted processes would be unable to send messages while trusted processes would be unaffected (fig. 2).

Fenix access control mechanisms are more appropriate to enforce dynamic access control policies than Mach mechanisms. If access rights had changed, the reference monitor would check all further requests according to new access rights, so all messages delivered by the kernel would be secure. Thread that had sent a message would exist in inactive state until the message is delivered to receiver, so no requests from dead processes are possible.

Fenix Operating System Architecture

Fenix operating system is a secure microkernel UNIX system. Fenix is supposed to be a secure server-oriented platform rather than a mainstream OS for user workstations. Fenix includes advanced security services and a limited number of functional components and user applications. A vital feature of Fenix is flexible access control mechanisms to enforce a wide range of security policies. This feature is implemented by strong separation of data processing functions and security mechanisms and by usage of mathematical models of trusted system components. The reference monitor enforces strong Bell-LaPadula policy combined with high granularity discretionary access control policy based on role hierarchy. The combination of mandatory and discretionary models provides both the strength of the former and the flexibility of the latter.

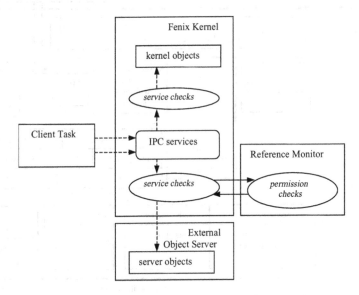

Fig. 2. Fenix Access Control Mechanisms

The separation of data processing and security functions provides invariance of data processing components about security policy. Mathematical models of system components are oriented to reflect interactions in operating system in terms of access control model. System component models provide unification and adequacy of access control mechanisms.

Fenix operating system consists of microkernel, basic services, security services, system services and a set of application programs and administrative applications, operating in Fenix operating system and MS Windows environment (fig. 3).

The microkernel manages system and application processes, performs message exchange, controls address spaces and other system resources allocation, controls hardware platform. It includes basic security mechanisms also. Any Fenix operating system component is accessed by message exchange mechanism.

The basic services, the security services and the system services are trusted, all applications are untrusted. Basic services provide access to hardware system resources and organize system objects hierarchy. Security services enforce security policy. System services manage operating system resources and provide them to application processes. Operating system resources are presented as objects (files, sockets etc) of various system services. System services support unified access interfaces for objects, while security services enforce security policy for operations upon these objects.

Compatibility with POSIX interfaces is provided by special program library. Fenix applications use this library to convert POSIX functions calls into appropriate messages addressed to system services.

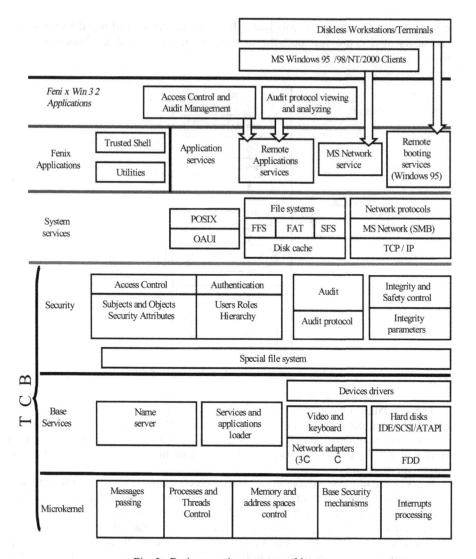

Fig. 3. Fenix operating system architecture

Most of Fenix applications are console administrative utilities, common network services (ftp, http etc.) and special services for remote administration.

Thus, operations of all Fenix operating system components are restricted to a small and relatively simple set of abstractions, which constitutes a compact and secure basis of the system.

Fenix Microkernel

Fenix operating system microkernel is up to modern technologies of operating system design and inherits the best features of Mach, MK++, VSTA, Chorus and other microkernels. Fenix microkernel is compact, simple and oriented to total control of system communications. Bringing a function, which manages system and application processes communications, out from microkernel allowed to make it invariant to security policy.

Fenix operating system microkernel has object-oriented architecture, implements powerful resource allocation mechanisms, provides compact program interface of system calls, sufficient for carrying out trusted communications between system applications and high-level program interfaces of operating system modules. Microkernel provides also all the tools, necessary for the implementation of such security services as access control, communication parties authentication and audit.

Fenix operating system microkernel program interface differs essentially from UNIX kernel interface. Fenix special-purpose library, used for system and applied applications design, provides compatibility with UNIX operating system standards.

Access control and testing it for correspondence with security policy are supported by operating system microkernel and special-purpose access control server, which mediates all the communications. Microkernel provides system and application processes isolation and assures communication mechanism uniqueness with the help of message passing mechanism, controlled by access control server. Access control server assures unconditional observance of security policy and access control rules.

Fenix architecture features provide direct microkernel and security services communications, protected from any contact with other operating system processes. Furthermore, there are trusted communication channels between system processes, protected from any contact with application processes.

Protected file system of Fenix operating system inherits recent advances in this area, including special clusterization and data aggregation algorithms, for more speed and capacity. Combination of careful data update methods with recovery utility provides integrity and safety of information storage. File system architecture allows to introduce cryptographic file protection services and service integrity checking structures.

Models of the Fenix Kernel and Servers

The Fenix kernel provides the limited set of well-defined system functions accessible to OS processes. Each function impacts the state of no more than two processes, namely active process (Pa), referring to kernel function, and passive process (Pp). Tab. 1 describes the set theoretic definition for kernel IPC system functions. (AttrSet function is available for authentication server only, MessageCheck function is available for reference monitor only).

The process is considered as a combination of virtual address space (AS) and attributes (AT):

$$P = \{AS, AT\}.$$

Table 1. Microkernel model

Message sending	{Pa', Pp'} = MessageSend(Pa, Pp), ASa' = ASa, ASp'=ASp.
Message receipt.	{Pa', Pp'} = MessageReceive(Pa, Pp) ASp'= ASp, $m: m˷ ASp, m˷ ASa' and m˅ ASa.
Message reply.	{Pa', Pp'} = MessageReply(Pa, Pp) ASa'= ASa, $m: m˷ ASa, m˷ ASp' and m˅ ASp.
Message check.	{Pa', Pp'} = MessageCheck(Pa, Pp) ASp'= ASp, $m: m˷ ASp, m˷ ASa' and m˅ ASa.
Address space increase.	{Pa'} = PageMap(Pa), ASa' ASa, '$ d: d˷ ASa', d˅ Asa
Address space decrease.	{Pa'} =PageUnmap(Pa), ASa' ASa.
Process creation.	{Pa', Pp'} = Create(Pa), Pa' = Pa, Pp = {0}, Pp'={ASa,{idp', typea, uida}}.
Process deletion.	{Pa'} = Delete (Pa), Pa'= {˅ }.
Process attribute setting.	{Pa', Pp'} = AttrSet(Pa, Pp, uid), uidp'= uid

Process address space is a set of memory pages available to the process. Process attributes include identifier, type, and user ID:

$$AT = \{pid, type, uid\}.$$

Processes are classified as trusted (OS servers) and untrusted. Untrusted processes act on behalf of computer system subjects, all the operations executed are managed by microkernel and OS security services. Trusted processes provide system functionality and security and are controlled by microkernel only.

Trusted processes are subdivided into functional servers, hardware drivers and security services.

Each functional server supports a portion of OS subjects and handles them when it is requested by other processes. The type and unique system identifier are assigned to each of the objects. The type of the object specifies the set of object operations. Functional server can use objects, supported by other servers. Hardware drivers have supplementary access to system hardware resources.

Trusted servers, as a part of security services, provide OS security (see fig. 4). Reference monitor and authentication server are critically important security services.

Reference monitor implements a basis for access control subsystem and audit subsystem and specifies a set of supported security models. Microkernel allows reference monitor to control all requests addressed to functional servers. Reference monitor manages subjects and objects security attributes, which are defined by security model. Security attributes server and subjects server are allocated to store security attributes. Moreover reference monitor takes audit of requests addressed to functional servers. Audit server is allocated to store audit protocol.

Authentication server provides a basis for authentication subsystem and changes user IDs for untrusted processes once they present correct authentication parameters.

Computer system subjects are classified as common users and pseudo-users. Common users are real users and system administrators. Pseudo-users are different untrusted system services.

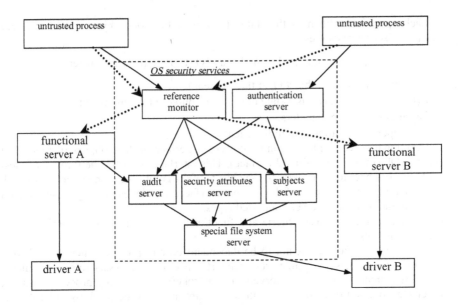

Fig. 4. Security services

Security service models represent special process managing functions, performed by reference monitor and authentication server.

The main operation, performed by reference monitor, is a single message check:

$$r = Permit(ATa, ATp, m),$$

r = TRUE, if a message m transfer from Pa to Pp is consistent with security policy,

r = FALSE, otherwise,

where Pa is a sender, Pp is a receiver, m is a message (m, Pa).

MessageCheck kernel function is intended to transfer messages to be checked from microkernel to reference monitor.

The role of authentication server is represented by user ID change operation:

$$\{Pa'\} = Authenticate(ATa, auth).$$

uida' = uid: auth = {uid, pwd}, if authentication was successful,

where Pa is a proccess that demands the authentication procedure, auth is an object, presented to Pa and containing an account name and authentication attributes (auth , Pa).

Authentication request is formed as a message addressed to the authentication server.

The model of functional server f is defined as a set of legal operations, dealing with objects and server resources (see fig. 5):

$$A = \{a_1, .., a_x\}.$$

Each operation influences the state of no more than one object and of arbitrary amount of server resources:

$$\{\{O_x(t+1), R_m(t+1)\}, Y_m\} = a_m(\{O_x(t), R_m(t)\}, Pa_m, x), \text{ where:}$$

- $O_k(t)$ is a state of the object O_k at the moment t,
- $R(t) = \{R_1(t), ..., R_k(t)\}$ is a state of server resources at the moment t,
- Pa_m are operation parameters,
- Y_m is operation output, sent to the client.

Functional servers enforce a unified access interface to their objects of functional servers (Object Access Unified Interface). This interface specifies the formats of messages used as operation requests to the objects of different types. The interface of each functional server is a subset of the unified interface, it includes messages used to gain access to the objects of server-supported types.

Microkernel correctness criteria and functional server correctness criteria are defined for models presented.

Microkernel efficiency is not to be dependent on any process states, microkernel functioning is to be safe against process interference. It is necessary to reserve microkernel resources for functional servers and microkernel itself. Any process' requests, unrelated to additional resource allocation, are to be processed by microkernel whether or not there are free system resources.

Functional server correctness is a set of the following principles:
- server executes well-defined set of operations, dealing with objects and resources;
- all the operations, involving change of object state, are performed on the client's request only;
- the set of operation-influenced system objects and resources, state changes of these objects and resources and parameter-dependent operation output are defined for each operation;
- server does not perform several operations on the same object simultaneously;
- server does not proceed to the new request, without having enough system resources to process all the requests available, including the new one;
  operation output is not sent to a client until request processing is completed.

Security Model Implementation

The main task of access control subsystem is to enforce security model adequately on the set of OS subjects and objects. For security architecture proposed and models of microkernel and trusted components considered, the following propositions are formally proved, substantiating the adequacy of access control.

Proposition 1. (Sufficient condition for security model implementation).
   Given OS architecture it is possible to provide system functioning consistent with security model, implemented by reference monitor, provided that microkernel and all the trusted components perform well.

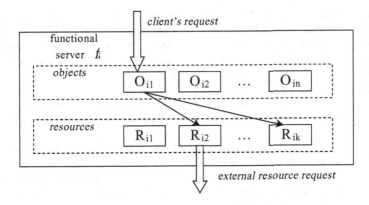

Fig. 5. Functional server model

The basis for the proof is unbypassability of security mechanisms, implemented by microkernel and reference monitor. So additional restrictions are imposed on system architecture. Hierarchies of functional servers and OS objects are defined, responsible for the interrelationship of servers and associated objects. Functional servers have exclusive access to the objects they use. Untrusted processes have access to the highest-level functional servers only. Authentication is allowed to pseudo-user processes only.

The ability to represent security model by reference monitor fully conforms to the requirement for flexibility of security model set.

Proposition 2. (Localization of security violation impacts).

Microkernel checks functional servers' calls with an granularity of OS object, and all the objects are used by functional servers exclusively; if so, it is possible to create such server/object hierarchy that any incorrectness in functional server f can produce confidentiality and integrity violations concerning server f objects and those servers, which use server f objects, only.

Servers and the microkernel are accessible to untrusted processes badware. Violation of microkernel or server correctness can cause the total system failure or OS objects security violation. To minimize the impact of functional servers failures the hierarchy of functional servers and objects is maintained. This method for violation impact localization makes it possible, for example, to exclude the influence of network components failure upon the file system security or upon the security service integrity.

These propositions substantiate the design of the Fenix microkernel, high-level functional servers and security services.

Resume

Fenix operating system is a secure special-purpose system used as a basis for secure information processing systems, compliant to Russia national information security requirements and standards. Microkernel concept, client-server technology and object-oriented approach substantiate the Fenix operating system architecture. Combination of these technologies is the best solution for modern operating systems, it has advantages over traditional operating system architecture because of more portability, compactness and effective resource utilization.

Using of the declared principles and proposed theory assures that all information flows in Fenix operating system are controlled by protection mechanisms. All intercommunications are restricted to a strict set of mechanisms controlled at the basic level.

The Fenix architecture provides unified flexible mechanism for access control and various security models implementation. Faults impact localization mechanism guarantees the consistence of objects managed by trusted servers. For the purpose of unification the access control mechanisms are localized, object-oriented programming had been used for system and security services implementation. The accordance of Object Access Unified Interface and access control models provides the adequacy of security policy implementation.

References

1. Trusted Information Systems, Inc. Trusted Mach System Architecture (October 1995)
2. Open Software Foundation, Inc. MK++ Kernel Executive Summary (November 1995)
3. Secure Computing Corporation. DTOS General System Security and Assurability Assessment Report. (June 1997)
4. Secure Computing Corporation. DTOS Lessons Learned Report. (June 1997)

Generalized Oblivious Transfer Protocols Based on Noisy Channels

Valeri Korjik[1] and Kirill Morozov[2]

[1] Section of Telecommunications, IPN CINVESTAV, AV. IPN No. 2508 ESQ Ticoman,
Col. San Pedro, Zacatenco, C.P. 07000, Mexico D.F., Mexico
Fax: 5747-7088, Tel.: 5747-3770
vkorjik@mail.cinvestav.mx
[2] Telecommunications Security Department, State University of Telecommunications,
Moika 65, St. Petersburg, 191186, Russia
Fax: 312-10-78, Tel.: 315-83-74
kirill@fem.sut.ru

Abstract. The main cryptographic primitives (Bit Commitment (BC) and Oblivious Transfer (OT) protocols) based on noisy channels have been considered in [1] for asymptotic case. Non-asymptotic behavior of BC protocol has been demonstrated in [2]. The current paper provides stricter asymptotic conditions on Binary Symmetric Channel (BSC) to be feasible OT protocol proposed in [1]. We also generalize this protocol using different encoding and decoding methods that require to regain formulas for Renyi entropy. Non-asymptotic case (finite length of blocks transmitted between parties) is also presented. Some examples are given to demonstrate that these protocols are in fact reliable and information-theoretically secure. We also discuss the problem

– how to extend $\binom{2}{1}$-OT protocol to $\binom{L}{1}$-OT protocol and how to arrange BSC

connecting parties. Both BC and OT protocols can be used as components of more complex and more important for practice protocols like "Digital cash", "Secure election" or "Distance bounding".

1   Introduction

The simplest of cryptographic protocols that are sufficient to accomplish many complex protocols can be called cryptographic primitives. One of such primitives is so called Oblivious Transfer ($\binom{2}{1}$-OT). According to this primitive, one party Alice has two secret strings b_0 and b_1, and another party Bob wants to learn b_c, $c = 0, 1$ for a secret bit c of his choice. Alice is willing to collaborate provided that Bob does not learn any information about $b_{\bar{c}}$ and Bob will only participate if Alice cannot obtain information about c. (We note that $\binom{2}{1}$-OT protocol is a particular case of

V.I. Gorodetski et al. (Eds.): MMM-ACNS 2001, LNCS 2052, pp. 219–229, 2001.

($\frac{L}{1}$)-OT protocol, where Alice has L secret strings and Bob wants to learn only one of them but in a manner to be completely unknown for Alice which of L secrets Bob receives.)

The algorithms to perform ($\frac{2}{1}$)-OT protocol (OT for brevity) have been considered in [3]. To be secure they require limitation on parties' computing power, so they are computational secure. This paper (following the idea of [1]) considers a scenario where both Alice and Bob have no limitation on their computing power. It would be impossible to accomplish this protocol without another assumption. Such an extra assumption we make is that Alice and Bob are connected by Binary Symmetric Channel (BSC$_j$) with bit error probability j .

In [1] the following OT information-theoretically secure protocol based on BSC$_j$ connecting parties was proposed. Alice and Bob agree on a binary linear code C that has constructive algorithm to correct maximum possible number of errors. It is also assumed that there exist noiseless channels to exchange messages between Alice and Bob. After this initialization phase Alice and Bob have to perform the following base protocol:

1. Alice picks randomly 2n bits x_i, i = 1, 2, K , 2n, repeats twice each of them and sends 4n-bit string to Bob over BSC$_j$.

2. Bob accepts each received pair if and only if it is either 00 or 11, otherwise he rejects it. The accepted pairs are transformed in 2n-bit string x¢, i = 1, 2, K , 2n, following the trivial decision rule: 00 fi 0, 11 fi 1.

3. Bob selects the desired bit string b_c, where c = 0 or 1 and picks two disjoint subsets I_0, I_1 of the set (1, 2, ..., 2n), satisfying the conditions: $|I_0| = |I_1| = n$, I_c contains only the numbers of positions corresponding to accepted bits in the step 2 of protocol.

4. Bob sends I_0 and I_1 to Alice over noiseless channel.

5. Alice computes the check strings (or syndromes) s_0 , s_1 to original n-bit substrings x_{I_0} , x_{I_1} using the known check matrix of the code C. Then Alice sends these check strings to Bob over noiseless channel.

6. Alice picks random n-bit string m, computes $\hat{b}_0 = b_0 \ ^\frown h_m(x_{I_0})$, $\hat{b}_1 = b_1 \ ^\frown h_m(x_{I_1})$, where "$^\frown$" is bitwise mod 2 addition, $h_m(K)$ is a hash function taken from universal$_2$ class [4] given known m and sends to Bob m, $\hat{b}_0$ and $\hat{b}_1$ over noiseless channel.

7. Bob corrects errors on x¢$_c$ using s_c , recovers x_{I_c} and computes the desired secret $b_c = \hat{b}_c \ ^\frown h_m(x_{I_c})$.

If Alice is honest (that means that she follows to this protocol) and the channel connecting parties is in fact BSC$_j$ without memory, then Alice has no chances to distinguish which of two subsets I_0 or I_1 corresponds to c. It is more complex to prove that Bob does not learn any information about $b_{\bar{c}}$ and on the other hand that Bob following honestly to the protocol given above is able to learn b_c with high probability. Solution to these problems will be considered in the next Sections.

The paper is organized as follows. In Section 2 the condition on the bit error probability j to provide both reliability and security of OT protocol is presented. We also give there an optimization of this parameter to provide more efficient protocol. Section 3 contains several generalizations of the base protocol and a description of their asymptotic behavior. In Section 4 we give the main non-asymptotic formulas to estimate reliability and security of the base protocol and consider also the problem how to optimize its parameters. In Section 5 we summarize the main results and consider possible transformations of $\binom{2}{1}$-OT protocols to $\binom{L}{1}$-OT protocol and discuss an arrangement of BSC_j between parties.

2   Feasibility of Base OT Protocol in Asymptotic Case

First of all we note that if Alice and Bob run the base protocol described above the following probabilities are true:

$P(\text{accept}) = e = j^2 + (1 - j^2) = (\text{the probability to accept any bit by Bob})$

$$P(x \not{c}, x \mid \text{accept}) = j^2/e = \tag{1}$$

$$= (\text{the probability to be error in any bit accepted by Bob})$$

In asymptotic case (n fi $¥$) we have the following sufficient conditions to get at least n bits accepted by Bob (it allows him to form the subset I_c, $|I_c| = n$) and to correct all errors in x_{I_c} using check string s_c, respectively:

$$e \ddagger 1/2 \tag{2}$$

$$r \sim nH(j^2/e), \tag{3}$$

where r is the length of check strings s_0 and s_1, $H(...)$ is the entropy function [5]. (It is easy to see that the inequality (2) is trivial because it has solution $0 £ j £ 1/2$).

The main theorem of privacy amplification [6] says that with high probability the amount of Shannon's information I_0, leaking to Bob about the secret string $b_{\bar{c}}$ is bounded by the following inequality

$$I_0 £ 2^{-(n-1-r-I_c)}/\ln 2, \tag{4}$$

where $1 = a n$ $(0 < a < 1)$ is the length of the secret strings b_0 and b_1, I_c is Renyi (or collision) information obtained by Bob about the string $x_{I_{\bar{c}}}$.

The amount of collision information in any of erased bits is zero, while the amount of collision information containing in the string of accepted bits can be expressed asymptotically as follows [6]:

$$I_c \sim n_a \left(1 - H\left(j^2/e\right)\right),$$ (5)

where n_a is the number of bits accepted by Bob.

Taking into account that asymptotically $n_a \sim ne$ and substituting (3) and (5) into (4) we obtain the following sufficient condition to be exponential decreasing of information about $b_{\overline{c}}$, as $n \to \infty$:

$$f_a(j) = 1 - a - H\left(j^2/e\right) - e\left(1 - H\left(j^2/e\right)\right) > 0$$ (6)

In a particular case of one-bit secret strings b_0 and b_1 ($a = 0$), we get from (6) the following condition to be secure OT protocol

$$f_0(j) = 1 - H\left(j^2/e\right) - e\left(1 - H\left(j^2/e\right)\right) > 0$$ (7)

The solution to the last inequality with respect to j shows that it will be true for every j within the interval (0, 1/2). In [1] only the case $l = 1$ has been considered and one of the open questions mentioned there was to find an efficient algorithm for $\binom{2}{1}$-OT using BSC_j for values j above 0.1982. We have proved it for the same as in [1] algorithm but taking into account the collision information in accepted bits of the string $x_{I_{\overline{c}}}$. The intervals for possible bit error probabilities j providing the secure OT protocol for different secret rates a are presented in Table 1.

Table 1. The intervals of bit error probabilities j providing the secure OT protocol

a	0	10^{-3}	$5 \cdot 10^{-3}$	10^{-2}	0.05	0.1	0.2	0.215
j	< 0.5	< 0.487 > 0.0005	< 0.470 >0.003	< 0.458 > 0.006	< 0.403 > 0.026	< 0.355 > 0.054	< 0.248 > 0.142	< 0.211 > 0.176

It is worth noting that for $a > 10^{-3}$ the bit error probability has to be sensibly restricted not only from above but also from below. OT protocol does not work at all for BSC_j, $0 \le j \le 1/2$ if $a > 0.217$.

We can see from (4) and (6) that the more is the function $f_a(j)$, the more efficient is OT protocol. Thus if the parties can establish BSC_j with different bit error probabilities j, they can optimize j to provide a maximum of $f_a(j)$. For example, if $a = 0$, then $f_0(j)$ reaches the maximum for $j \approx 0.2$.

3   Generalizations of Base OT Protocol

The algorithm to accomplish OT protocol proposed in [1] and described in Section 1 is very natural but not the only one possible to solve the same problem. Let us consider some possible generalizations of it.

3.1   Algorithm with a-Multiple Repetition of Bits x_i , $i = 1, 2, K , 2n$

It is very natural to generalize algorithm given in [1] if Alice repeats a times each of 2n random chosen bits x_i, $i = 1, 2, K , 2n$ and sends them to Bob over BSC_j . Bob accepts each received a-bit string if and only if the number of zeroes or ones in this string is at least b, b < a. This algorithm results in changing the probabilities (1) in the following manner

$$P(accept) = e\cancel{c} = \sum_{\substack{i\ (0,a-b)\\i\ (b,a)}} \begin{pmatrix} a \\ i \end{pmatrix} j^i (1-j)^{a-i} \tag{8}$$

$$P(x\cancel{c}, x \mid accept) = \frac{1}{e\cancel{c}} \sum_{i=b}^{a} \begin{pmatrix} a \\ i \end{pmatrix} j^i (1-j)^{a-i} \tag{9}$$

In asymptotic case this algorithm results in the following sufficient conditions to receive by Bob at least n accepted bits in the string $x\cancel{c}$, $i = 1, 2, K , 2n$

$$e\cancel{c}\ddagger 1/2 \tag{10}$$

and to correct all errors in x_{1_c} using the check string s_c

$$r \sim nH \left\{ \frac{1}{\sum e\cancel{c}} \sum_{i=b}^{a} \begin{pmatrix} a \\ i \end{pmatrix} j^i (1-j)^{a-i} \right\} \tag{11}$$

The amount of collision information in any bit received by Bob depends on the Hamming weight of a-bit string corresponding to this bit. It can be shown that the average collision information obtained by Bob about the string x_{1_c} in asymptotic case is the following

$$I_c \sim n \left\{ 1 - \frac{1}{\sum} \sum_{i=0}^{a} \begin{pmatrix} a \\ i \end{pmatrix} \left(j^i (1-j)^{a-i} + j^{a-i} (1-j)^i \right) \right.$$

$$H \left. \frac{j^i (1-j)^{a-i}}{\sum j^i (1-j)^{a-i} + j^{a-i} (1-j)^i} \right\} \tag{12}$$

Now if we substitute (11) and (12) for r and I_c in (4), respectively, we can find the optimal bit error probability j that maximizes the exponent in (4) or functions

$f_a(j)$ and $f_0(j)$ similar to those given in (6), (7), respectively. Unfortunately, the inequality (10) is not as trivial as it was for (2). So we have to maximize $f_a(j)$, $f_0(j)$ over the set j satisfying (10).

The numerical calculations show that the function $f_0(j)$ by (7) reaches the maximum 0.21 for $j = 0.2$, while similar function for generalized algorithm (a>2, b<a) reaches the maximum 0.285 for $j = 0.327$, if a = 8, b = 6, that is a slightly better result.

3.2   Algorithm with Arbitrary Linear Binary (N, K)-Code (K < n) Used to Send K-Bit Substrings of the String x_i, i = 1, 2, K , 2n

Next step to generalize the base OT protocol is to replace a-multiple repetition of bits x_i, i = 1, 2, K , 2n by the use of some binary linear (N, K) code V known for both parties and having maximum possible minimum code distance D.

Without the loss of generality let us take n = Kb , where b is some integer. Alice and Bob also agree on $q = 2^K$-ary ($\tilde{N}, \tilde{K}$)-Reed Solomon (RS) code, where $\tilde{K} = b$. Then the base OT protocol has to be changed to the following one:

1. Alice picks at random 2n bits x_i, i = 1, 2, K , 2n , encodes this string by blocks of (N, K)-code and sends these blocks y_j, j = 1, 2, K , 2b over BSC_j .

2. Bob receives the noisy versions of all blocks, corrects at most t < [(D – 1)/2] errors on every block and detects errors using (N, K)-code. He accepts blocks with undetected errors and rejects blocks with detected errors.

3. Bob selects the desired bit string b_c , where c = 0 or 1 and picks two disjoint subsets I_0, I_1 of the set (1, 2, ..., 2 b), satisfying the conditions: $|I_0| = |I_1| = b$, I_c – contains only the numbers of blocks accepted by Bob.

4. Bob sends both I_0 and I_1 to Alice over noiseless channel.

5. Alice computes the q-ary check strings s_0 and s_1 to y_{I_0}, y_{I_1} respectively using ($\tilde{N}, \tilde{K}$)-RS code. She sends then both s_0 and s_1 to Bob over noiseless channel.

6. Alice picks random n-bit string m, computes $\hat{b}_0 = b_0 \bar{} h_m(x_{I_0})$, $\hat{b}_1 = b_1 \bar{} h_m(x_{I_1})$, where x_{I_0}, x_{I_1} are the information symbols of blocks corresponding to subsets I_0 and I_1 respectively and sends $\hat{b}_0$, $\hat{b}_1$, and m to Bob over noiseless channel.

7. Bob corrects errors on x_{I_c} using s_c, recovers x_{I_c} and computes the desired secret $b_c = \hat{b}_c \bar{} h_m(x_{I_c})$.

It is easy to show that in asymptotic case (n fi ¥) the following sufficient condition has to be hold for the number $\tilde{R} = \tilde{N} - \tilde{K}$ of q-ary check symbols of

$(\tilde{N}, \tilde{K})$-RS code transmitted over noiseless channel to correct all errors in the string x_{I_c}

$$\tilde{R} \sim -\frac{b}{L} P_{ue} \log_2 \frac{P_{ue}}{q-1} + (1 - P_{ue}) \log_2 (1 - P_{ue}) \Big/ \log_2 q,\qquad (13)$$

where P_{ue} is the probability of undetected error obtained after a completion of error correcting and detecting procedures by (N, K)-linear binary code V. Because each of q-ary symbols can be represented by $K = \log_2 q$ binary symbols we automatically get the following number of binary check bits

$$R \sim -K\tilde{K}\frac{P_{ue}}{L} \log_2 \frac{P_{ue}}{q-1} + (1 - P_{ue}) \log_2 (1 - P_{ue}) \Big/ \log_2 q =$$

$$= -\tilde{K}\frac{P_{ue}}{L} \log_2 \frac{P_{ue}}{q-1} + (1 - P_{ue}) \log_2 (1 - P_{ue})\qquad (14)$$

The sufficient condition to receive by Bob at least b accepted (without detected errors) code blocks of (N, K)-code among all $2b$ code blocks can be expressed as follows

$$\sum_{i=0}^{t} \binom{N}{i} j^{i} (1-j)^{N-i} > 1/2$$

The bottleneck of this algorithm is the finding the collision information obtained by Bob about the string $x_{I_{\bar{c}}}$ taking into account that this substring has been transmitted by code blocks of the code V.

If we assume that the amount of collision information coincides with the amount of Shannon information in asymptotic case with large probability, then we can use the results of [7] to express I_c as follows

$$I_c \sim \frac{b}{L} K - NH(j) - \sum_{j=1}^{2^{N-K}} P(G_j) \log_2 P(G_j),\qquad (15)$$

where $H(...)$ is the entropy function,

$$P(G_j) = \sum_{i=0}^{n} A_{ij} j^{i} (1-j)^{n-i},$$

A_{ij} – is the number of words of weight i in the j-th coset of the standard decompositions V_N / V (weight distribution of cosets).

Substituting (14) for r and (15) for I_c into exponent of (4) and dividing the result by n we get the following function (similar to (6)) that should be maximized over $j \in (0, 1/2)$

$$f_a(j) = 1 - a - \frac{1}{K}\left[\sum_L P_{ue}\log_2\frac{P_{ue}}{q-1} + (1-P_{ue})\log_2(1-P_{ue})\right] -$$

$$- \frac{K - NH(j) - \sum_{j=1}^{2^{N-K}} P(G_j)\log_2 P(G_j)}{L} \right\} \tag{16}$$

We note that the probability P_{ue} can be found for the chosen (N, K)-code as follows [8]

$$P_{ue} = (1-j)^N A_c^{(t)}\left[\frac{j}{1-j}\right] \tag{17}$$

where $A_c^{(t)}(z) = \sum_{j=D-t}^{N} A_{t,j}z^j$,

$$A_{t,j} = \sum_{i=j-t}^{j+t} A_i N_t(i,j),$$

t is the multiplicity of errors correcting by the code V, providing $t \leq [(D-1)/2]$, where D is the minimum distance of the code V,

$$N_t(i,j) = \sum_{g=max(0,i-j)}^{[(t+i-j)/2]}\binom{N-i}{g+j-i}\binom{i}{g},$$

A_i is the number of code words of weight i (weight distribution function) of the code V. We note that $A_i = A_{i1}$.

It is seen from (16) and (17) that OT protocol optimization requires the knowledge of weight distribution of cosets for chosen code V. Unfortunately this distribution is known only for a limited classes of linear binary codes [8].

4   Non-asymptotic Case

Let us consider the base protocol [1]. The requirement that the number of accepted bits in the string x_i', $i = 1, 2, K, 2n$ has to be at least n is not necessary for Bob to receive the desired secret b_c because he can correct both errors and erasures in the chosen substring x_{I_c} of length n using check string s_c of the code C. If the minimum code distance of this code is d then it is capable to correct both t errors and t' erasures when the following condition holds [9]

$$2t + t' \leq d - 1 \tag{18}$$

It is easy to show that if Bob corrects both erasures and errors in some n-bit substring $x_{I_\varepsilon}^i$ of 2n-bit string x^i, $i = 1, 2, K, 2n$, then the probability to recover x_{I_ε} correctly can be upper bounded as follows

$$P_c \ddagger \sum_{i=n-d+1}^{n-1} \binom{2n}{i} e^i (1-e)^{2n-i} \sum_{j=0}^{[(d-1-(n-i))/2]} \binom{i}{j} \left[\frac{j^2}{e}\right]^j \left[1 - \frac{j^2}{e}\right]^{i-j} +$$

$$+ \sum_{i=n}^{2n} \binom{2n}{i} e^i (1-e)^{2n-i} \sum_{j=0}^{[(d-1)/2]} \binom{n}{j} \left[\frac{j^2}{e}\right]^j \left[1 - \frac{j^2}{e}\right]^{n-j} \quad (19)$$

The inequality (4) can be used to estimate the amount of Shannon's information I_0 leaking to Bob about the string $b_{\tilde{\varepsilon}}$, where $I_c = g I_{c_1}$, g is the number of accepted bits in x_{I_ε} and I_{c_1} is the amount of collision information obtained by Bob about each of these accepted bits. Because the execution of the base protocol results for Bob in $BSC_{j\phi}$ with $j\phi = j^2/e$ we get for I_{c_1} [6]

$$I_{c_1} = 1 + \log_2 \left[\left(1 - j^2/e\right)^2 + \left(j^2/e\right)^2 \right] \quad (20)$$

To provide the guaranteed security of OT protocol we have to design it under the conservative assumption that Bob can be a dishonest party and distribute the accepted (non-erased) bits equally between substring $x_{I_c}^i$, $x_{I_{\tilde{\varepsilon}}}^i$ thinking to extract at least some information from both secrets b_0 and b_1. If we want to prevent this attack then one can find I_c from (4) for given value I_0, where r is the number of check symbols for chosen $(n+r, n)$-code C and estimate the probability of risk to be accepted by Bob at least $2g$ bits on the string x^i, $i = 1, 2, K, 2n$, where $g = I_c/I_{c_1}$, as follows

$$P_r = \sum_{i=2g}^{2n} \binom{2n}{i} e^i (1-e)^{2n-i} \quad (21)$$

The problem how to select the main parameters of OT base protocol in non-asymptotic case can be solved by the following algorithm:

1. Fix P_r, $\tilde{I}_0$, P_c, n, l, and j .
2. Find d from (19) satisfying P_c in the point 1 given j and n.
3. Find r using the bound for BCH codes given n and d.
4. Find g from (21) given n, j and P_r.
5. Find I_{c_1} from (20) and then $I_c = g I_{c_1}$.
6. Find I_0 from (4) given n, l, r, and I_c .
7. If I_0 that was found in the point 6 occurs at most equal to $\tilde{I}_0$ given in the point 1, then decrease n and repeat the points 2-6. Otherwise optimize j to provide a

positive result. In the case if no one j gives $I_0 \pounds \tilde{I}_0$ then increase n and repeat the points 2-6.

Example. Let us take $l = 20$, $P_c = 0.9999$, $P_r = 10^{-4}$, $I_0 = 10^{-10}$ bit. Then we can evaluate (following algorithm above) the minimal possible $n = 3840$ that results in $r = 255$, $d = 43$ and optimal chosen $j = 0.0453$.

We can see from this example that OT protocol "works", in fact, because it provides both reliability ($P_c \ddagger 0.9999$) and security ($I_0 \pounds 10^{-10}$, $P_r \pounds 10^{-4}$). Unfortunately the length of the string x has to be significant ($n = 3840$) but this is the feature of this base OT algorithm.

5   Discussion of the Main Results and Some Open Problems

Our contribution (in comparison with paper [1]) is the following:
ú extension of OT protocol to multibit secret strings,
ú tighter bounds on the bit error probabilities that makes the base OT protocol more feasible,
ú generalizations of base OT protocol,
ú performance evaluation of OT protocol in a non-asymptotic case.

We showed that base OT protocol practically "works" (in non-asymptotic case), although it requires a significant bit string to be sent over BSC_j . This protocol is the remarkable example of the combination of both codes and cryptography. In fact, on the one hand OT protocol does not work without the use of error correcting codes and on the other hand OT based on noisy channel is typical cryptographic protocol that is in addition information-theoretically secure.

We do not consider in the current paper how to prevent attacks on OT protocol, that can be initiated by dishonest Alice, who deviates from base OT protocol to find out which of two secrets Bob wants. (The solution to this problem was given in [1] and it also holds for our extensions.)

To extend $\binom{2}{1}$-OT protocol to the case $\binom{L}{1}$-OT protocol, there exist different possibilities to proceed. The simplest way is to arrange a special dichotomous procedure that reduces several $\binom{2}{1}$-OT protocols to one $\binom{L}{1}$-protocol. (The complexity of such $\binom{L}{1}$-OT protocol is $(L-1)$ times more than complexity of $\binom{2}{1}$-OT protocol). Next way is to use base protocol [1] for L secret strings initially. It results in harder conditions than (6) and (7) to be feasible $\binom{L}{1}$-OT protocol in BSC_j . We are going to publish these results later and to specify which of $\binom{L}{1}$-protocol versions is the best.

Finally the main problem that has to be solved for practical implementation of OT-protocol (as well as for other protocols based on noisy channels) is an arrangement of

BSC_j between parties. This problem has already been discussed in [2]. We can add only that there is some progress in our investigations to arrange BSC_j as quantum channel with low intensity of random polarized photons.

References

1. Crepeau, C.: Efficient Cryptographic Protocols Based on Noisy Channels. Proc. Eurocrypt'97. Lecture Notes in Computer Science. 1233, 306–317
2. Korjik, V., Morozov, K.: Non-asymptotic bit commitment protocol based on noisy channels, 20 Biennal Symposium on Communications Department of Electrical and Computing Engineering Queen's University, Proc. Kingston, Canada (2000) 74–78
3. Schneier, B.: Applied Cryptography. John Wiley (1996)
4. Stinson, D.R.: Universal hashing and authentication codes. Advances in Cryptology, Proc. of Crypto' 91. Lecture Notes in Computer Science, Vol. 576. (1999) 74–85
5. Feinstein, A.: Foundations of information theory. New-York, McGraw Hill (1958)
6. Bennett, C., Brassard, G., Crepeau, C. and Maurer, U. M.: Generalized privacy amplification. IEEE Trans. on IT, Vol. 41. 6 (Nov. 1995) 1915–1923
7. Korjik, V., Yakovlev, V.: Capacity of communication channel with inner random coding. Problems of Information Transmission, Vol. 28. 4 (1992) 317–325
8. Klove, T., Korjik, V.: Error Detecting Codes. Kluwer (1995)
9. Mac Williams, F.J. and Sloan, N.J.A.: The theory of error-correcting codes. New-York: North-Holland (1977)

Controlled Operations as a Cryptographic Pri   mitive

Boris V. Izotov, Alexander A. Moldovyan, and Nick A. Moldovyan

Specialized Center of Program Systems "'SPECTR'",
Kantemirovskaya str., 10, St. Petersburg 197342, Russia;
ph./fax.7-812-2453743, spectr@vicom.ru

Abstract. Controlled two-place operations (CTPO) are introduced as a new cryptographic primitive for the block ciphers. Design criteria, structure, some good cryptographic properties of the CTPO are considered. There are proposed CTPO representing single nonlinear operations on the operands of relatively large length. It is shown that CTPO in combination with the controlled permutations (CP) can be efficiently used to construct fast block ciphers. Three different cryptoschemes based on data-dependent two-place operations and data-dependent permutations are presented. A feature of the cryptoschemes is the use of the CTPO and CP for construction of some mechanisms of the internal key scheduling which consists in data-dependent transformation of the round keys.

1   Introduction

Cryptographic transformation is one of the basic mechanisms used in computer security systems. The design of ciphers based on data-dependent operations appears to be a new and perspective direction in applied cryptography. Data-dependent rotations (DDR) have been used for the first time by W. Becker [1] and later by W. Madryga [2] and R. Rivest [3]. After Rivest's proposing the RC5 cipher based on extensive use of DDR this cryptographic primitive gained recognition in last years. It has been shown that mixed use of DDR with some other simple operations is an effective way to thwart linear cryptanalysis. Recently DDR have been used in new ciphers RC6 [4] and MARS [5].

DDR can be interpreted as a particular case of controlled permutations (CP) $P_{n/m}$. CP can be easy performed with interconnection networks (IN) [6,7]. A variant of the symmetric cryptosystem based on IN and Boolean functions is presented in [8]. Another cryptographic application of IN is presented in [9]. The CP operations are very quick and inexpensive in hardware. They are used in fast hardware encryption methods [10,11]. Searching other types of the controlled operations with good cryptographic properties is also of great interest for different applications.

Present paper introduces controlled two-place operations (CTPO) as a new cryptographic primitive. In the second section the design criteria, structure, and some algebraic and probabilistic properties of the CTPO are considered. Several cryptoschemes based on both CTPO and CP operations are presented in the third section. A feature of these cryptoschemes is the use of the internal key scheduling based on the data-dependent operations.

V.I. Gorodetski et al. (Eds.): MMM-ACNS 2001, LNCS 2052, pp. 230–241, 2001.
© Springer-Verlag Berlin Heidelberg 2001

2   Design of the CTPO

2.1   Basic Notions

Definition 1. Under controlled two-place operation we mean some transformation $Y = Q_V(X,A)$ above two operands $\{X,A\}$, $Z^{(n)}$, which depends on control vector V, $Z^{(m)}$ ($m \updownarrow n$), where $Z^{(n)}$ is a set of n-dimensional binary vectors $Z=(z_1,z_2, \dots ,z_n)$, " i, $\{1,..,n\}$ z_i, $\{0,1\}$ ($\% Z^{(n)}\% \bar{\sigma} 2^n$).

In cryptoschemes the control vector V is supposed to be dependent on key K and/or transformed data T: $V=V(K, T)$. Operands X, A correspond to data blocks and/or to round keys.

Definition 2. Transformation $Y = Q_V(X,A)$ corresponding to concrete value of the control vector V is called CTPO-modification (or modification Q_V) and defines a mapping $Z^{(n)} \times Z^{(n)} fi Z^{(n)}$, where " $\times$" denotes direct product of two sets.

Definition 3. Two different modifications Q_V and $Q_{V^{\ell}}(V^{\ell},, V)$ are called equivalent if " X,A $Q_V(X,A) = Q_{V^{\ell}}(X,A)$.

Definition 4. A modification Q_V is called unique if all other modifications are not equivalent to Q_V.

Definition 5. CTPO is called complete if all 2^m modifications are unique.

Definition 6. Let $m = n$. CTPO is called effectively complete if for all pairs $\{X,A\}$ and " $\{V,V^{\ell}, V\}$, $Z^{(n)} Q_V(X,A)$ „ $Q_{V^{\ell}}(X,A)$.

If $m > n$, then CTPO is not effectively complete, since the mapping $\{V\} \cdot \{Y\}$ ($Z^{(m)} \cdot Z^{(n)}$) is not one-to-one.

Since for fixed value V CTPO defines a mapping $Z^{(n)} \times Z^{(n)} fi Z^{(n)}$, for arbitrary modification Q_V there exists a set of different pairs $\{\{X,A\},\{X^{\ell},A^{\ell}\}\}$ satisfying the equation $Q_V(X,A) = Q_V(X^{\ell},A^{\ell})$. This case we shall call a collision on two operands X and A.

Definition 7. Let for some modification Q_V and some fixed A there exist two different values X and X^{ℓ}, X such that $Q_V(X,A) = Q_V(X^{\ell},A)$. This is a collision on one operand X.

Definition 8. A transformation $Q_V(X,A)$ is called invertible over X if for all fixed $\{A,V\}$ there exists a single-valued transformation Q_V^{-1} such that " X: $\{Y = Q_V(X,A)\}$ $\{X = Q_V^{-1}(Y,A)\}$.

In general the transformation $Y = Q_V(X,A)$ defines a mapping $\{X\} \times \{A\} \times \{V\} fi \{Y\}$ ($Z^{(n)} \times Z^{(n)} \times Z^{(m)} fi Z^{(n)}$) which induces the following mappings:

- $Y = p_A^{(V)}(X)$: ($Z^{(n)} fi Z^{(n)}$) if the pair $\{A,V\}$ is fixed;
- $Y = h_X^{(V)}(A)$: ($Z^{(n)} fi Z^{(n)}$) if the pair $\{X,V\}$ is fixed;
- $Y = \mu_{X,A}(V)$: ($Z^{(m)} \cdot Z^{(n)}$) if the pair $\{X,A\}$ is fixed.

The transformations Q_V, $p_A^{(V)}$, $h_x^{(V)}$ and $\mu_{X,A}$ are interdependent and each of them reflects special properties of the CTPO. Below we shall use different representations of the CTPO.

2.2 General Model of the CTPO

Let consider a substitution $p_A^{(V)}$ as a basic representation of the CTPO:

$$p_A^{(V)}(X) = \{\, y_1, y_2, \ldots, y_n \,\} = \{ f_1^{(V)}(X,A), \ldots, f_n^{(V)}(X,A) \}. \tag{1}$$

where x_i fi y_i, $y_i = f_i^{(V)}(X, A)$ are some Boolean functions depending on binary vectors X, A, V and represented as algebraic normal form polynomials.

Representation of the CTPO in terms of substitutions $p_A^{(V)}$ is called general CTPO model. Considering this model independently of the operand A, we obtain a general model of the controlled one-place operation (controlled substitution). Thus, model (1) is a general model of the controlled operation.

Definition 9. Each CTPO corresponding to some model is called realization of the model.

Practical use of the realizations of the model (1) as cryptographic primitives is connected with certain restrictions caused by peculiarities of the concrete encryption schemes and by hardware implementation. An interesting case is the use of CTPO in traditional Feistel cryptoscheme instead of the bit-wise exclusive-or operation "$^-$". In similar cryptographic applications CTPO should have some special properties, for example, the following:

1. Mappings $p_A^{(V)}$ should be one-to-one (bijective).
2. Control mechanism of the CTPO should provide large number of different modifications.
3. Mapping inverse of $p_A^{(V)}$ should be easy to perform.
4. Number of collisions for each CTPO-modification $Y = Q_V(X,A)$ should be not large.
5. Mapping $p_A^{(V)}$ should posses certain algebraic and probabilistic properties.
6. Functional structure of the CTPO should be simple to provide cheap and fast hardware and/or software implementation.

Let us consider the property of the invertibility for detailing of the general model (1) in correspondence with the special requirements indicated above. According to definition 8 in terms of the general model this mean that for each substitution $p_A^{(V)}$ there exists respective inverse substitution $p_A^{-1\,(V)}$, i.e. if " $i_\_$ {1,..,n} $y_i = f_i^{(V)}(X,A)$, then " $i_\_$ {1,..,n} there exists such one-to-one function $f_i^{-1\,(V)}$ that $x_i = f_i^{-1\,(V)}(Y,A)$.

Statement 1. If the CTPO-modification is invertible over X, then mapping $p_A^{(V)}$ is bijective and the related transformation $Y = Q_V(X,A)$ has no collision on X.

Proof. Suppose the statement is false. Then there exist such X, $X'_,,$ X and Y that $Y = p_A^{(V)}(X)$ and $Y = p_A^{(V)}(X')$. According to the condition for $p_A^{(V)}$ there exists the inverse substitution $p_A^{-1\,(V)}$, therefore $p_A^{-1\,(V)}(Y) = X$ and $p_A^{-1\,(V)}(Y) = X'$, where $X'_,,$ X. This is impossible, since $p_A^{-1\,(V)}$ is a one-to-one mapping.

Definition 10. Let the invertible over X transformation Q_V depends on one parameter more, which is called invertibility parameter (denoted as e) and which takes on two values: 0 (direct transformation) and 1 (inverse transformation). If for arbitrary fixed $\{A,V\}$ and $"X$ the equations $Y=Q_V(X,A)=Q_{V,0}(X,A)$ and $X=Q^{-1}_V(Y,A)=Q_{V,1}(Y,A)$ are hold, then the transformation $Q_{V,e}$ is called parametrically invertible over X.

In terms of the general model (1) this mean that direct and inverse substitutions are symmetric, they being connected by the condition:

$p_A^{(V)}=p_A^{(V,0)}$ and $p^{-1\,(V)}_A=p_A^{(V,1)}$ i.e." $i_ {1,..,n}$ $y_i=f_i^{(V,0)}(X,A)$, $x_i=f_i^{(V,1)}(Y,A)$.

Thus, the general model of the parametrically invertible CTPO has the following form:

$p_A^{(V,e)}=\{f_1^{(V,e)}(X,A),..., f_n^{(V,e)}(X,A)\}$.

Idea of the parametric invertibility is very important while designing cryptographic primitives, since it allows one to realize data encryption and data decryption without change of functional scheme, i.e. with the same electronic circuit.

2.3   Sequential Model of the CTPO

If in the general model (1) " $i_ {1,..,n}$ $f_i^{(V)}(X,A) = f_i^{(V)}(x_1, x_2,..., x_i, A)$, then we get sequential model of the CTPO:

$$p_A^{(V)} = \{f_1^{(V)}(x_1, A),\ f_2^{(V)}(x_1, x_2, A),\ ...\ ,\ f_n^{(V)}(x_1, x_2,..., x_n, A)\}. \qquad (2)$$

In this model some properties of the CTPO which are important for the respective cryptographic applications can be easy satisfied. To study properties of the sequential model (2) we shall use the following notations:

$(x_1, x_2,\ ...\ x_i)=X^{(i)}$, $(a_1, a_2,...,a_i)=A^{(i)}$, $(y_1, y_2,...,y_i)=Y^{(i)}$ $(A^{(i)},X^{(i)},Y^{(i)},\ Z^{(i)},\ X=X^{(n)},\ A=A^{(n)},\ Y=Y^{(n)})$.

Using this notation the parametrically invertible sequential model (2) can be written in the form:

$p_A^{(V,e)} = \{f_1^{(V,e)}(x_1, A),\ f_2^{(V,e)}(X^{(2)}, A),\ ...\ ,\ f_n^{(V,e)}(X^{(n)}, A)\}$.

Statement 2. For sequential model (2) the sufficient condition of the invertibility of the mapping $p_A^{(V)}$ is satisfaction of the relation:

$$"\,i_ {1,..,n}\ f_i^{(V)}(X^{(i)}, A)=x_i\ ^-\ j_i^{(V)}(x_1, x_2,\ ...\ x_{i-1}, A)=x_i\ ^-\ j_i^{(V)}(X^{(i-1)}, A), \qquad (3)$$

where x_0 is the initial condition.

Proof. Let Y is arbitrary n-bit vector. Let us construct such X that $Y=p_A^{(V)}(X)$:

$x_1= y_1\ ^-\ j_1^{(V)}(x_0, A)$, $x_2= y_2\ ^-\ j_2^{(V)}(x_0, x_1, A)$, $...$, $x_n=y_n\ ^-\ j_n^{(V)}(X^{(n-1)}, A)$.

For this $X= \{x_1,\ ...\ x_n\}$ the equation $Y = p_A^{(V)}(X)$ is satisfied. Suppose there exists such $X^{\not=}$,, X that $Y=p_A^{(V)}(X\not=)$. Then $y_1= x_1^{\not=}\ ^-\ j_1^{(V)}(x_0, A)$ and $y_1= x_1^-\ j_1^{(V)}(x_0, A)$, hence $x_1^{\not=}= x_1$. Considering all y_i, where i sequentially change from 2 to n, we get $x_i^{\not=}=x_i$ for all $i_ {1,..,n}$, i.e. $X^{\not=} X$. Thus, these algorithm of the construction X corresponding to the given Y defines the inverse substitution $p^{-1\,(V)}_A$.

Corollary 1. If the condition (3) is satisfied, then in accordance with Statement 1 all modifications of the transformation $Y = Q_V(X,A)$ have no collisions on X.

Corollary 2. If $\forall i \in \{1,..,n\}$ the relation

$$f_i^{(V)}(X^{(i)}, A)= x_i^- a_i^- j_i^{(V)}(x_1, x_2, ..., x_{i-1}; a_1, a_2, ..., a_{i-1})= x_i^- a_i^- j_i^{(V)}(X^{(i-1)}, A^{(i-1)})$$

is satisfied, then the transformation $Y = Q_V(X,A) = p_A^{(V)}(X)$ is invertible separately over X and over A, hence this transformation is bijective and has no collisions separately on X and on A.

These corollaries can be easy proved considering CTPO as substitution $Y = h_x^{(V)}(A)$ and using Statements 1 and 2.

Corollary 3. If control mechanism of the CTPO has the form:
$\forall i \in \{1,..,n\}$ $f^{(V)}(X, A)= v_i^- j_i^{\phi\phi(i-1)}(X, A)$, then the CTPO is effectively complete in accordance with Definition 6.

The proof can be obtained from Statements 2 and 1 considering the CTPO as substitution $Y=\mu_{X,A}(V)$. This induce the following:

Corollary 4. If $\forall i \in \{1,..,n\}$ the expression
$f_i^{(V)}(X^{(i)}, A)= x_i^- a_i^- v_i^- j_i^{\phi\phi(i-1)}(X^{(i-1)}, A^{(i-1)})$ is satisfied, then the CTPO is effectively complete and all modifications Q_V are invertible separately over X and over A, hence the modifications are bijective and have no collisions separately on X and on A.

Statement 3. [12] To define parametric invertibility of the substitution $p_A^{(V)}$ in sequential model (3) of the CTPO it is sufficient to use sequentially from $i =1$ to n the following functions:

$$\forall i \in \{1,.., n\} \quad y_i = f_i^{(V,e)}(X^{(i)}, A) = x_i^- j_i^{(V)}(W_e^{(i-1)}, A), \tag{4}$$

where x_0 is the initial condition, $W_e^{(i-1)}$ is a delta-function $d_i^{(e)}$ of the form:

$$W_e^{(i-1)} = d_i^{(e)}(X^{(i-1)}, Y^{(i-1)}) = \begin{matrix} ; \\ < \end{matrix} \quad \begin{matrix} = \\ = \end{matrix} \quad \begin{matrix} [& [& [\\ \backslash & \backslash & \backslash \end{matrix} \quad \begin{matrix} LI \ H= \ , \\ LI \ H= \ . \end{matrix}$$

Proof. Indeed, from conditions of Statement 3 for $e = 0$ we have:

$$y_i = f_i^{(V,0)}(X^{(i)}, A) = x_i^- j_i^{(V)}(d_i^{(0)}(X^{(i-1)}, Y^{(i-1)}), A)=x_i^- j_i^{(V)}(X^{(i-1)}, A). \tag{5}$$

Then in expression (4) we substitute y_i for x_i and replace $e = 0$ by $e = 1$:
$$x_i^{\phi}= f_i^{(V,1)}(Y^{(i-1)}, A) = y_i^- j_i^{(V)}(d_i^{(1)}\{(Y^{(i-1)}, (x_1^{\phi}, x_2^{\phi}, ... x_{i-1}^{\phi}\}, A) =$$

$$y_i^- j_i^{(V)}(x_1^{\phi}, x_2^{\phi}, ... x_{i-1}^{\phi}, A) = y_i^- j_i^{(V)}(X^{\phi(i-1)}, A). \tag{6}$$

Selecting sequentially in (6) values i from 1 to n and using (5), we obtain:
$$x_1^{\phi}= y_1^- j_1^{(V)}(x_0, A) = (x_1^- j_1^{(V)}(x_0, A))^- j_1^{(V)}(x_0, A) = x_1,$$
$$x_2^{\phi}= y_2^- j_1^{(V)}(x_0, x_1^{\phi}, A) = y_2^- j_1^{(V)}(x_0,x_1,A) = (x_2^- j_1^{(V)}(x_0,x_1,A))^- j_1^{(V)}(x_0,x_1,A) = x_2,$$
and so on to $x_n^{\phi} = x_n$, i.e. for substitution $p_A^{(V)}$ we have constructed parametrically inverse substitution $p_A^{-1}{}^{(V)}$: $(Y= p_A^{(V,0)}(X), X= p_A^{(V,1)}(Y))$.

A feature of this transformation is the necessity of the hardware implementation of the "switch" $d^{(e)}(X^{(i-1)}, Y^{(i-1)})$ with feedback loop. A feature of this realization of the model (3) is that the direct transformation is very fast, since it is performed in parallel for all i. On contrary, the inverse transformation is executed sequentially for i from 1 to n causing significantly larger time delay.

In some ciphers it is not necessary to perform inverse transformations while deciphering, although the transformation only should be bijective. In such cases the use of the direct transformation corresponding to model (3) is preferable due to its low time delay. It is used as Q-box in cryptoschemes described in section 3.

Other variant of the sequential model (2) has the form of the recursive CTPO model of the type:

$$p_A^{(V,e)}(X) = \{f_1^{(V)}(x_1, u_0, A), f_2^{(V)}(x_2, u_0, u_1, A),...,f_n^{(V)}(x_n, u_0, u_1, \ldots, u_{n-1}, A)\}, \qquad (7)$$

where $"i_\smile \{1,.., n-1\}\ u_i = Y_i^{(V,e)}(x_i, u_0, u_1, \ldots, u_{i-1}, A)$, $u_0 = const$ - initial condition, $\{f_i^{(V)}\}$ are output functions, $\{Y_i^{(V,e)}\}$ are transition functions.

To provide invertibility condition in accordance with Statement 2 it is sufficient to consider the realizations of the model (8) with the following output functions:

$$" i_\smile \{1,..,n\}\ f_i^{(V)}(x_i, u_0, u_1, \ldots, u_{i-1}, A) = f_i^{(V)}(x_i, U^{(i-1)}, A) = x_i^- j_i^{(V)}(U^{(i-1)}, A). \qquad (8)$$

If the functions

$$y_i = f_i(x_i, U^{(i-1)}, A) = x_i^- j_i(u_{i-1}, A) = x_i^- j_i(u_{i-1}, a_i) = x_i^- (a_i^- u_{i-1}),$$
$$u_i = Y_i^{(e)}(x_i, U^{(i-1)}, A) = Y_i^{(e)}(x_i, u_{i-1}, a_i) = (x_i^- e)(a_i^- u_{i-1})^- (a_i u_{i-1}),$$

which do not depend on vector V are selected as output and transition functions in the recursive model (8), then we get the realization of the modulo 2^n addition operation for operands • and A. Natural complication of the modulo 2^n addition is connected with the parametrically invertible realizations of the recursive model (8), known as elementary controlled adders [12] which are defined by the following conditions:

$$y_i = f_i^{(v_i)}(x_i, u_{i-1}, a_i) = x_i^- j_i^{(v_i)}(u_{i-1}, a_i) = x_i^- [a_i^- (u_{i-1} * v_i)],$$
$$u_i = Y_i^{(v_i,e)}(x_i, u_{i-1}, a_i) = (x_i^- e)[a_i^- (u_{i-1} * v_i)]^- [a_i (u_{i-1} * v_i)], \text{ where } "*" \text{ is some}$$

Boolean operation.

Selecting other more complex functions $\{j_i^{(V)}\}$ and $\{Y_i^{(V,e)}\}$ in model (8) one can get different CTPO having special cryptographic properties. Necessary and sufficient conditions of the parametrical invertibility of the CTPO belonging to the recursive model (8) are the following relations between output and transition functions [12]:

$$"i_\smile \{1,.., n-1\}$$
$$u_i = Y_i^{(V,e)}(x_i, U^{(i-1)}, A) = [ej_i^{(V)}(U^{(i-1)}, A)^- x_i]L_i^{(V)}(U^{(i-1)}, A)^- L_i^{e(V)}(U^{(i-1)}, A),$$

where $L_i^{(V)}(U^{(i-1)}, A)$ and $L_i^{e(V)}(U^{(i-1)}, A)$ are arbitrary generator polynomials.

A feature of this realization of the model (8) is that the direct and inverse transformations are executed sequentially for i from 1 to n causing some time delay. If functions $\{j_i^{(V)}\}$ and $\{Y_i^{(V,e)}\}$ are unificated and have low order of the algebraic normal form, the implementation complexity of this realization does not significantly exceed that of modulo 2^n addition.

For cryptographic applications it is interesting to use CTPO with output and transition functions having the following properties [17,18]: correlation immunity, high non-linearity (bent or other functions), correspondence to propagation criteria of high order and degree.

2.4   CTPO and Security against Differential and Linear Cryptanalysis

Justification of the use of CTPO in different cryptoschemes is connected with the problem how CTPO influence the security against differential (DCA) [13,14] and linear (LCA) cryptanalysis [15,16]. Let us enter the following notations:

$K=\{K_1,..., K_r\}$ is the set of keys and $X,\ Z^{(2n)}$ is the plain text block. We shall consider an iterative r-round cipher having traditional Feistel cryptoscheme with 2n-bit input. Let the encryption is described by the following transformation:

1. For $i = 1$ to $i = r-1$ do
   $\{X_R(i) = X_L(i-1),\ X_L(i) = F(X(i-1), K_i) =$
   $F(X_L(i-1), K_i)^- X_R(i-1),\ X(i) = X_L(i)\|X_R(i)$; where $K_i,\ Z^{(n)}$ is the current round key, $F(X_L, K_i)$ is the round function $(Z^{(n)} \times Z^{(n)} \text{fi } Z^{(n)})$, "$\|$" denotes the concatenation operation, $i,\ \{1, ...,r\}$, $X(i),\ Z^{(2n)}$, $\{X_L(i), X_R(i)\},\ Z^{(n)}$, $X(0)= X_L(0)\|X_R(0)$ is input data block}.

2. For $i = r$ do
   $\{Y_R = F(X_L(r), K_i)^- X_R(r),\ Y_L=X_L(r)\ Y=Y_L\|Y_R$; where $Y,\ Z^{(2n)}$ is the output of the encryption function, i.e the ciphertext block}.

Replacing the operation "$-$" by an invertible CTPO of the model (3) or (8) we obtain a new encryption function based on the same round function F. The differences between two input blocks X and X* and respective output blocks Y and Y* are defined as: $DX=DX_L\|DX_R=X^- X^*$, $DY=DY_L\|DY_R=Y^- Y^*$.

Let us compare the probability distributions of the one-round differences corresponding to source ($\Pr_, D</D;$) and modified ($\Pr^p D</D;$) encryption functions, where probability is considered over the set of all vectors $X,\ Z^{(2n)}$ and $K,\ Z^{(n)}$. For source encryption function we have the following distribution:

$$\Pr_, D</D, = \Pr_,) X . ^-) X^- D, . ^- X^- X^- D, = D</D, =$$
$$\Pr^{(-)}_{,\ L} D^- D = D \Big/ D_{1},$$

where $D) =) X . ^-) X^- D;$ . . For modified encryption function, in which substitution $p_A^{(V,e)}$ is used instead of "$-$" we have:

$$3 U^{p)} DY\ DX) =$$

$$\Pr^{p)}(\pi^{(\ ,\ )} - \pi^{(\ ,)} - D ^- D = D \Big/ D). \qquad (9)$$

Let the substitution $p_A^{(V,e)}$ is a CTPO realization of the sequential model (3) (note that analogous consideration is also valid for the output functions of model (8)) which is represented in the form:

$$\pi^{(\ )} ;) ; . = ; ^-) ; . ^- j^{(\ ,\ )}, \qquad (10)$$

where $V=V(X,K)$ and $j = j$;) ; . =
^{j} ;) ; . ... $j^{(\ ,)}$;) ; . ` .
Then from (9) and (10) we obtain the distribution:

$$3 U^p\ D\ /D) = \Pr^p_{,\ L} D^- D ^- Dj^{(\ ,)} = D \Big/ D_{1}, \qquad (11)$$

where $\Delta \dot{j} = \dot{j}$ ' ;) ; . $^- \dot{j}$ $^{-4}$ ' ; $^-$ D;) ; . $^-$ D) and
DV = V(X, K) $^-$ V(X$^-$D X, K). Hence the properties of the distribution
$3U^p$ D< /D;) unlike distribution $3U$ D /D) depend directly on the choice of
the vector function $\dot{j}$ ' .

If components of vector $\dot{j}$ ' are a bent functions [17,18], then each component
of the vector function $D\dot{j}$ $^{(V,e)}$ is a balanced function independently of the values
D) D; DQGΔ9 . Thus, for the modified encryption function from (11) we
have:

$$3U^p \text{ D< /D;) =} \Pr_{"D', ()} (D\dot{j}^{(\cdot)} = D \text{ }^- D \text{ '})\Pr^- (D \text{ '/D)}$$

This means that the use of bent-function-based substitution $p_A^{(V,e)}$ instead of the "$^-$"
operation can average the source probabilities so that
$\underset{D,D}{PD[} 3U^p$ D< /D;)$< \underset{D,D}{PD[} 3U$ D< /D;) ensuring strengthening against DCA.
To obtain good evaluations it is necessary to design such CTPO operations which
provide independence or minimal interdependence between components of the vector
function $D\dot{j}$ $^{(V,e)}$.

Use of the invertible CTPO instead of the operation "$^-$" makes the cryptoscheme
to be also more secure against LCA [15,16]. Let F = F$_-$ (X,K) denotes the round
function of the source Feistel cipher, then according to (10) Feistel-like cipher with
invertible CTPO can be represented as ordinary Feistel cipher with modified round
function $F_p(X,K)=F_- (X_L,K)^- \dot{j}$;)$_-$; . . If the components of the vector
function $\dot{j}$ are, for example, bent functions with maximal non-linearity or another
functions with high non-linearity, then one can configure these functions in such way
that N(F$_p$) > N(F$_-$), where N(F) = $\underset{(\,...,\,\lambda\,(+)}{min}$ $|\{; |) (;)_{,,} Z^- Z[\,^- ...^- Z[\,\}|$. This
means that the respective use of the substitution $p_A^{(V,e)}$ instead of the "$^-$" operation
strengthens significantly the initial cryptoscheme against LCA.
More precise estimations just as for DCA and LCA can be obtained on the basis of
the analysis the concrete design of both the round function and the CTPO realization.

3   Cryptoschemes Based on Controlled Operations

An interesting strategy of the design of the block ciphers is oriented to extensive use
of the controlled operations. As cryptographic primitives the CP and CTPO have the
following peculiarities:
1. It is easy to construct CP-box and CTPO-box operations representing a single
   nonlinear operation above data blocks of large size (32, 64 or 128 bit).

2. CP- and CTPO-boxes realize relatively large number of different modifications (for example, 2^{32} for $P_{16/32}$-box).
3. Avalanche effect spreads mainly via the use of the data bits as control ones while performing controlled operations. In the case of the CP-box permutations this is the only mechanism causing avalanche effect. In the case of the CTPO both the input bits and the control bits influence the output bits.
4. Round keys can be combined with data via using them as a part of the control vector V. This is a way to define key-dependent fixed permutations and key-dependent two-place operations.
5. CP and CTPO operations give the possibility to use so called internal key scheduling (IKS) realized as data-dependent transformation of the round keys [11]. Operations corresponding to IKS can be performed in parallel with some other operations corresponding to data ciphering. In this case IKS introduces no additional time delay keeping high encryption speed in the case of frequent change of keys.

A general structure of the iterative cryptoscheme is shown in Fig.1a, where F is the one-round encryption function. Input data block $X_L \| X_R$ is represented as concatenation of two 32-bit subblocks X_L and X_R. The iterative cryptoschemes uses r = 16 encryption rounds and 256-bit secrete key, which is represented as a set of 32-bit round keys K_1, K_2, ..., K_8. This cryptoscheme executes encryption or decryption procedure depending on the order of the use of keys. While enciphering the i-th key is used in i-th and (i+8)-th rounds. While deciphering the i-th key is used in (9-i)-th and (17-i)-th rounds. Three different ciphers based on controlled operations correspond to the round functions presented in Fig.1b, 1c, and 1d. In these round functions there are used left (<<<11) and right (>>>11) rotations by 11 bits. Fixed rotations improve data diffusion being obviously cheap in hardware and introducing no time delay. Transformation of keys with $P_{32/96}$-box operation in Fig.1b and 1c and with Q-box operation in Fig.1d represents two variants of IKS. The extension box E transforms 32-bit input into 96-bit output. $P_{32/96}$-boxes can be designed analogously to that in [9].

To estimate numerically the avalanche effect one can use the following properties of the controlled operations. On the average one bit of the control vector influences statistically one output bit of the CP-box. One input bit of the Q-box influences on the average two bits of the output. In it easy to see that one bit of the data subblock X_L affects statistically tree output bits of the $P_{32/96}$-box. In one round single input bit of the left subblock influences on the average G»4.5 (Fig.1b), G»6.5 (Fig.1c), and G»4.5 (Fig.1d) bits of the right one.

After three rounds every input bit influence statistically all output bits. The ciphers have been implemented in software and tested. Statistic examination have shown that eight rounds sufficed to get uniform correlation between input and output bits.

The ciphers suites well to hardware implementation. Fixed rotations and extension box are implemented with simple connections. The cell count for the implementation of one round is about 1000 nand gates. Using controlled operations (CP-boxes and CTPO) we have elaborated a fast cryptochip with about 100,000 transistors, the internal encryption speed about 1Gbit/s being obtained with the 1.2 micron technology.

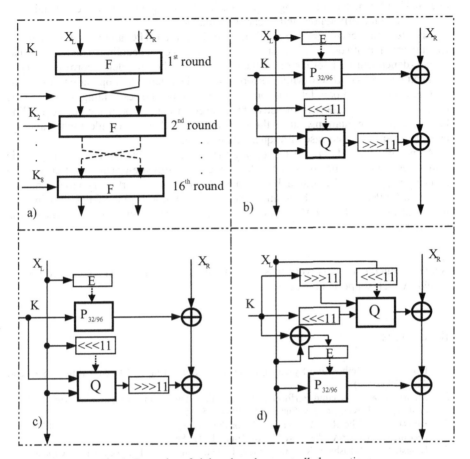

Fig. 1. Examples of ciphers based on controlled operations

Preliminary analysis shows the described ciphers are secure against differential, linear and other known attacks, although much more work on cryptanalysis is to be done. The aim of present paper is to focus on the controlled operations as a source of cryptographic strength and rapidity.

4   Conclusion

One can easy construct some CTPO operations with good cryptographic properties and with large number of different CTPO-modifications ($\ddagger 2^n$). The CTPO operations represent a single nonlinear operations above the data blocks of relatively large length. These controlled operations in combination with CP can be efficiently used to construct fast ciphers based on both the key-dependent and the data-dependent operations. The fact that current modifications of the CTPO are not predetermined is one of significant factors strengthening such ciphers against known cryptanalytical attacks including differential and linear cryptanalysis.

An interesting cryptographic application of both the CTPO and the CP is their use for construction of some mechanisms of the internal key scheduling. The IKS consists in data-dependent transformation of the round keys. The time delay corresponding to IKS can be reduced to zero while performing the data-dependent operations above the current round key in parallel with some other operations corresponding directly to the data encryption. The IKS is attractive to be used instead of ordinary key scheduling based on precomputations, since the ciphers with IKS can provide higher average encryption speed in the case of frequent change of keys.

At present the CTPO and CP operations can be considered as a cryptographic primitives which are perspective for designing fast hardware-oriented fast encryption algorithms (> 2 Gbit/s), although chip makers can easy support encryption technique based on CTPO and/or CP by adding respective instructions to the CPU allowing one to design fast (> 400 Mbit/s for Pentium-like CPU) and secure software encryption algorithms.

Acknowledgement. This work was carried out as a part of the AFRL funded project #1994P which supported the authors.

References

1. Becker, W.: Method and System for Machine Enciphering and Deciphering. U.S.Patent. 4157454 (1979)
2. Madryga, W.E.: A High Performance Encryption Algorithm. Computer security: a global challenge, Elsevier Science Publishers (1984) 557–570
3. Rivest, R.L.: The RC5 Encryption Algorithm. Fast Software Encryption – FSE'94 Proceedings. Springer-Verlag LNCS. Vol. 1008 (1995) 86–96
4. Rivest, R.L., Robshaw, M.J.B., Sidney, R. and Yin, Y.L.: The RC6 Block Cipher. Proceedings of the 1st Advanced Encryption Standard Candidate Conference. Venture, California (Aug. 20-22, 1998) (http://www.nist.gov/aes)
5. Burwick, C., Coppersmith, D., D'Avingnon, E., Gennaro, R., Halevi, Sh., Jutla, Ch., Matyas, Jr.S.M., O'Connor, L., Peyravian, M., Safford, D., and Zunic, N.: MARS – a Candidate Cipher for AES. Proceedings of 1st Advanced Encryption Standard Candidate Conference. Venture, California (Aug. 20–22, 1998)
6. Benes, V.E.: Mathematical Theory of Connecting Networks and Telephone Traffic, Academic Press, New York (1965)
7. Waksman, A.A.: Permutation Network. Journal of the ACM, Vol. 15. 1 (1968) 159–163
8. Portsa, M.: On the Use of Interconnection Networks in Cryptography. Advances in Cryptology –EUROCRYT'91 Proceedings. Springer Verlag LNCS, Vol. 547. (1991) 302–315
9. Goots, N.D., Moldovyan, A.A., Moldovyan, A.A.: Fast Encryptuion Algorithm SPECTR-H64. International Workshop Mathematical Methods, Models and Architectures for Computer Network Security – MMM-2001 Proceedings. Springer Verlag LNCS. This vol. (2001)
10. Moldovyan, A.A., Moldovyan, N.A. and Moldovyanu, P.A.: A Method of the Block Cryptographical Transformation of the Binary Information. Russian patent №2141729. Bull. no 32 (1999)

11. Maslovsky, V.M., Moldovyan, A.A., and Moldovyan, N.A.: A Method of the Block Encryption of Discrete Data. Russian patent•2140710. Bull. no 30 (1999)
12. Goots, N.D., Izotov, B.V., Moldovyan, A.A., Moldovyan, N.A.: Design of the Controlled Two-place Operations for Fast Flexible Cryptosystems, Security of the Information Technologies,•., MIPhI, 4 (2000, in Russian)
13. Biham, E., and Shamir, A.: Differential Cryptanalysis of the Data Encryption Standard, Springer Verlag (1993)
14. Nyberg, K.: Differentially Uniform Mappings for Cryptography. Advances in Cryptology –EUROCRYT'93 Proceedings. Springer Verlag LNCS. Vol. 765 (1994) 55–64
15. Matsui, M.: Linear Cryptoanalysis Method for DES Cipher. Advances in Cryptology – EUROCRYT'93 Proceedings. Springer Verlag LNCS. Vol. 765 (1994) 386–397
16. Nyberg, K.: Linear Approximations of Block Ciphers. Advances in Cryptology – EUROCRYT'94 Proceedings. Springer Verlag LNCS. Vol. 950 (1994) 139–144
17. Nyberg, K.: Constructions of Bent Functions and Difference Sets. Advances in Cryptology –EUROCRYT'90 Proceedings. Springer Verlag LNCS. Vol. 473 (1991) 151–160
18. Kurosawa, K., Satoh T.: Desigh of SAC/PC(l) of Order k Boolean Functions and Three Other Cryptographic Criteria. Advances in Cryptology – EUROCRYT'97 Proceedings. Springer Verlag LNCS (1998) 434–449
19. Nyberg, K., Knudsen, L.: Provable Security Against a Differential Attack. Advances in Cryptology – CRYPTO'92 Proceedings. Springer Verlag LNCS. Vol. 740 (1994) 566–574
20. Lai, X., Massey, J.L.: Markov Ciphers and Differential Cryptanalysis. Advances in Cryptology– EUROCRYT'91 Proceedings. Springer Verlag LNCS. Vol. 547 (1992) 17–38

Key Distribution Protocol Based on Noisy Channel and Error Detecting Codes

Viktor Yakovlev[1], Valery Korjik[2], and Alexander Sinuk[1]

[1]State University of Telecommunications
St. Petersburg, Russia
viyak@robotek.ru
[2]Section of Telecommunications IPN Cinvestav, Mexico D.F., Mexico
vkorjik@mail.cinvestav.mx

Abstract. Secret key agreement based on noisy channel connecting parties and on public discussion has been considered in [1-4] for asymptotic case. Extension of the information-theoretically secure key sharing concept to non- asymptotic case was given in [5]. In the last paper several channel transform protocols (corresponding to different algorithms of public discussion) were presented. Unfortunately the efficiency of these protocols was very low in comparison with asymptotic key capacity found in [1]. The reason of this was that these protocols do not use a redundancy efficiently. One of the considered protocols (the so-called advantage to the main channel primitive) exploits a repetition of binary symbols only. It may be much better to use linear error detecting codes, that is just a subject of consideration in the current paper. We regain the main formulas to compute Renyi entropy which is necessary to bound the information about the final key leaking to an eavesdropper after execution of such modified protocol. The use of this protocol causes an increase in the key-rate by several times, that is very important in practical implementations of key sharing procedures.

1 Introduction

It is well known that key distribution problem is one of the most important and delicate problems in information security of computer networks.

Many protocols have been suggested to solve it [6]. But we consider a specific scenarios when the main difference between legal and illegal parties is only that illegal parties are passive. It means that they can eavesdrop any information transmitted between legal parties but cannot change this information.

If we assume the existence of a noisy channel between legal parties and also the existence of noisy wire-tap channel (which is not necessarily inferior to the main channel) then we result in a scenario known as "secret key agreement by public discussion" considered in [1-4].

Let us repeat the setting up a problem for this case. One of legal parties (say Alice) can transmit to another legal party (say Bob) any information through the main binary

V.I. Gorodetski et al. (Eds.): MMM-ACNS 2001, LNCS 2052, pp. 242–250, 2001.

symmetric channel (BSC) where the bit error probability is S , whereas an eavesdropper (say Eve) can receive this message through independent wire-tap BSC where the bit error probability is S .Thereafter both Alice and Bob exchange messages with one to another through noiseless and public channels. The final goal of the key sharing protocol between legal parties is to generate the key reliable and secure enough. In fact, the final key .   computed by Alice may be different from the final key .   computed by Bob. Thus the first goal of public discussion is to agree .   and .   with very high probability, that is 3 .   ,,   3   , where 3   is as small as desired. The secrecy of this protocol can be described by Shannon information about the final key leaking to Eve. So the second goal of the protocol is to provide a very little eavesdropper's information about   .   (.   ), that is , . 8 $\overline{=}$, where U is the total Eve's knowledge including the information received by her on BSC, noiseless channels of public discussion between Alice and Bob and the full knowledge of protocol and key computing algorithm.

   We can define the key-rate 5   as the ratio of the final key length O (in bits) to the length $\hat{I}$ of the bit string ; $\hat{I}$ transmitted by Alice to Bob through BSC, that is

$$5 \ = \frac{O}{\hat{I}} \ .$$

This scenario is shown in Fig. 1

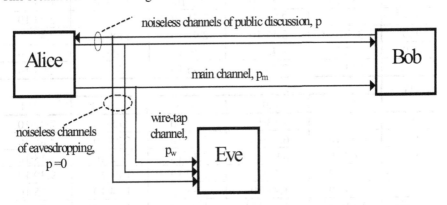

Fig. 1. Scenario of key sharing by public discussion in the presence of passive eavesdropper

   Key-capacity & can be defined as the maximal possible key-rate for every 3 >   and , >   and sufficiently large $\hat{I}$ given S , and S . The key-capacity in our case is [1]

$$\& \ = K \ S - \ K \ S \ , \tag{1}$$

where $S = S + S - S S$ and $K S = - \mathfrak{O}RJ S - S ORJ - S -$ the entropy function.

It is worth noting that $\& >$ even in the case $S \ddagger S$, that is, when the main channel is a inferior to wire-tap channel.

It has been shown in [1] that such capacity is achieved by using only a single forward (backward) transmission as public discussion between legal parties. Key-capacities calculated by (1) for different pairs of $(S , S .)$ probabilities are presented in Table 1.

Table 1. Key-capacities for different probabilities in the main channel (S) and in the wire-tap channel (S)

P_w	P_m				
	0.01	0.02	0.03	0.04	0.05
0.01	$5.95 \cdot 10^{-2}$	$5.09 \cdot 10^{-2}$	$4.51 \cdot 10^{-2}$	$4.07 \cdot 10^{-2}$	$3.71 \cdot 10^{-2}$
0.02	$1.12 \cdot 10^{-1}$	$9.72 \cdot 10^{-2}$	$8.69 \cdot 10^{-2}$	$7.88 \cdot 10^{-2}$	$7.2 \cdot 10^{-2}$
0.03	$1.59 \cdot 10^{-1}$	$1.4 \cdot 10^{-1}$	$1.26 \cdot 10^{-1}$	$1.15 \cdot 10^{-1}$	$1.05 \cdot 10^{-1}$
0.04	$2.02 \cdot 10^{-1}$	$1.8 \cdot 10^{-1}$	$1.63 \cdot 10^{-1}$	$1.49 \cdot 10^{-1}$	$1.37 \cdot 10^{-1}$
0.05	$2.43 \cdot 10^{-1}$	$2.17 \cdot 10^{-1}$	$1.97 \cdot 10^{-1}$	$1.81 \cdot 10^{-1}$	$1.67 \cdot 10^{-1}$
0.06	$2.81 \cdot 10^{-1}$	$2.52 \cdot 10^{-1}$	$2.3 \cdot 10^{-1}$	$2.11 \cdot 10^{-1}$	$1.95 \cdot 10^{-1}$
0.07	$3.16 \cdot 10^{-1}$	$2.86 \cdot 10^{-1}$	$2.61 \cdot 10^{-1}$	$2.4 \cdot 10^{-1}$	$2.23 \cdot 10^{-1}$
0.08	$3.5 \cdot 10^{-1}$	$3.17 \cdot 10^{-1}$	$2.91 \cdot 10^{-1}$	$2.68 \cdot 10^{-1}$	$2.49 \cdot 10^{-1}$
0.09	$3.82 \cdot 10^{-1}$	$3.48 \cdot 10^{-1}$	$3.19 \cdot 10^{-1}$	$2.95 \cdot 10^{-1}$	$2.74 \cdot 10^{-1}$
0.1	$4.13 \cdot 10^{-1}$	$3.76 \cdot 10^{-1}$	$3.46 \cdot 10^{-1}$	$3.21 \cdot 10^{-1}$	$2.98 \cdot 10^{-1}$
P_w	P_m				
	0.06	0.07	0.08	0.09	0.1
0.01	$3.4 \cdot 10^{-2}$	$3.13 \cdot 10^{-2}$	$2.89 \cdot 10^{-2}$	$2.68 \cdot 10^{-2}$	$2.49 \cdot 10^{-2}$
0.02	$6.62 \cdot 10^{-2}$	$6.11 \cdot 10^{-2}$	$5.66 \cdot 10^{-2}$	$5.25 \cdot 10^{-2}$	$4.88 \cdot 10^{-2}$
0.03	$9.69 \cdot 10^{-2}$	$8.96 \cdot 10^{-2}$	$8.31 \cdot 10^{-2}$	$7.72 \cdot 10^{-2}$	$7.18 \cdot 10^{-2}$
0.04	$1.26 \cdot 10^{-1}$	$1.17 \cdot 10^{-1}$	$1.09 \cdot 10^{-1}$	$1.01 \cdot 10^{-1}$	$9.39 \cdot 10^{-2}$
0.05	$1.54 \cdot 10^{-1}$	$1.43 \cdot 10^{-1}$	$1.33 \cdot 10^{-1}$	$1.24 \cdot 10^{-1}$	$1.15 \cdot 10^{-1}$
0.06	$1.81 \cdot 10^{-1}$	$1.68 \cdot 10^{-1}$	$1.56 \cdot 10^{-1}$	$1.46 \cdot 10^{-1}$	$1.36 \cdot 10^{-1}$
0.07	$2.06 \cdot 10^{-1}$	$1.92 \cdot 10^{-1}$	$1.79 \cdot 10^{-1}$	$1.67 \cdot 10^{-1}$	$1.56 \cdot 10^{-1}$
0.08	$2.31 \cdot 10^{-1}$	$2.15 \cdot 10^{-1}$	$2.01 \cdot 10^{-1}$	$1.87 \cdot 10^{-1}$	$1.75 \cdot 10^{-1}$
0.09	$2.55 \cdot 10^{-1}$	$2.37 \cdot 10^{-1}$	$2.21 \cdot 10^{-1}$	$2.07 \cdot 10^{-1}$	$1.93 \cdot 10^{-1}$
0.1	$2.77 \cdot 10^{-1}$	$2.59 \cdot 10^{-1}$	$2.42 \cdot 10^{-1}$	$2.26 \cdot 10^{-1}$	$2.11 \cdot 10^{-1}$

2   Constructive Methods of Key Sharing

It is commonly to perform three main procedures in order to share secure reliable keys between legal users under the condition that S $\ddagger$ S :

ú   Channel transform (CT) (to converge the condition S $\ddagger$ S to the condition S $<$ S);

ú   Forward error correction (FEC) (to provide reliable key-sharing, e.g. 3 . „ $\mathbf{1}$);

ú   Privacy amplification (PA) to provide the condition , . 8 £ .

There are several CT (or primitives in other words) satisfying the requirement given above but we consider only one of them because it has been proved in [5] that this primitive has advantage in the key-rate against other known primitives.

It can be called the advantage to the main channel primitive (AMC). In this protocol each of binary symbols chosen by Alice at random is repeated Q times and transmitted to Bob over the main BSC. Bob accepts each of groups consisting of Q symbols if and if they are all ones or zeros only. Thereafter he takes evident decision about the information symbol corresponding to the accepted group, otherwise Bob rejects these groups completely (say corresponding information symbols). His decision about the rejection of the corresponding groups is transmitted to Alice over a noiseless public channel. Both Alice and Bob store the information symbols which were not rejected. This protocol has been considered in [5] but it showed still minor efficiency.

Let us extend this protocol and try to use some binary linear Q N-code $\&$ having good error detection capability [7]. Then Alice divides bit string ; [1] on blocks of length N and encodes each of these blocks by code words of the code $\&$. At the receiving side Bob accepts each of code blocks if and only if he does not detect error on this block by code $\&$ known for him. Thereafter he stores the information bits of the accepted blocks and sends his decision about the rejection of some blocks to Alice. Both Alice and Bob store the information symbols of code blocks that were not rejected. We will call this protocol modified AMC(MAMC).

FEC procedure can be done if Alice computes the check symbols of some constructive linear code to information bit string stored by parties after the completion of protocol and sends this check string to Bob over a noiseless public channel. It is easy to prove (following the idea of time sharing) that the probability 3 of a disagreement between the strings stored by parties upon completion CT may be provided as small as desired for large information bit string length 1 if the number of binary check bits transmitted over noiseless channels satisfies inequality

$$U \ddagger 1 - \& , \qquad (2)$$

where $\&$ is the capacity of virtual $T = {}^{N}$-ary channel obtained between parties after the execution of CT protocol.

Capacity of channel corresponding to AMC protocol can be trivially found as follows

$$\&_{\$0\&} = - K \overset{a}{\$}, \tag{3}$$

where K is the entropy function, $\overset{a}{\$}_P = \dfrac{S_P^Q}{S_P^Q + - S_P{}^Q}.$

Capacity of virtual channel obtained after MAMCtransform was derived in [8]

$$\overset{a}{\&}_P N = N \frac{\overset{Q}{\$}_M S_P^M - S_P{}^{Q-M}}{S_{DF}} - ORJ \frac{S_P^M - S_P{}^{Q-M}}{S_{DF}}, \tag{4}$$

where

$$S_{DF} = - S_P{}^{Q+} \$ S - S^-, \tag{5}$$

$\$_M$ – the number of code words of the code $\&$ having Hamming weight M

PA procedure means that legal parties use the shared string (after the completion CT and FEC) as the input to some hash function chosen at random by one of them from the class of universal hash functions [9] and transmitted to another party over a noiseless public channel. The output of this hash function is the desired final key shared by Alice and Bob. The main theorem of privacy amplification [9,10] says that the amount of Shannon's information , leaking to Eve is bounded by the following inequality

$$, £ - 1 - \Theta - U W OQ \text{ (bits)}, \tag{6}$$

where 1 length of the bit string stored by legal parties after the completion of CT and FEC procedures,

O– is the length of final key,

W– is the Renyi information obtained by Eve from her eavesdropping over wire-tap channel,

U – the length of check string used in FEC procedure (these symbols Eve obtains from noiseless channels of public discussion).

It has been found in [5] that for AMC transform the Renyi information is upper bounded as follows

$$W£ \quad 1 - 5, \tag{7}$$

were

$$5 = \overset{Q}{\underset{= ŁGł}{}} 3 G 5 G,$$

$$5 G = - ORJ \frac{S - S^- + S^- - S}{S - S^- + S^- - S},$$

$$3 G = S_Z^G - S_Z{}^{Q-G} + S_Z{}^Q G - S_Z{}^G.$$

The key rate for AMC transform is

$$5 = \frac{\partial S}{1Q},\tag{8}$$

where

$$\partial_{DF} = - S_P{}^Q + S_P^Q .$$

The efficiency of this key-sharing protocol can be estimated as a ratio between the key capacity and the key rate

$$h = \frac{\&}{5}.\tag{9}$$

If the final key length O and allowable information , leaking to Eve about final key are given, we can optimize AMC transform with respect to parameter Q. The values of coefficients h by (9) and corresponding to them optimal parameter Q for different pairs S , S and $O = , , = \cdot$ bit are presented in Table 2.

Table 2. Performance coefficients h and optimal block length Q obtained for AMC transform, different probabilities S , S and $O = , , = \cdot$ bit

p_w	p_m					
	0.05	0.06	0.07	0.08	0.09	0.1
0.01	$3.97 \cdot 10^2$	$5.58 \cdot 10^2$	$2.26 \cdot 10^3$	$2.84 \cdot 10^4$	$1.06 \cdot 10^6$	$7.05 \cdot 10^6$
	4	4	4	6	8	8
0.02	$1.23 \cdot 10^1$	$6.57 \cdot 10^1$	$1.75 \cdot 10^2$	$2.1 \cdot 10^2$	$3.12 \cdot 10^2$	$1.22 \cdot 10^3$
	2	2	4	4	4	4
0.03	6.97	9.62	$1.94 \cdot 10^1$	$1.03 \cdot 10^2$	$1.13 \cdot 10^2$	$1.35 \cdot 10^2$
	2	2	2	4	4	4
0.04	5.59	6.56	8.61	$1.44 \cdot 10^1$	$7.22 \cdot 10^1$	$7.72 \cdot 10^1$
	2	2	2	2	4	4
0.05	4.92	5.42	6.32	8.1	$1.26 \cdot 10^1$	$3.91 \cdot 10^1$
	2	2	2	2	4	4

We can see that efficiency of AMC transform is very low if $S > S$. It means that this transform is very far from optimal one and we can try to improve it using MAMC transform.

3   Derivation of Renyi Information for MAMC Transform

Theorem . The amount of Renyi information that gains Eve from each block of $Q N$ linear binary code C over wire-tap channel with bit error probability S can be found as follows:

$$, _5 N = N+ \quad 3 \, \&ORJ \quad \S_{\mathbb{ME}}^{Q} \frac{\O{S_Z}^M - S_P}{\mathbb{E} \quad 3 \, \&_L} \quad \text{œ} \quad , \quad (10)$$

where

$$P(C_i) = \sum_{\bar{z}, C_i} P(z_\cdot C_i) = \sum_{j=0}^{n} A_{ji} p_w^{j} (1 - p_w)^{n-j} ,$$

A_{ji} – the number of words of Hamming's weight j in the i-th coset of linear binary $Q N$-code C.

Proof. Because the information N bit strings are uniformly distributed we get
$$W= N \quad 5 ; \; =, \quad (11)$$
where $5 ; <$ is the conditional Renyi entropy of each block $[$ of $Q N$-linear binary code received by Eve as $=$ over wire-tap channel with bit error probability S .

We can write a posteriori probability $S [\;]$ using Bayes formula
$$P(x/z) = \frac{P(z,x)}{P(z)} = \frac{P(x) \; P(z/x)}{P(z)} . \quad (12)$$
Assume that $] _\cdot \&$, where $\&$ is the i-th coset of linear code $\&$. Then we get
$$S] [= 3 H_{1[} , \quad (13)$$
where $H_{1[} =]^- [_\cdot \&$ and $3 H_{1[}$ is considered as the probability of error pattern $H_{1[}$ in the wire-tap channel, $^-$ is bitwise modulo two sum.

Because all code words are equally likely we get
$$3] = \sum_{[\cdot \&} 3 [3] [= ^{-N} \sum_{[\cdot \&} 3] [= 3^N \&_L , \quad (14)$$
where $3 \& = H.$

Substituting (14) and (13) into (12), we obtain
$$S [\;] = \frac{3 H_{1[}}{3 \&_L} . \quad (15)$$
By definition given in [9] we get for $] _\cdot \&$ from (15)

$$5 \; = =] = - ORJ \quad \S [\;] \;] \;] = - ORJ \; \$ \; \frac{Ø S - S ø}{Œ \quad 3 \& \quad œ}{Œ \quad \quad \wp} \quad (16)$$

In contrast to model with original BSC, this channel transform results (as we can see from (16)) in dependence of Renyi entropy from the coset to which belongs the received block] . Therefore we have to find the Renyi entropy averaged on all binary blocks] of length Q. It gives the relation

$$5 \; = = S] 5 [= =] = \quad (17)$$

$$3 \&_L ORJ \quad \sum_{L=} \quad \$ \quad \frac{Ø S_Z^M - S_P^{Q M}}{Œ \quad 3 \&_L \quad œ}{Œ \quad \quad \wp} .$$

Substituting (17) into (11) we get (10) that is just the statement of theorem. o

Now we can use (6) and (10) to find the amount of Shannon's information , leaking to Eve after execution of MAMC protocol.

It is easy to see that for MAMC protocol the key rate is

$$5 = \frac{OS \; N}{1Q} . \quad (18)$$

In Table 3 we present efficiency h of MAMC protocol where 5 is calculated by (18) under the conditions $O= , , = ^- $ bit.

Table 3. Performance coefficients h and optimal error detecting codes of MAMC protocol obtained for different probabilities S , S and $O= , , = ^-$ bit

S	S					
	0.05	0.06	0.07	0.08	0.09	0.1
0.01	109	218	-	-	8.4· 10^4	1.3· 10^6
.	(8,4)	(8,4)			(32,16)	(16,05)
0.02	-	21	43	-	-	-
		(15,11)	(7,4)			
0.03	-	-	-	18.6	34.6	48.2
				(7,4)	(8,4)	(8,4)
0.04	-	-	-	-	13.0	18.8
					(7,4)	(7,4)
0.05	-	-	-	-	-	10.5
						(7,4)

Remark. Spaces in Table mean that it is unknown which codes provide better results than AMC protocol for corresponding probabilities S , S .

Because code spectra and, moreover, coset spectra are known for some classes of codes only [7] we have to select namely these codes.

Therefore the notations (7,4), (8,4), (15,11) in Table 3 mean the parameters of the corresponding Hamming codes, while (32,16), (10,5) are the parameters Reed-Muller codes.

We may compare Table 2 and 3 and conclude that the use of MAMC protocol instead of AMC protocol results in an increase in efficiency at least in several times.

4   Conclusions

The problem of information-theoretically secure key sharing based on an existence of both noisy and noiseless channel can be put into practice if the key rate (given high level of security and reliability) is close enough to the key capacities.

It is worth noting here that small key-rates may be tolerable for practice, in contrast to the code rate for error correcting codes, because we need to share small key sizes, that are enough for further cryptographic applications (as a rule they are at most 64-256 bits).

It is no problem to provide the key rate close to the key capacity for good situation when $S \ll S$. But in the worst of considered cases ($S =$, $S =$) we have got several orders difference between the key capacity and the key rate. Nevertheless MAMC protocol is one order better than AMC and several orders better than other known protocols. We believe that key rate can be further increased if so called challenge-response protocol (CR in [5]) is followed by MAMC protocol. We intend to present our results in this direction in the future.

References

1. Ahlswede, R., Csiszar, I.: Common Randomness in Information Theory and Cryptography-Part 1: Secret Sharing. IEEE Trans. on IT, Vol. 39. 4 (1993) 1121–1132
2. Maurer, U.: Secret Key Agreement by Public Discussion Based on Common Information. IEEE Trans. on IT. Vol. 39. (May 1993) 733–742
3. Maurer, U.: Protocols for Secret Key Agreement by Public Discussion Based on Common Information. Advances in Cryptology – CRYPTO '92, Lecture Notes in Computer Science, Vol. 740. Springer–Verlag, Berlin (1993) 461–470
4. Maurer, U.: Linking Information Reconciliation and Privacy Amplification. J. Cryptology. 10 (1997) 97–110
5. Korjik, V., Yakovlev, V., Sinuk, A.: Performance Evaluation of different Secret Key Agreement Protocols Based on Noisy Channels. ICCI'2000, Kuwait
6. Scheier, B.: Applied Cryptography. J. Wiley (1994)
7. Klove, T., Korjik, V.: Error Detecting Codes. Kluwer (1995)
8. Korjik, V., Yakovlev, V.: Capacity of Communication Channel with Inner Random Coding. Problems of Information Transmission, Vol. 28. 4 (1992) 317–325
9. Bennett, C., Brassard, G., Crepeau, C., Maurer U.: Generalized Privacy Amplification. IEEE Trans. on Inform. Theory. Vol. 41. 6 (1995) 1915–1923
10. Korjik, V., Morales Luna, G, Balakirsky, V.: Enhanced Privacy Amplification Theorem for Noisy Main Channel (submitted for Crypto'2001)

Dynamic Group Key Management Protocol

Ghassan Chaddoud [1], Isabelle Chrisment [2], and Andr´e Scha [1]

[1] LORIA - University of Nancy I
[2] LORIA - University of Nancy II
Campus Scienti que - BP239
54506 Vandoeuvre-Les-Nancy - FRANCE
{chaddoud,ichris,scha }@loria.fr

Abstract. If multicast communication appears as the most e cient way to send data to a group of participants, it presents also more vulnerabilities to attacks and requires services such as authentication, integrity and con dentiality to transport data securely. In this paper, after introducing the research work related to securing multicast communication, we present the protocol Baal [1] as a solution to the scalability problems of key management in dynamic multicast group and show how Baal resolves the user's revocation problem. This protocol is based on distributed group key management by local controllers within sub-networks.

1 Introduction

The main multicast security problem is often considered as key distribution and management problem. Indeed, in multicast communications the group security association or GSA cannot directly be negotiable, i.e. group members cannot interact among them for creating a GSA because the number of participants in the group may be very signi cant. [13,8] propose that only one entity, for example, the group manager or the group controller, GC, chooses a GSA. Then, this entity distributes the GSA to the group members by means of secure tunnels. The most important parameter in GSA is the symetric key used to encrypt group data. This key is called multicast key or group key, K_{grp}, and ensures the con dentiality to group communications. The GC creates and distributes the key, in a secure manner, to the di erent members of the group. Multicast messages sent by a group member and encrypted with K_{grp}, only shared among group members, can be received and decrypted by all group members owning K_{grp}. In this way, using this key ensures group access control as well, because only entities having K_{grp} can receive and decrypt group tra c. Therefore, the group key allows con dentiality, access control, group authentication, and integrity.

There are many proposals to manage group key and secure group communications. Unfortunatly, neither one of these proposals can scale to large groups or groups with highly dynamic memberships, like groups extended over a wold wide

---

[1] In the Canaanite's mythology, Baal is a generic semetic name meaning Master. He is the storm's god victorious against Mot, death's god.

V.I. Gorodetski et al. (Eds.): MMM-ACNS 2001, LNCS 2052, pp. 251–262, 2001.

network (e.g. Internet), where joins and evictions of entities happen frequently. A such solution should provide [17] :

— minimal time for secure group con guration;
— con guration tra c as reduced as possible;
— dynamic group, i.e. join or eviction member is possible at any time;
— independancy of routing protocols and of unicast security protocols (IKE [21], SSL [19]);
— con dentiality, integrity and authentication of group data;
— decentralized group key management.

In this paper, we present the protocol   Baal as a solution to the scalability problem of group key management. In Section 2, we describe background related to securing group communications. Section 3 describes brie y our proposal. Section 4 presents in detail the algorithms of our approach. Section 5 analyses its performance and compares it with another propositions. Finally, in Section 6, we conclude our work.

2   Background

Multicast protocols exhibit two types of scalability failures summarized by   Mittra   [2]:

{ 1 a ects n: this failure occurs when a group member a ects all the other members.
— 1 does not equal n: this failure occurs when a protocol has to deal with each member separately.

In a secure multicast communication where membership is dynamic (dynamism can depend on the security policy adopted by group membership or on multicast applications), member eviction or join is possible at any time. In order to ensure a forward (backward) secrecy, the group key must be replaced by another one after each leaving (joining) of a member. In multicast key management protocols, the joining of a new member exhibits only the rst failure type and the leaving of a member presents both types. According to   Canetti   [1] the second scalability failure is known as revocation problem

Until now, there are some proposals to secure group communication, but an all-encompassing solution for multicast security is di cult, even impossible. The approaches, used for the construction and the distribution of the group key, can be either theoretic or pragmatic.

The theoric approaches encompass solutions based on the information theory [3], hybrid approaches [4], and Di e-Hellmann key exchange approaches [10, 11]. The storage space, computation and exchanged message number increase linearly with the group size (participants number) for join or eviction of a member. These approaches are very di cult even impossible to be implemented and the security for some of them is not yet proven.

As for the pragmatic category, it contains SKDC (Single Key Distribution Center) methods [16,15,9] and hierarchical approaches [9,5,12,7]. SKDCs scale linearly with group size and su er from the centralization of group management. They do not resolve the two scalability failures, but they are suitable for small discussion group. The hierarchical approaches scale logarithmically in group size. We can distinguish between two types of hierarchical approaches : the approaches requiring trusted routers such as SMKD (Scalable Multicast Key Distribution) [14] and the approaches which do not require trusted intermediate nodes. This is the case, for example, of LKH (Logical Key Hierarchy) [9,5] and OFT (One-way function Tree) [12,7].

During group initialization, the approach SKDC is more e cient than the hierarchical approaches. Moreover, it requires a storage space less signi cant than others. On the other hand, the hierarchical approaches are more e cient for dynamic groups ; because they distribute the computational cost of re-keying among the whole group. The hierarchical approaches try to resolve the second scalablity failure   1 does not equal n   by changing an   O(n) problem to an O(log(n)) problem where   n denotes group size. We note here that neither one of these distribution group key approaches resolve the failure   1 a ects n   .

In the literature, we nd two strategies for multicast security management : centralized management where only one entity controls the group security [16, 9,5] and decentralized management [2,6,8] where many entities participate in group management. The second strategy resolves the failure   1 a ects n   by dividing the multicast group into sub-groups. Each sub-group, managed by a local controller, has its own key. The sub-groups are linked by intermediate agents for building a virtual group. The intermediate agent role is to translate multicast data di used by a member within its sub- group to all members of the virtual group. Consequently, this strategy ts better dynamic groups. But, it is less e ecient for di usion of group data which undergoes encryption/decryption operations by the intermediate agents. On the other hand, the rst strategy is more e cient for data di usion because it uses only one key shared between group members.

It would be better to set up a protocol which ensures the decentralized management with only one key. In the following section, we present a new group key management protocol, called   Baal , which is a solution for scalability problems, in particularly for the   revocation problem , again 1 does not equal n   .

3   Baal : Overview

Baal uses a groupe controller, GC, and local controllers,LCs, in order to con gure and manage secure dynamic groups on Internet. A single group key is used at any time to cipher the group tra c. When a member is evicted, its local controller generates a new group key and multicasts it to the rest of the LCs and to group members upstream, but not downstream. It unicasts the key to its own local members under their respective public keys, but omits the one being evicted.

The speci cation of our approach was motivated by the fact that in specialized conferences, there may be many participants coming from the same specialized establishment (laboratory, university, society, ...), which has the same research domain than the conference. Participants number can vary from one to many dozens. Therefore, the group controller could deliver only one copy of the group key to local controllers, which, in their turn, distribute it to participants in their establishments.

We de ne the new parameter   , participation coe cient   , as the average of participant number of an establishment, or the average of the number of participants managed by a local controller.

Baal de nes three entities : group controller, local controller and group member.

- The group controller , GC, may be an organizer or a chairman of a conference and has the right to create secure groups on Internet. It has a list, called PL (Participants List), of future participants. GC creates the group key and distributes it to group members via local controllers. We assume that the group controller creates the PL list by means of other ways (e-mail, fax, phone, post, ...).
- A local controller , LC, per sub-network, is delegated by the group controller. It receives the group key and distributes it to group members in its own sub-network. At any time of the group life, a local controller can play the role of the group controller if there is any change on the state of the group in its sub-network. It can create and distribute a new group key, accept or refuse a new member in the group, and notify any change in the group to the other group controllers. LCs must be trusted entities and cannot be group members.
- Group member is a member of the participants list, PL, or any host joining the group later.

In a simpli ed scheme of our approach, we consider the following hypothesis:

- an establishment consists of only one sub-network or LAN;
- we have one local controller per sub-network;
- there are no multihomed LANs or hosts;
- a sub-network is linked to Internet by an IGMPv3-capable multicast router and the local controller is this router.

Then, in an extended scheme we plan to have one local controller for many sub- networks. In this case a local controller can be a border router within routing domain, a gateway within an autonomous system, a home agent in IP mobility, or an independent machine dedicated to security management.

In order to assure that all group members use, at the same time, the same group key, joins and evictions, must be bu ered until a break point. But, if group members receive many group keys sent by di erent controllers at a break point, which key would be chosen? A solution can consist of the de nition of a new parameter, called the priority number, for each controller. In this case, group members would chose the one coming from the controller with the highest priority, i.e. the smallest priority number. This parameter can be managed by the group controller having the priority number of zero. To do this, the GC de nes a counter, p _count, initialized to zero. p _count will be incremented, before the delegation of a new local controller, and attributed to this controller.

4   Baal : Algorithms

We now present further in detail the three main algorithms of   Baal : group initialization, junction of a new entity and eviction of an entity.

4.1   Group Initialization

The group controller starts the process of the group initialization by creating the group key K_{grp}. For simpli cation purposes, we assume that every controller can securely generate cryptographic keys. Then, the group controller communicates the key K_{grp} to group members via local controllers. A local controller communicates its public key to the group controller by presenting itself as a candidate local controller.

The group inilialization is composed of invitation and key distribution steps. The invitation step is dedicated to invite PL members and local controllers to participate in group communication. Key distribution step sends securely the key and the identity of the group to group members and local controllers.

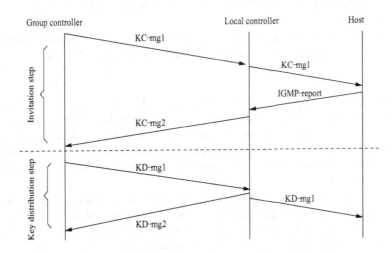

Fig. 1.   Group initialization

Invitation step.   In order to protect plain-text messages during this step, the sender de nes a token in the message. A token forms an essential part of the authentication process of exchanged messages and assists the receiver to verify the sender identity. We adopt the token de nition given in [14] where a token contains recipient's unique identity, a timestamp and a peudo-random number.

Figure 1 shows the message exchange during this step. The message KC_mg1 is sent by the GC to PL members. This message is signed by the GC which includes its public key and a signed token. The candidate local controller of the receiver authenticates the message and, if successful, stores the GC'public key before forwarding the message to the receiver which, in its turn, authenticates the sender. If a receiver accepts to participate to the group communications then, it stores the GC's public key and acknowledges this message by an IGMPv3-report involving its signed token and public key. We suppose that IGMPv3 [20] de nes a message type which indicates the presence of a signed token.

At the receipt of the IGMPv3-report, the candidate local controller authenticates the sender and, if successful, sends to the GC a signed KC_mg2 containing its signed token and public key and those of the host. The sender presents itself as a candidate local controller within a domain, a sub-network. The GC authenticates the candidate local controller and the host. If the authentication is successful and if the sender is accepted as delegated controller, the GC stores LC's public key, adds it to the local controllers list, LCL, and attributes to it a priority number .

When a local controller receives many KC_mg1 messages (the number of messages equals the number of participants within its sub-network,   i.e,   the participation coe cient), that implies that there are   participants in the group coming from the same sub-network. In this case, the candidate local controller replies by sending only one message KC_mg2 when it receives the rst IGMP-report of a host.

The number of messages used to achieve this step is   n unicast messages of the type KC_mg1 and n=   unicasts messages of the type KC_mg2.

Key distribution step.   With the messages KD_mg1 described in Figure 1, the GC distributes K_{grp}, group identity, its identity and the priority number of the receiver, the local controller. The messages KD_mg1 and KD_mg2 must be encrypted because they contain the group key and the identity of the sender. The controller uses the public key asymetric cryptographic technic to encrypt these messages.

At the receipt of the KD_mg1 message, a local controller decryptes it, and then acknowledges it by the message KD_mg2 encrypted with the GC'public key. The message KD_mg2 contains only the identity of the local controller and thus of the group. Finally the group key is forwarded to the local members.

This step is accomplished by 2 .(n=) unicast inter-domain messages of type KD_mg1 and KD_mg2 and for each LC,   unicast intra-domain messages of type KD_mg1. But, this last ones are not a scalability concern.

Until now, a local controller do not have all informations needed to take part in the group key management. They need the LCL list and their public keys. The only entity holding all these informations is the GC. So, the GC multicasts these information to the local controllers.

4.2   New Member Join

The entity which wants to join the group sends an IGMP group membership report to its local router. It includes in this report its signed token. When the IGMP- report is received, the local router authenticates the entity's token. If successful, it waits for the next break point. Then, two cases may occur:

1. The local router is a delegated controller,   i.e, it owns the group key K_{grp} and the LCL list. Then it generates a new group key K_{grp} ', unicasts a message to the entity, containing the new group key K_{grp} ' encrypted with the entity's public key, and multicasts the same massage, encrypted with the group key, to all the other group members and controllers. The local controller includes in these messages its priority number.
2. The local router is not yet delegated. So, it must negotiate with the GC to obtain the permission to participate in the management of the group key. The local controller starts the negotiation session by the message KC_mg2 (g. 1) [2]. If the key K_{grp} has to be replaced by another, the GC creates a new key K_{grp} '. Then, it multicasts this new key and the new controller's identity and its public key, in a message encrypted with K_{grp}, to group controllers and members. Next, it uses KD_mg1 to send the new group key and a new priority number, the updated list LCL to the new controller. Finally, the controller distributes the new key to its new group member. The local controller includes in these messages its priority number. The messages exchanged during the addition of a member are one multicast message and one unicast's.

4.3   Group Member Eviction

To prevent the evicted entity (ex-member) from further accessing to future group data, the group must be re-keyed. The local controller creates a new key K_{grp} ' , encrypts it with the group key K_{grp} and multicasts it with its priority number to group controllers and members, upstream, which are capable to decrypt and extract it. Since the local controller is on the multicast group distribution tree, it must not forward this message to members within its sub-network. It unicasts the new group key to them under their respective public keys, but not to the evicted one. The priority number of the local controller must be included in these messages.

The local controller, which is a local router or an IGMPv3 [20] capable multi-cast router, can bene t from the IGMPv3's new functionnality, source ltering,

---

[2] The messages KC_mg2, KD_mg1, and KD_mg2 are the same of these used during the group initialization

that is, the ability for a system to report interest in receiving packets only from speci c source address, or from all but speci c source addresse, sent to a particular multicast address. This functionnality can permit to the local controller to multicast a message involving the group key, to the group members except those behind him. In this case, with IGMPv3-capable multicast routers as local routers, group re-keying is faisable with a unique interdomain multicast message and   unicast intra-domain messages.

4.4   Security and Secrecy of Baal

This section focuses on the security of the group key and data secrecy.

Group key security.   Both messages of the invitation step are provided with signed tokens for anti- replay. They do not contain any sensitive security information about the group key. Since the messages of key distribution step contain the identity and the key of the group, they will be encrypted with receiver's public key. Consequently, eavesdroppers cannot retrieve the group key in the group initialization.

An evicted member cannot obtain the new group key because its LC, which proceed to change the key, unicasts the new key, downstream, to members under their public keys, but, not to the evicted one. With the assumption (hypothesis of our simpli ed scheme) that the evicted member is only linked to one sub-network, evicted members cannot retrieve the new group key.

Similarly, the group key is replaced by a new one after the joining of a new member. Its LC multicasts the new key to all group members under the old group key and unicasts it to the new member encrypted with the public key of the member. So, new members cannot retrieve the old group key.

Group data secrecy.   Group data con dentiality is assured ; only entities, group members, owning the group key can decrypt group data. Members receiving many group key by di erent rekeying messages, choose the key coming from the controller with the highest priority and start to use it at the next break point. By doing this, all group members use, at the same time, the same group key and new (evicted) members cannot access old (future) group data because they cannot retrieve the old (new) group keys. Consequently, forward and backward secrecy is assured to group tra c.

5   Performance Evaluation and Comparative Analysis

In this section, we analyse our approach and compare it with others, especially, those which resolve the scalability problems.

In the following we denote k cryptographic key size, h height of hierarchical tree and equal to   log(n) for balanced trees,   participation coe cient and   n the group size. So, n=   stands for the local controller number.

At the time of group con guration, unicast transmission size required to distribute the group key is equal to (n=   ).k. K$_{grp}$ distribution within sub-networks
is not a scalability concern. Consequently, as long as   is signi cant transmission
size is reduced.

Transmssion size required to rekey group members after member eviction
or member join, is equal to   k, i.e, only one multicast inter-domain message is
needed for rekeying the group.

We note that storage space is equal to   k for a member and (n=   +1) .k for a
controller.

Table 1 summarizes the size of transmission required for initiating and rekeying the secured group, and the space requirements for   Baal , SKDC, LKH and
OFT. We note that   Baal carries out less transmission than the other approaches.
Also, it requires the smallest storage space for a controller. This proves that the
e ciency of our approach depends on the value of   at the step of group key
distribution. Figure 2 shows the number of packets sent by the group controller
according to group size. We observe that as long as   becomes important the
number of sent packets approximate the group size. On the other hand, for
equal to 1, the number of packets is about twice the group size.

Table 1.   Comparison of Baal with SKDC, LKH et OFT

	Baal	SKDC	LKH	OFT
transmission size (init.)	(n=).k	nk	2nk + h	2nk + h
transmission size (rekey)	k	nk	2hk + h	hk + h
manager storage space	(n=).k	nk	2nk	2nk
membrer storage space	:k	2k	hk	hk

When we Compare   Baal with the hierarchical approaches [2,5,12] ensuring
forward and backward secrecy, we note that those last ones resolve the scalability
problems by means of the hierarchy of keys or systems. Indeed, they attempt to
change a O(n) problem into a   O(log(n)) problem, i.e, the number of messages
needed to rekey the group after a join or an eviction of a member is of the order
of h which equals to log(n). On the other hand,   Baal change the   O(n) problem
into a O(1) problem. The second raw of Table 1 depicts this comparison. In
order to rekey the group, of Baal , after a join or an eviction of a member, a
local controller needs only one multicast message to distribute upstream the
new key and   unicast message downstream. But, these unicast messages are
intra-domain, consequentlly, they are not a scalability concern.

Table 2 recapitulates this comparison and shows that by having local controllers within sub-networks, our approaches resolves the scalability problem.

From Table 1 we note that packet number, required to initiate the secured
group with a key, is equal to   n= . Again,   = n/pkt _n, where pkt_n is the
packet number. For a xed value of   n and di erent values of   we draw the

Table 2.   Comparaison of Baal with theoric and pragmatic approaches

	Complexity	scalability problems
theoric approaches	O(n)	no
SKDC	O(n)	no
hierarchical approaches	O(log(n))	yes
Baal	O(1)	yes

graph (g. 3) representing   pkt_n/n vs.   . This graph shows the asymptotic relationship between packet number/group size and participation coe cient.

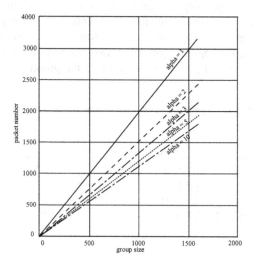

Fig. 2.   Transmitted packet vs. group size for di erent alpha

6   Conclusion

This paper has presented a new protocol for key management in dynamic multicast group,   Baal . It o ers a scalable solution to the scalability problem, user revocation, based on the delegation of key managers per sub-networks where group members exist. This protocol is independent of routing or unicast security protocols and assures decentralized group key management .

Comparative analysis have showed that, in group con guration step,   Baal performs less packet transmission, needs less data to con gure groups. It needs a unicast transmission size of (n=).k in order to distribute group keys (key size equals to k) and initiate secure groups. The previous hierarchical approaches

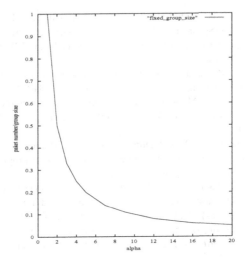

Fig. 3.   Packet number/group size vs.

change a $O(n)$ problem into a $O(\log(n))$ problem, but Baal changes this problem into $O(1)$. This improvement is due to the delegation of the group key
management to local controllers. In the simpli ed scheme, a local controller is
an IGMPv3-capable multicast router and bene ts from the new functionnality
of IGMPv3 [20], source ltering, which allows a local controller to multicast
the rekeying message to all group members except those within its sub-network.
Therefore, Baal changes $O(n)$ into $O(1)$, A controller needs only one multicast
message to distribute the new group key to members out of its sub-network and
unicast messages to distribute it to members within its sub-network.

We have also seen that as long as   , the coe cient of participation, is signi cant, the transmission size is small and consequently   Baal is e cient. In the
simpli ed scheme, the de nition of   is limited within a sub-network, so in our
further works, we envisage to expand it to cover a set of sub-networks,   i.e. to
have a local controller for many sub-networks, for example, for an autonomous
system or routage domains. We will use NS (Network Simulator) in order to
validate our propostion and compare it with another hierarchical approaches.

References

1. Canetti, R. and Pinkas, B. and Garay, J. and Micciancio, D. and Noar, M. and Itkis,
   G.: Multicast Security: A Taxonomy and E cient Authentication. Internet draft
   (work in progress): draft-canetti-secure-multicast-taxonomy- 00.txt (May 1998)
2. Mittra, S.: Iolus: A Framework for Scalable Secure Multicasting, in proceedings of
   ACM/SIGCOMM'97 , Cannes, FRANCE (September 1997)

3. Blundo, C. and Santis, A. and Herzberg, A. and Kutten, S. and Vaccaro, U. and Yung, M.: Perfectly-Secure Key Distribution for Dynamic Conferences, Advances in Cryptology: proceedings of Crypto92, E. F. Brickell, Ed., LNCS 740, Springer-Verlag (1992) 471–486

4. Fiat, A. and Noar, M.: Broadcast Encryption, Advances in Cryptology: proceedings of Crypto93, D. R. Stison, Ed. , LNCS 773, Springer-Verlag. 480–491

5. Wong, C. and Gouda, M. and Lam, S.: Secure Group Communications using Key Graphs, in proceedings of ACM SIGCOMM'98, Vancouver, British Columbia (Septembre 1998)

6. Hardjono, T. and Cain, B. and Doraswamy, N.: A Framework for Group Key Management for Multicast Security, Internet draft (work in progress): draft-ietf-ipsec-gkmframework-01.txt (August, 1999)

7. Balenson, D. and McGrew, D. and Sherman, A.: Key Management for Large Dynamic Groups: One-way Function Trees and Amortized Initialization, Internet draft (work in progress): draft-balenson-groupkeymgmt-oft-00.txt (February 1999)

8. Hardjono, T. and Cain, B. and Monga, I.: Intra-Domain Group Key Management Protocol, Internet draft (work in progress): draft-ietf-ipsec-intragkm-00.txt (August 1999)

9. Wallner, D. and Harder, E. and Agee, R.: Key Management for Multicast: Issues and Architecture, Internet draft (work in progress): draft-wallner-key-arch-01.txt, (September 1998)

10. Steiner, M. and Tsudik, G. and Wainder, M.: De e-Hellmen Key Distribution Extended to Group Communication, 3rd ACM conference on Computer and Communication Security, New Delhi India (March 1996)

11. Burmester, M. and Desmedt, Y.G.: E cient and Secure Conference-Key Distribution, Secure Protocol, M. Lomas, Ed. , LNCS 1189, Springer-Verlag (1997) 119–130

12. McGrew, David A. and Sherman, Alan T.: Key Establishment in Large Dynamic Groups using One-way Function Trees, TIS Labs at Network Associates, Inc. Gleenwood, Maryland (1998)

13. Atkinson, R. and Kent, S.: Security Architecture for the Internet Protocol, Request For Comments rfc-2401, Network Working Group (November 1998)

14. Ballardie, T.: Scalable Multicast Key Distribution, Request For Comments rfc-1949, Network Working Group (May 1996)

15. Harney, H. and Mucknhirn, C.: Group Key Management Protocol (GKMP) Spec-i cation, Request For Comments rfc-2093, Network Working Group (July 1997)

16. Harney, H. and Mucknhirn, C.: Group Key Management Protocol (GKMP) Architecture, Request For Comments rfc-2093, Network Working Group (July 1997)

17. Chaddoud, G. and Chrisment, I. and Scha , A.: Secure Multicaasting Survey, in proceedings of SEC2000, the 15th Internatioanl Conference on Information Security, Beijing China (August 2000)

18. Chaddoud, G. and Chrisment, I. and Scha , A.: Baal : Securisation des communications de groupes dynamiques, in proceedings of CFIP'2000 : Colloque Francphone sur l'Ingenierie des Protocoles, Toulouse France (October 2000)

19. Freier, A. and Karlton, P. and Kocher, P. : The SSL Protocol Version 3.0, ftp://ftp.netscape.com/pub/review/ssl-spec.tar.Z, March 1996.

20. Cain, B. and Deering, S. and Kouvelas I. and Thyagarajan, A.: Internet Group Management Protocol, Version 3, Internet draft (work in progress): draft-ietf-idmr-igmp-04.txt (June 2000)

21. Harkins, D. and Carrel, D.: The Internet Key Exchange (IKE), Request For Comments rfc-2104, Network Working Group (November 1998)

SVD-Based Approach to Transparent Embedding Data into Digital Images

Vladimir I. Gorodetski[1], Leonard J. Popyack[2], Vladimir Samoilov[1], and Victor A. Skormin[3]

[1] St. Petersburg Institute for Informatics and Automation, St. Petersburg, 199178, Russia
[2] AFRL/IF, 525 Brooks Rd., Rome, NY 13441-4505, USA
[3] Watson School, Binghamton University, Binghamton, NY 13902, USA
[1]{gor, samovl}@mail.iias.spb.su,
[2]Leonard.Popuack@rl.af.mil,
[3]vskormin@binghamton.edu

Abstract. A new approach to transparent embedding of data into digital images is proposed. It provides a high rate of the embedded data and is robust to common and some intentional distortions. The developed technique employs properties of the singular value decomposition (SVD) of a digital image. According to these properties each singular value (SV) specifies the luminance of the SVD image layer, whereas the respective pair of singular vectors specifies image geometry. Therefore slight variations of SVs cannot affect the visual perception of the cover image. The proposed approach is based on embedding a bit of data through slight modifications of SVs of a small block of the segmented covers. The approach is robust because it supposes to embed extra data into low bands of covers in a distributed way. The size of small blocks is used as an attribute to achieve a tradeoff between the embedded data rate and robustness. An advantage of the approach is that it is blind. Simulation has proved its robustness to JPEG up to 40%. The approach can be used both for hidden communication and watermarking.

1   Introduction

Transparent hiding data into digital images called "digital image steganography" (DIS) presents an effective way for secret communication, watermarking and other applications. Although DIS is a quite new field of research, development of the Internet digital media and practical needs stimulated its recent rapid development. Steganography by itself aims to conceal the very existence of the fact of communication. Combined with encryption, steganography provides a higher level of secrecy. A DIS technique is required to support a high rate of the hidden data, assuring invisibility of the hidden data, and robustness to common and some types of intentional distortions. While these requirements are conflicting, a rational tradeoff must be found for any particular application requirement [1].

A number of techniques for DIS and watermarking has been developed during the last five years. There is not one superior technique. Each technique has each own

V.I. Gorodetski et al. (Eds.): MMM-ACNS 2001, LNCS 2052, pp. 263–274, 2001.

264   V.I. Gorodetski et al.

merits and flaws and the overall evaluation of a technique depends on the application. The overviews of DIS techniques are given in [7], [12], [13], [14]. [18], and [24].

In general, the proposed techniques can be classified as follows:

ú Techniques that utilize a spatial domain. To insert data into an image, they use a selected subset of the image pixels[1] using a bit-wise approach ([3], [4], [6], [16], [17], [19], [25], [26], etc.).

ú Transform-based techniques, that operate with images represented by a finite set of orthogonal or bi-orthogonal functions called "basis functions" ([5], [15], [20], [21], [23], [28], [30], etc.). Examples are Discrete Cosine Transform, and Wavelet transform.

ú Fractal-based techniques that construct a "fractal code" of an image in a way that allows to encode both the cover and the hidden images ([22]).

It should be noticed that most developed techniques are aimed at solving the watermarking problem under the major requirement of robustness to a wide range of distortions, but not the high rate of embedded data. That is why many existing approaches do not pay noteworthy attention to the development of techniques capable of embedding high volumes of data transparently and robustly. Indeed, the only well-known approach that could be used to embed an image into the cover image is so-called LSBs approach. One of such techniques is proposed in [8]). It uses segmentation of the image to be embedded into blocks of size 8ô8, applying DCT transform to each block, quantizing and encoding its coefficients and embedding each coded block into one or into two LSBs of the cover image. Unfortunately, such and similar techniques cannot provide robustness and are highly sensitive to many common distortions, image format transformations and lossy compression like JPEG.

However, many military and industrial applications, such as hidden communication (HC) and hidden transmission of digital images call for transparent and robust embedding of high volumes of data. Examples are transmission of top-secret projects, industry secret, plans of covert operations [12], etc. An important aspect of HC is the necessity to support survivability of the transmitted information.

In this paper a new approach to transparent embedding of data into digital images is proposed. It can be classified as a "Transform-based" because it uses Singular Value Decomposition (SVD) transform of digital images. It presents a rational tradeoff between the embedded data rate and robustness to common and some intentional distortions. The developed technique uses properties of SVD of a digital image. According to these properties each singular value (SV) specifies the luminance (energy) of the SVD image layer, whereas the respective pair of singular vectors specifies the image geometry. Therefore, slight variations of SVs cannot affect the visual perception of the quality of the cover image. The proposed approach is based on the embedding of data through slight modifications of SVs of a small block of the segmented covers. The approach is potentially robust because it supposingly embeds extra data into low bands of a cover image in a distributed way. The size of small blocks is used as an attribute to achieve a tradeoff between the embedded data rate and robustness. An additional advantage of the approach is that it is blind.

The rest of the paper is organized as follows. In Section 2 the concept of the proposed approach to hiding data into digital images using SVD is explained. In Section 3 the developed techniques of data hiding is described and the results of a

---

[1] For example, using masking effect, or using a pseudo-random seeding.

simulation-based study, focused on the robustness issue as well as on embedded data rate, are outlined. The developed technique is illustrated by several examples. In conclusion a general assessment of the paper results is given.

2   Singular Value Decomposition of Digital Images

A digital image in bitmap format is specified by a môn matrix $A = \{a_{i,j}\}_{m,n}$. If an image is represented in RGB format then it is specified by three such matrices A_R, A_G and A_B.

An arbitrary matrix A of size môn can be represented by its SVD ([11]) in the form

$$A = X L Y^T = \sum_{i=1}^{i=r} l_i X_i Y_i^T \qquad (1)$$

where X, Y are orthogonal môm and nôn matrices respectively, $X_1, X_2, ..., X_m$ and $Y_1, Y_2, ..., Y_n$ are their columns, Λ is diagonal matrix with non-negative elements, and $r \pounds \min\{m, n\}$ is the rank of the matrix A. Diagonal terms $l_1, l_2, ..., l_r$ of matrix Λ are called singular values (SV) of the matrix A and r is the total number of non-zero singular values. Columns of the matrices X, Y are called left and right singular vectors of the matrix A respectively. Singular values $l_1, l_2, ..., l_r$ can be calculated as $l_i = \sqrt{\mu_i}$ i=1,2,...,r, where μ_i is an eigenvalue of the matrix AA^T, or $A^T A$. The left singular vector X_i, i = U is equal to the eigenvector of the matrix AA^T corresponding to μ_i. Similarly, the right singular vector Y_i, L= U is equal to the eigenvector of the matrix $A^T A$ that corresponds to its eigenvalue μ_i. If an image is given in RGB format then it is represented by three SVDs in the form (1).

Thus, SVD of an image decomposes the respective matrix into layers $l_1 X_1 Y_1^T$, $l_2 X_2 Y_2^T, ..., l_r X_r Y_r^T$. As a rule, SVs are enumerated in descending mode, i.e. if $l_i > l_j$ then i<j, and l_1 is the maximal SV.

SVD possesses several interesting properties ([11]; two of them are utilized below to achieve invisible and robust hiding of extra data in digital images.

The first property is that each SV specifies the luminance (energy) of the SVD image layer, whereas the respective pair of singular vectors specifies an image "geometry". It was discovered that slight variations of SVs do not affect visual perception of the quality of the cover image. This property is used to embed a bit of data through slight modifications of SVs of a small block of a segmented cover.

The second property is that without the loss of image quality an image could be represented by so-called Truncated SVD (TSVD), i.e. by the sum

$$A_s = \sum_{i=1}^{i=s} l_i X_i Y_i^T , \; s<<r. \qquad (2)$$

instead of sum (1) ([10]). In other words, a TSVD of an image can be used for its compressed representation. SVD-based image compression was proposed in [2]. Later a number of approaches that combine SVD image transform with other transforms was developed. However, the major attention was paid to the development of a lossyless SVD image compression and its combinations with other transforms. For example, to code images, in [29] SVD transform is combined with Vector Quantization approach, in [27] a combination of SVD and Karhunen-Loeve transform is used to develop a hybrid compression. In [9] SVD transform is combined with wavelet transform. In [10] a format for SVD-based lossy compression of digital images was proposed. This format provides the rate of compression close to 2 bit/pixel while preserving the appropriate quality of the restored image. This format is used below to solve the task of robust embedding a digital image in a cover image.

3   Novel Approach and Techniques for Hiding Data into Digital Images

The following are the techniques that were developed to utilize the first of the two aforementioned properties of SVD image representation.

3.1   Technique 1

In brief, the first proposed technique is as follows. A cover image represented in 24 bpp (RGB) format is segmented into blocks of size sôs[1] and SVDs for each such a block and for each matrix of Red, Green and Blue layers are computed. Each block of every color layer is used to embed a bit of data. In the Technique 1, a bit of data is embedded through a slight modification of the largest singular value of the block. The implemented and explored algorithm of modification is described below.

Let B(k,l) be a block, where k is the block number and l {Red, Green, Blue}. Let the largest SV of the block B(k,l) of size sôs be l_1^k. Let b be a bit of data to be embedded into this block. The embedding algorithm is as follows:

For each pair (k, l)
1. Choose the quantization step, d, of the largest singular value of the block. The value d may be different depending on the layer of color.[2]
2. Compute integer number S such that $l_1^k(l)$=Sôd+ d , d <S.
3. Embed bit b of data as follows:

   If S is the odd number then
       if b=1 then S is not changed

---

[1] The developed software implements a particular case of this technique when s=4, generally, this number could be extended. Note that an increase of k value results in the increased robustness of the technique and in the decreased volume of transparently embedded data.

[2] The value of d is selected on the basis of statistical exploration of correlation between distortion caused by JREG compression of various percentages and the probability of a bit recovery. This exploration must be made for each color. The respective results are given below in this section.

else
    if b=0 then S:=S+1.

If S is the even number then
    if b=1, then S:=S+1,
    else
    if b=0, then S is not changed.

4. Compute the modified singular value $\tilde{l}_1^k(l)$:

$$\tilde{l}_1^k(l)=d\hat{o}S+d/2.$$

5. Compute the matrix of the block having modified largest singular value:

$$\tilde{B}(k,l) = \tilde{l}_1^k(l)X_1(l)Y_1^T(l) + \sum_{i=2}^{4}l_i(l)X_i(l)Y_i^T(l)$$

6. Result: Matrix $\tilde{B}(k,l)$ containing embedded bit of data.

The major implementation concern of the above algorithm is how to choose the quantization step d. It is obvious that the increase of the value d leads both to more robust data hiding and to less transparency of the embedded data. To explore this dependency quantitatively, the respective simulation was performed for several cover images segmented into blocks of size 4ô4 pixels and several images to be embedded. The latter were segmented into small blocks of size 8ô8 and compressed using TSVD proposed in [10] and mentioned in Section 2. The results are given in Fig. 1a (for Red layer), Fig. 1.b (for Green layer) and Fig. 1.c (for Blue layer). The plots illustrate the correlation between the percentage of JPEG compression and percentage of the correctly recovered blocks of the embedded image. Based on this result the following

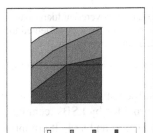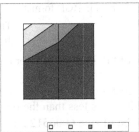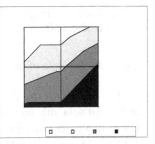

Fig. 1. The plots of dependencies between the step of quantization d of the largest singular value of a cover image and survivability of the embedded image after JPEG compression for red (left), green (center) and blue (right) color layers of the cover image. These results correspond to the case when cover image is segmented into blocks of size 4ô4

values of quantization steps within the Red, Green and Blue layers were chosen:[1]

$$d(red)=46, d(green)=22 \text{ and } d(blue)=52.$$

---

[1] Note that the appropriate choice is specific for every way of cover images segmentation.

It should be noticed that the quantization step value d might be used as a component of the secret key providing the restricted access to the hidden information. The extraction is very simple. Let $\widetilde{B}(k,l)$ be a block with an embedded bit of data.

For each pair (k, l) do

    1. Compute the largest singular value $\widetilde{l}_1^k(l)$.

    2. Compute $\widetilde{l}_1^k(l)/d=S+d/2$.

    3. If S is even number then the embedded bit value is 0 otherwise it is 1.

The simulation indicated that this way of embedding data into a cover image is robust against 100% JPEG compression.

Fig. 2.b. Image to be transmitted. It is gray and is of size 240ô120

Fig. 2.a. Cover image. It is presented in RGB format and is of size 600ô512 pixels

Fig. 2.c. Recovered hidden image. The stego-image was subjected to JPEG compression

Fig. 2. An example of the use of the Technique 1 for embedding image into cover one

This approach to data embedding provides a sufficiently high rate of the embedded data although its bit rate is less than the one provided by LSBs techniques. For example, let cover image be of size 600ô512 (see Fig. 2.a) and s=4. It comprises 150ô128 blocks of size 4ô4 in each color. Therefore this technique makes it possible to embed up to 57600 bits. The picture to be embedded (see Fig. 2.b) is of size 240ô120 and segmented into 200 blocks of size 12ô12. Due to TSVD compression (see Section 2), each such block is represented into blocks of 12ô12 using segmentation. Its length is 288 bits, that is why the total size of the image of Fig. 2b in TSVD format is equal to 56000 bits. Hence, it is possible to embed it into image depicted in Fig. 2.a.

This technique was subjected to several experimental studies. An example of such a study is given in Fig. 2. The cover image (Fig. 2a) containing embedded image (Fig. 2b) was distorted by JPEG compression and then transformed back in BMP format. The hidden image extracted from the JPEG distorted stego-image is depicted in Fig. 2.c. One can see that the quality of the reconstructed image is not excellent but is very satisfactory. Notice that the bit rate of the TSVD image to be embedded can be increased using a Haffman-like compression of the TSVD files for each block.

Embedding and recovery procedures could be equipped with a secret key to seed blocks thus providing additional level of security. To improve the survivability of the hidden data during transmission it is possible to transmit the same image several times using various covers. It should be noticed that if the image to be hidden is of very large size then its transmission could be implemented using several cover images.

3.2   Technique 2

This technique uses a different approach to embed data into a cover image. Notice that the type of embedded binary file is irrelevant. A bit of information is embedded into a block of the segmented cover image. The block size could be chosen arbitrarily.

Let cover image be represented in a 24 bit (RGB) format. Data can be embedded independently into each RGB layer of the cover image or, optionally, in specified layer(s). Let the size of blocks of the cover image segmentation be môk pixels, A be the matrix of a block of the covers corresponding to a color layer from Red, Green or Blue. Let a bit b to be embedded into block A of a layer.
The algorithm of the embedding procedure is as follows:

1. Compute singular value decomposition of matrix A. Let $V' =[l_1, l_2, ..., l_r]$ be the vector of singular values of matrix A ordered in decreasing mode, X_i and Y_i are singular vectors of matrix A and $i=1,2,...,r$, where r is the rank of matrix A.

2. Compute Euclidean norm of vector V', $Norm(V')=\sqrt{\sum_{i=1}^{r}(v_i')^2}$, where v_i', $i=1,2,...,r$, are the components of vector V'.

3. Select the value of Delta that is the step of quantization of the Euclidean norm of vector V'.

   Remark 1. The appropriate value of Delta depends on the color layer used for embedding a bit of data and has been chosen for each layer through simulation. Notice that the value of Delta can play the role of a component of the secret key restricting access to the hidden data. One more way to increase the secrecy is to use an uneven quantization of $Norm(V')$.

4. Compute the integer $N=[Norm(V')/Delta]$, where [*] is the integer part of the quotient of division.

5. Embed bit b according to the following algorithm:

If b=1 then
{if N is odd then $\tilde{N}$ =N+1 else $\tilde{N}$ =N}
else (if b=0)
{if N is even then $\tilde{N}$ =N else $\tilde{N}$ =N+1}}

6. Compute the modified norm of the vector of the singular values:

$$Norm(\tilde{V}')= \tilde{N} \, ô Delta+(Delta/2).$$

7. Compute the modified vector of the singular values:

$$\widetilde{V}' = V' \, ô(\text{Norm}(\widetilde{V}')/\text{Norm}(V')).$$

8. Compute the modified matrix of the block in which bit b is embedded:

$$\widetilde{A} = \sum_{i=1}^{r} \widetilde{l}_i X_i Y_i^{T}.$$

9. End of the embedding procedure.

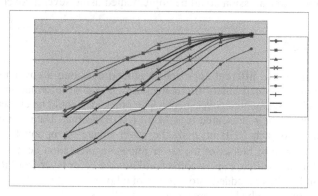

Fig. 3. Dependencies between degree of JPEG compression (horizontal) and probability of the watermark presence (vertical) for various value Delta. (Averaged over a set of images)

The described algorithms must be applied to each block of the covers in which a bit of data is to be embedded.

The embedding procedure may provide restricted access to the hidden data via using seeding of the binary string to be embedded in pseudo random mode or via transpositions of the lines and columns [10].

The extraction task is simpler then the embedding. Let $\widetilde{A}$ be the matrix of block containing hidden bit b of data, which must be extracted.

1. Compute the singular value decomposition of block $\widetilde{A}$. Let $\widetilde{V}'$ be vector of singular values ordered in decreasing fashion, X_i and Y_i are singular vectors of matrix $\widetilde{A}$ and i=1,2,...,r, where r is the rank of matrix $\widetilde{A}$.

   Remark 2. SVs and singular vectors of the same block $\widetilde{A}$ of the modified cover image can be additionally modified during transmission or changed intentionally. For simplicity, we denote them in the same way as they were denoted in the embedding procedure.

2. Compute the Euclidean norm of vector $\widetilde{V}'$, i.e. Norm $\widetilde{V}' = \sqrt{\sum_{i=1}^{r}(\widetilde{v}'_i)^2}$.

3. Compute $\widetilde{N}$ =[Norm$\widetilde{V}'$ /Delta], where [*] is the integer part of the quotient of division.

4. Compute the value of the hidden bit b:

$$\{\text{If } \widetilde{N} \text{ is even then b=1 else b=0}\}$$

5. End of extraction procedure.

The described embedding (extraction) algorithm must be applied to each block of the covers (stego-image), in which a bit of data is embedded. The extracting procedure may require the knowledge of the secret key if it has been applied in embedding.

This technique was investigated statistically from several points of view. The first task was to explore the optimality of Delta values in the trade-off between the robustness of the hidden data and its visibility. It was established that the appropriate value of Delta depends on the color layer. In Fig. 3 the results of the simulation-based investigation of the aforementioned dependencies are given. One can see that optimal choices of the of Delta given color layer are close to the followings:

Delta(Red)=40, Delta(Green)=24, Delta (Blue)=48.

Given such values of Delta, Technique 2 proves robustness to JPEG compression up to degree 40% provided that the value of the watermark presence probability is (0.7–0.8).

A special effort was made to study the relationship between the degree of JPEG

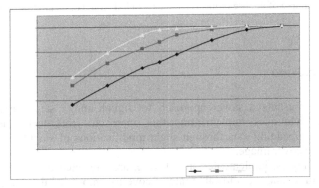

Fig. 4. Dependency between Degree of JPEG compression (horizontal axis) and Probability of the watermark presenc e for various degrees of redundancy

compression, redundancy of the embedded watermark and the probability of the watermark presence. In this study redundancy is understood as the number of embedded copies of the watermark. The results are displayed in Fig. 4. One can see that the redundancy presents an additional way to increase the robustness of Technique 2.

In Fig. 5 an example of using Technique 2 for embedding emf-file into a digital image is given. Left-hand image (Fig. 5a) corresponds to the cover image (in gray scale) with the embedded emf-file. The right-hand image depicts the cover image with extracted data that indicates, for example, to a pilot the route to follow that was hidden into transmitted image.

Fig. 5.a. Cover image with embedded emf-   Fig. 5.b. Extracted route drawn over the cover
file representing an aircraft route   image

Fig. 5. An example of the use of the Technique 2 to hide route of an aircraft into the map

4   Conclusion

This paper presents a novel approach to digital image steganography and two
alternative techniques implementing this approach. Both techniques utilize the
concept of embedding data through slight modifications of singular values of small
blocks of the cover image.

The techniques, implemented in software, are subjected to statistical analysis of
their robustness and the allowable rate of covertly embedded data. In particular, it is
shown the developed approach for data hiding is robust to JPEG compression up to
40%. It is blind, i.e. extraction of hidden data does not require the presence of the
original image. Embedding and recovery procedures can be equipped with a password
and secret key to seed blocks thus providing restricted assesses to the hidden data.
There exist several additional attributes resulting in additional security assurance.

The analyses indicate that the developed techniques are suitable for hidden
communication that calls for high rate and appropriate survivability of the embedded
data, and invisibility of the hidden image, and for watermarking where the main
requirement is robustness to common and some intentional distortions.

The approach is demonstrated by several examples of practical applications.

Acknowledgment. The authors are grateful to the Air Force Research Laboratory at
Rome NY for funding this research.

References

1. Anderson, R., Petitcolas, F.A.P.: On the Limits of Steganography. In: IEEE Journal of Selected Areas of Communications, Vol. 16(4) (1998) 474-481
2. Andrews, H.C., Patterson, C.L.: Singular Value Decomposition (SVD) for Image Coding. In: IEEE Transaction on Communication. Vol. 24 (1976) 425-432
3. Bender, W., Gruhl, D., Morimoto, N., and Lu, A.: Techniques for Data Hiding. In: IBM System Journal, Vol. 35 (3&4) (1996)
4. Bruyndonckx, O., Quisquater, J.J., and Macq, B.: Spatial Method for Copyright Labeling of Digital Images. In: Proceedings of IEEE Workshop on nonlinear signal and image Processing, Greece (1995)
5. Burget, S., Koch, E., and Zhao, J.: A Novel Method for Copyright Labeling Digitized Image Data. Technical Report, Fraunhofer Institute for Computer Graphics, Germany (1994)
6. Chen B., and Wornell, G.W.: Dither Modulation: A new Approach to Digital Watermarking and Information Embedding. In: Ping Wah Wong and E.J.Delp (eds). Vol. 3657, San Jose, CA, USA (1999)
7. Cox, I.J. and Miller, M.: A Review on Watermarking and the Importance of Perceptual Modeling. In: Proceedings of the Conference on Electronic Imaging (1997)
8. Fridrich, J., Goljan, M.: Protection of Digital Images using self-embedding. In: Proceedings of the Second International Scientific Conference in the Republic of Kaazakhstan "Information Technologies and Control", Almaty, Kazakhstan (1999) 302-311
9. Fukutomi, T., Tahara, O., Okamoto, N., Minami, T.: Encoding of Still Pictures by a Wavelet Transform and Singular Value Decomposition. In: Proceedings of the 1999 IEEE Canadian Conference on Electrical and Computer Engineering. Alberta (1999) 18-23
10. Gorodetski, V., Skormin, V., Popyack, L.: Singular Value Decomposition Approach to Digi-tal Image Lossy Compression. In: Proceedings of the 4-th World Conference "Systems, Cybernetics and Informatics-2000" (SCI-2000), Orlando, USA (2000)
11. Horn, R.A. and Johnson, C.R.: Matrix Analysis. Cambridge University Press (1988)
12. Johnson, N., Jagodia, S.: Exploring Steganography: Seeing the Unseen. Computer, February (1998) 26-34
13. Johnson, N., Duric, Z., Jajodia, S.: Information Hiding. Steganography and Watermarking–Attacks and Countermeasures. Kluwer Academic Pub. (2000)
14. Katzenbeisser, S., Petitcolas F.A.P. (eds): Information Hiding Techniques for Staganography and Digital Watermarking. Artech House Books (2000)
15. Kundur, D. and Hatzinakos, D.: A robust Digital Watermarking Method using Wavelet-based Fusion. In: Proceedings of the International Conference on Image Processing, Vol. 1, USA, IEEE (1997)
16. Machado, R.: EZ Stego. Http://www.stego.com.
17. Matsui, K. and Tanaka, K.: Video Steganography: How to Embed a Signature in a Picture. In: Proceedings of IMA Intellectual property, Vol. 1(1) (1999) 187-206.
18. Petitcolas, F.A.P., Anderson, R.J. and Kuhn, M.: Information Hiding–A Survey. In: Proceedings of the IEEE, Special Issue on Protection of Multimedia Content, Vol. 87(7) (1999) 1062-1078
19. Pitas, I.: A method for signature casting on digital images. In: Proceedings of the International Conference on Image Processing (ICIP'96) (1996)
20. Piva, A, Barni, .M., Bartoloni, E., and Cappellini, V.: DCT-based Watermarking Recovering without Restoring to the Uncorrupted Original Image In: Proceedings of the International Conference on Image Processing (ICIP), Vol. 1, IEEE (1997)
21. Podilchuck, C.I. and Zeng, W.: Perceptual Watermarking of Still Images. In: Proceedings of the Workshop on Multimedia Signal Processing, Princeton, NJ, USA (1997)

22. Puate J. and Jordan F.: Using Fractal Compression Scheme to Embed a Digital Signature into an Image. In: Proceedings of SPIE, Video Techniques and Software for Full-Service Network, Vol. 2915, Boston, MA, USA (1996) 108-118

23. Smith, J.R. and Comiskey, B.O.: Modulation and Information Hiding in Images. In: Lecture Notes in Computer Science; Vol. 1174, Springer Verlag (1996) 207-226

24. Swanson, M.D., Kobayashi, M. and Tewfic, A.H.: Multimedia Data Embedding and Watermarking Technologies. In: Proceedings of the IEEE, Vol. 86(6) (1998) 1064-1087

25. Tanaka, K., Nakamura, Y. and Mitsui, K.: Embedding the Attribute Information into a Dithered Image. In: Systems and Computers in Japan, 21(7) (1990)

26. van Schydel, R., Tirkel, A. and Osborne, C.: A digital Watermarking. In: Proceedings of ICASSP, Piscataway, NJ, IEEE, Vol. II (1994) 86-90

27. Waldemar, P., Ramstad T.: Hibrid KLT-SVD Image Compression. In: Proceedings IEEE International Conference on Acoustics, Speech, and Signal Processing, Germany, IEEE Computer Society Press, Vol. 4 (1997) 2713-2716

28. Xia, X.G., Boncelet, C.G. and Arce, G.R.: A Multi-resolution Watermark for Digital Images. In: Proceedings of International Conference on Image Processing (ICIP), Vol.1, USA, IEEE (1997)

29. Yang, J-F., Lu ,C-L.: Combined Techniques of Singular Value Decomposition and Vector Quantization for Image Coding. In: IEEE Transaction on Image Processing, Vol. 4 (8) (1995) 1141-1146

30. Zhu, B., Tewfic, A.H. and Gerec, O.: Low Bit Rate Near Transparent Image Coding. In: Proceedings of International Conference on Wavelet Applications for Dual Use, Vol. 2491, Orlando, FL (1995) 173-184

Fast Encryption Algorithm Spectr-H64

Nick D. Goots, Alexander A. Moldovyan, and Nick A. Moldovyan

Specialized Center of Program Systems "SPECTR",
Kantemirovskaya str., 10, St. Petersburg 197342, Russia
ph./fax.7-812-2453743, spectr@vicom.ru

Abstract. This paper describes a fast hardware-oriented 64-bit block cipher SPECTR-H64 based on combination of the data-dependent permutations and data-dependent transformation of subkeys.

Introduction

Design of ciphers based on data-dependent operations appears to be a new and perspective direction in applied cryptography. Using data-dependent rotations R. Rivest designed extremely simple cryptosystem RC5 [1]. Different studies [2-3] have provided a good understanding of how RC5's structure and data-dependent rotations contribute to its security. It have been shown that the mixed use of data-dependent rotations and some other simple operations is a very effective way of thwarting differential and linear cryptanalysis. Since several studies provide some theoretical attacks based on the fact that few bits in a register define selection of concrete modification of the current rotation operation, in AES candidates MARS [4] and RC6 [5] data-dependent rotations are combined with integer multiplication.

In the case of n-bit registers containing encrypted data subblocks there are available n different modifications of the rotation operation and consequently only $\log_2 n$ bits can be used directly as controlling ones. It is very attractive to use data-dependent permutations (DDP) for increasing the number of the controlling bits [6]. DDP are easy to be performed with permutation networks (PN) developed previously [7,8]. A PN can be used in the cryptosystems design as an operational block performing so called controlled permutations (CP). CP in the form of key-dependent permutations are used in cipher ICE [9], such use of CP has been shown [10] to be not very effective against differential cryptanalysis though. Different variants of the use of CP are proposed in patents [6,11]. In [6] it is proposed to encrypt data with DDP combining them with modulo 2^{32} addition and bitwise exclusive-OR (XOR) operations ($^-$). In [11] it is proposed to construct a block cipher based on data-dependent transformation of the round subkeys.

In present paper we consider a fast hardware-oriented 64-bit block cipher based on DDP performed on both the data and the round subkeys. Concrete type of the CP-box permutation $P_{n/m}$ is characterized by an ordered set $\{P_0, P_1, \ldots, P_{2^m-1}\}$, where all P_i, i = $0, 1, \ldots, 2^m-1$, are fixed permutations and each number i can be represented as a m-bit control vector $V(v_1, \ldots, v_m)$ such that $i = v_1 + 2v_2 + \ldots + 2^{m-1}v_m$ is hold. Therefore instead of P_i we can use notation P_V. Such fixed permutations we shall call the CP-

V.I. Gorodetski et al. (Eds.): MMM-ACNS 2001, LNCS 2052, pp. 275–286, 2001.
© Springer-Verlag Berlin Heidelberg 2001

modifications. Thus, the notation $Y = P_{n/m}(X,V)$ means $Y = P_V(X) = P_i(X)$. To construct CP boxes we use the approach proposed in [8] which is based on standard elementary CP-boxes $P_{2/1}$ (Fig. 6a). Each $P_{2/1}$-box is controlled by one bit v: $y_1 = x_{1+v}$ and $y_2 = x_{2-v}$.

CP-boxes can be constructed on the basis of the simple layered structure shown in Fig. 5-7. Two CP-boxes $P_{n/m}$ and $P^{-1}_{n/m}$ we shall call mutually inverse, if for all fixed values of the vector V the respective CP-modifications P_V and P^{-1}_V are mutually inverted.

1   Design of the Block Cipher SPECTR-H64

Our design strategy was oriented to the following objectives:
- Cryptoscheme should be iterative in structure.
- Cryptoscheme should be based on fast operations (CP-box permutations, XOR, fixed permutations, special fast nonlinear functions).
- Key scheduling should be very simple in order to provide high encryption speed in the case of frequent change of keys.

1.1 Notations

Let $\{0,1\}^n$ denotes the set of the binary vectors of length n, then $X \in \{0,1\}^n$ means $X = (x_1,...,x_n)$, where $x_i \in \{0,1\}$ and $i \in \{1,...,n\}$. Let x_1 is the least significant bit and x_n is the most significant bit. Let denote $X_{lo} = (x_1,...,x_{n/2})$ and $X_{hi} = (x_{n/2+1},...,x_n)$, then $X_{lo}, X_{hi} \in \{0,1\}^{n/2}$ and $X = (X_{lo}, X_{hi})$ (or $X = X_{lo}|X_{hi}$, where "|" denotes concatenation operation). Let $X, Y, A \in \{0,1\}^n$, $c \in \{0,1\}$. Let $X \wedge A$ denotes bitwise-AND operation ("$\wedge$"), and let denote cyclic rotation ">>>" as follows:

$$< = ; \qquad >>> \qquad \begin{array}{l} \backslash = [+ \\ \backslash = [+ . \end{array} \qquad \begin{array}{l} L= \qquad Q- N \\ L= Q- N+ \end{array} \qquad , \text{ where } 0 \pounds k < n.$$

1.2 General Encryption Scheme

General encryption scheme is defined by the following formulas: $C = E(M,K)$ and $M = D(C,K)$, where M is the plaintext, C is the ciphertext $(M,C \in \{0,1\}^{64})$, K is the secret key $(K \in \{0,1\}^{256})$, E is encryption function, and D is decryption function.

In the block cipher SPECTR-H64 encryption and decryption functions coincide $(E = D = F)$. Thus, transformation algorithm is defined by formula:

$$Y = F(X, Q^{(e)}),$$

where $Q^{(e)} = H(K,e)$ is the extended key (EK), the last being a function of the secret key $K = (K_1,...,K_8)$ and of the transformation mode parameter e (e = 0 defines encryption, e = 1 defines decryption), the input data block being $X = \begin{array}{l} 0 \quad H= \\ \& \quad H= \end{array}$.

In 12-round block cipher EK is represented as concatenation of 14 subkeys:

$$Q^{(e)} = (Q^{(e)}_{IT}, Q^{(e)}_1, \ldots, Q^{(e)}_{12}, Q^{(e)}_{FT}),$$

where $Q^{(e)}_{IT}, Q^{(e)}_{FT} \in \{0,1\}^{32}$ and $\forall\ j = 1, \ldots, 12\ Q^{(e)}_j = (Q^{(1,e)}_j, \ldots, Q^{(6,e)}_j)$, where $\forall\ h = 1, \ldots, 6$ $Q^{(h,e)}_{j} \in \{0,1\}^{32}$.

Thus, EK is a sequence of 74 32-bit binary vectors. Output value Y is the ciphertext C in the encryption mode or is a plaintext M in the decryption mode.

Description of the encryption algorithm (function F) is given in section 1.3 and procedure of the formation of EK $Q^{(e)} = H(K,e)$ is given in section 1.4.

1.3 Encryption Algorithm

The algorithm (function F) is designed as sequence of the following procedures: initial transformation IT, 12 rounds with procedure Crypt, and final transformation FT.

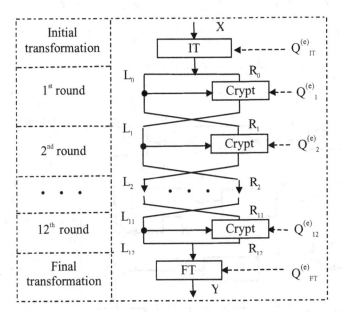

Fig. 1. General encryption scheme

Ciphering (Fig. 1) begins with the procedure IT.

$$Y = IT(X, Q^{(e)}_{IT}).$$

Then data block Y is divided into two blocks L_0 and R_0, i.e. $(L_0, R_0) = Y$, where $L_0, R_0 \in \{0,1\}^{32}$. After that 12 sequential rounds are performed with procedures Crypt in accordance with the formulas:

$$L_j = Crypt\,(R_{j-1}, L_{j-1}, Q^{(e)}_j); \qquad R_j = L_{j-1}, \qquad \text{where } j = 1, \ldots, 12.$$

The final transformation FT is executed after the 12-th round:

$$Y = FT(X, Q^{(e)}_{FT}),$$

where $X = (R_{12}, L_{12})$ is the output of 12-th round.

Initial transformation   IT. It has the following form: $Y = IT(X,A)$, where X,Y ⌐ $\{0,1\}^{64}$, A ⌐ $\{0,1\}^{32}$. Fig.3a shows the scheme of this transformation which defines transposition of each pair of bits $x_{2j-1}x_{2j}$ ($j = 1,...,32$) of X if $a_j = 1$, otherwise ($a_j = 0$) the bits are not transposed. After that each even bit is inverted.

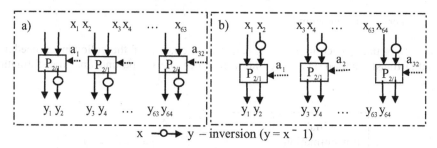

$$x \;\; —\!\!\!\!○\!\!\!\!→\;\; y \;-\; \text{inversion} \; (y = x^- 1)$$

Fig. 2. Initial transformation IT (a) and final transformation FT (b)

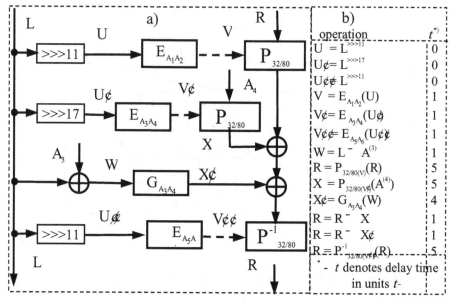

Fig. 3. Procedure Crypt (a) and time delays of individual operations (b)

Final transformation   FT. Procedure FT is inverse of the procedure IT. Transformation FT has the following form: $Y = FT(X,A)$, where X,Y $\{0,1\}^{64}$, A $\{0,1\}^{32}$. Initially (Fig. 3b) each even bit of the input block is inverted and then each pair of bits with indices 2j-1 and 2j (j = 1,...,32) is transposed, if a_j = 1, otherwise (a_j = 0) the bits are not transposed.

Procedure Crypt. This procedure has the form: $R = \text{Crypt}(R,L,A_1,A_2,...,A_6)$, where $L,R,A_1,A_2,...,A_6 \in \{0,1\}^{32}$. Thus, Crypt transforms data subblock R under control of the data subblock L and 192-bit subkey. It uses the following operations: cyclic rotation ">>>" by fixed amount, XOR operation "⁻", non-linear function G, data-dependent permutations $P_{32/80}$ and $P^{-1}_{32/80}$, and extension operation E over control vector. The sequence of the performed transformations is represented in Fig. 4.

Non-linear function G.   Justification of the function G is given in section 2. Formally this transformation can be represented as follows: $X\phi = G(W,A,B)$, where $X\phi,W,A,B \in \{0,1\}^{32}$. Realization of the function G is defined by the following expression:

$$X\phi = M_0 \bar{} \ M_1 \bar{} \ (M_2 \tilde{} \ A) \bar{} \ (M_2 \tilde{} \ M_5 \tilde{} \ B) \bar{} \ (M_3 \tilde{} \ M_5) \bar{} \ (M_4 \tilde{} \ B),$$

where binary vectors $M_0, M_1,..., M_5$ are expressed recursively through W as follows:

$$M_0 = (m_1^{(0)},m_2^{(0)},...,m_{32}^{(0)}) = (w_1, w_2,...,w_{32}) \quad \text{and} \quad "j = 1,...,5$$

$$M_j = (m_1^{(j)},m_2^{(j)},m_3^{(j)},...,m_{32}^{(j)}) = (1,m_1^{(j-1)},m_2^{(j-1)},...,m_{31}^{(j-1)}).$$

Taking into account that some operations in function G are performed in parallel it is easy to see that delay time of this function is about $4t\text{-}$.

Controlled permutation boxes P $_{32/80}$ and P $^{-1}_{32/80}$. CP-box transformation is represented in the following form: $Y = P_{32/80}(X,V)$, where $X, Y \in \{0,1\}^{32}$, $V \in \{0,1\}^{80}$, X is the input, V is the control vector. In necessary cases we use also the notation $P_{32/80(V)}(X)$. Detailed mathematical representation depends on concrete realization of the CP box $P_{32/80}$. In general, each output bit y_j (j = 1,...,32) is a Boolean function of the variables $\{x_i\}$ and $\{v_i\}$, i.e. $y_j = g_j(x_1,...,x_{32}; v_1,...,v_{80})$.

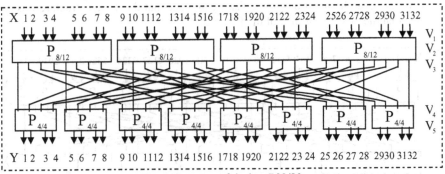

Fig. 4. Structure of the box P32/80

Construction scheme of the box $P_{32/80}$ is shown in Fig. 5. Box $P_{32/80}$ has 32-bit input, 32-bit output, and 80-bit control input. This CP box consists of five layers of 16 parallel elementary CP boxes $P_{2/1}$ with some fixed connection between layers. The first three layers are structurally combined in four boxes $P_{8/12}$ and the last two are combined in eight boxes $P_{4/4}$. Design schemes of the boxes $P_{2/1}$, $P_{4/4}$, and $P_{8/12}$ are shown in Fig. 6.

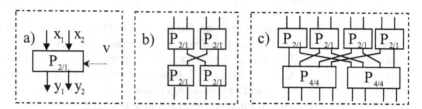

Fig. 5. Structure of the boxes P2/1 (a), P4/4 (b), and P8/12 (c)

In the box $P_{32/80}$ all elementary CP boxes are numbered consequently from left to right and from top to bottom. In accordance with such enumeration the j-th bit of vector V controls the elementary box $P_{2/1}^{(j)}$ and vector V can be represented as concatenation of the five vectors $V_1, V_2, V_3, V_4, V_5 \in \{0,1\}^{16}$, i.e. $V = (V_1|V_2|V_3|V_4|V_5)$. Two lower bits x_1 and x_2 of the transformed data subblock are the input of the first box $P_{2/1}^{(1)}$. Respectively, two higher bits x_{31} and x_{32} of the transformed data subblock are the input of the 16-th elementary box $P_{2/1}^{(16)}$. Respectively, the bits y_1 and y_2 of the value Y correspond to the box $P_{2/1}^{(65)}$. Bits y_{31} and y_{32} correspond to the box $P_{2/1}^{(80)}$.

Interconnection between the third and the fourth layers of the box $P_{32/80}$ has the form of the following involution:

$$I = (1)(2,9)(3,17)(4,25)(5)(6,13)(7,21)(8,29)(10)(11,18)(12,26)$$

$$(14)(15,22)(16,30)(19)(20,27)(23)(24,31)(28)(32)$$

One can remark that interconnection between boxes $P_{8/12}$ and $P_{4/4}$ is designed in accordance with the "each-to-each" principal, i.e. the i-th output digit of the j-th box $P_{8/12}$ is connected with the j-th input digit of the i-th box $P_{4/4}$. It is easy to see that boxes $P_{2/1}$, $P_{4/4}$, $P_{8/12}$ are the CP boxes of the first order. The CP box $P_{n/m}$ of the order h (0£ h £n) is a CP box which has at least one CP-modification P_V moving arbitrary given h bits in the given h output positions. Taking into account the design structure it is easy to see that the box $P_{32/80}$ is also of the first order.

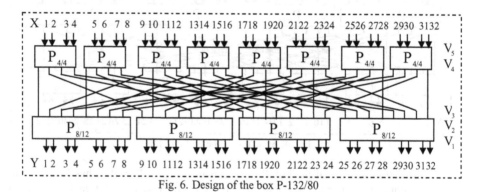

Fig. 6. Design of the box P-132/80

Formal representation of the inverse transformation $P^{-1}_{32/80}$ has the following form:

$$Y = P^{-1}_{32/80}(X,V), \qquad \text{where } X, Y \in \{0,1\}^{32}, V \in \{0,1\}^{80}.$$

Box $P^{-1}_{32/80}$ (Fig. 7) can be constructed analogously to the design of the box $P_{32/80}$ with exception that input and output are swapped.

If for CP-box $P_{32/80}$ enumeration of the elementary boxes $P_{2/1}$ is given from left to right and from top to bottom for $P^{-1}_{32/80}$ enumeration is given from left to right from bottom to top, then the same control vector $V=V_1|V_2|V_3|V_4|V_5$ defines a pair of respective inverse modifications $P_{32/80(V)}$ and $P^{-1}_{32/80(V)}$. Such design of the box $P^{-1}_{32/80}$ allows one to use the same extension box E to form control vector for both the box $P_{32/80}$ and the box $P^{-1}_{32/80}$. Structure of the CP boxes $P_{32/80}$ and $P^{-1}_{32/80}$ provides that superposition $P_{32/160} = P_{32/80} * P^{-1}_{32/80}$ is a CP box of the maximal order [12].

Extension box E. The extension box E is used to form an 80-bit control vector the given 32-bit value. Formal representation of the extension transformation is:

$$V = (V_1|V_2|V_3|V_4|V_5) = E(U,A,B) = E_{A|B}(U),$$

where $V \in \{0,1\}^{80}$, $V_1,V_2,V_3,V_4,V_5 \in \{0,1\}^{16}$, $U,A,B \in \{0,1\}^{32}$.

Actually the vectors V_1,V_2,V_3,V_4,V_5 are determined accordingly to the formulas:

$$V_1 = U_{hi}; \; V_2 = p((U^- A)_{hi}); \; V_5 = p((U^- A)_{lo}), \; V_3 = p\phi((U^- B)_{hi}); V_4 = p\phi((U^- B)_{lo}),$$

where fixed permutations p and $p\phi$ are the following:

$$p(Z) = Z_{hi}^{>>>1}|Z_{lo}^{>>>1} \qquad \text{and} \; p\phi(Z) = Z_{hi}^{>>>5}|Z_{lo}^{>>>5}.$$

Correspondence between control bits of both the vector V and the vector U for box $P_{32/80}$ and for $P^{-1}_{32/80}$ are given in Tables 2.

Table 1.

E																
V_1	17	18	19	20	21	22	23	24	25	26	27	28	29	30	31	32	U_{hi}
V_2	26¢	27¢	28¢	29¢	30¢	31¢	32¢	25¢	18¢	19¢	20¢	21¢	22¢	23¢	24¢	17¢	$(U^- A)_{hi}$
V_3	30¢	31¢	32¢	25¢	26¢	27¢	28¢	29¢	22¢	23¢	24¢	17¢	18¢	19¢	20¢	21¢	$(U^- B)_{hi}$
V_4	14¢	15¢	16¢	9¢	10¢	11¢	12¢	13¢	6¢	7¢	8¢	1¢	2¢	3¢	4¢	5¢	$(U^- B)_{lo}$
V_5	10¢	11¢	12¢	13¢	14¢	15¢	16¢	9¢	2¢	3¢	4¢	5¢	6¢	7¢	8¢	1¢	$(U^- A)_{lo}$

For example, in Table 2 the number 19 means that bit u_{19} controls the box $P^{(3)}_{2/1}$ (i.e. u_{19} is used as control bit v_3), 22¢ corresponds to value $u_{22}^- a_{22}$ used as control bit v_{29}, and 7¢ corresponds to value $u_7^- b_7$ used as control bit v_{58}.

Extension box is designed in such a way that lower 16 bits of the vector U take part in controlling the boxes $P_{4/4}$ only and the higher 16 bits control the boxes $P_{8/12}$ only. This provides the uniformity of the use of each bit of vector U in controlling the box $P_{32/80}$. This also provides avoiding repetition of arbitrary bit of U to be used twice in some set of five control bits defining the path of each input bit to each output position.

1.4 Procedure of the Formation of the Extended Key

Extended encryption key represents $Q^{(e)} = H(K,e)$ some sequence of 32-bit subkeys K_i $(i = 1,...,8)$, in which parameter e defines encryption (e = 0) or decryption (e = 1) mode. Sequences $Q^{(0)} = H(K,0)$ and $Q^{(1)} = H(K,1)$ are different in general, but partially they coincide. Respective modification of the extended key is performed with single-layer box $P_{256/1(e)}$ which is represented by four parallel boxes $P_{2\times32/1(e)}$. Four couples of subkeys K_i (Fig. 2), i.e. (K_1,K_2), (K_3,K_4), (K_5,K_6), and (K_7,K_8) are inputs of the corresponding boxes $P_{2\times32/1(e)}$. Each box $P_{2\times32/1(e)}$ represent 32 parallel elementary boxes $P_{2/1(e)}$.

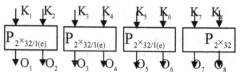

Fig. 7. Switching of subkeys

In Fig. 2 eight 32-bit outputs of four boxes $P_{2\times32/1(e)}$ are denoted as O_j (j = 1,...,8): $O_{2j-1}=K_{2j-1+e}$ and $O_{2j}=K_{2j-e}$ (j = 1,...,4). Extended key is specified in Table 1.

Table 2.

$Q^{(e)}_{IT}$	j =	1	2	3	4	5	6	7	8	9	10	11	12	$Q^{(e)}_{FT}$
O_1	$Q^{(1,e)}_j=$	K_1	K_8	K_5	K_4	K_1	K_6	K_7	K_4	K_2	K_6	K_5	K_3	O_2
	$Q^{(2,e)}_j=$	K_2	K_6	K_7	K_3	K_5	K_8	K_5	K_3	K_1	K_8	K_7	K_4	
	$Q^{(3,e)}_j=$	O_6	O_1	O_2	O_5	O_3	O_4	O_8	O_6	O_1	O_2	O_5		
	$Q^{(4,e)}_j=$	O_7	O_4	O_3	O_8	O_6	O_1	O_2	O_5	O_7	O_4	O_3	O_8	
	$Q^{(5,e)}_j=$	K_3	K_5	K_6	K_2	K_4	K_7	K_6	K_1	K_4	K_5	K_8	K_1	
	$Q^{(6,e)}_j=$	K_4	K_7	K_8	K_1	K_3	K_5	K_8	K_2	K_3	K_6	K_6	K_2	

Let r denotes the number of encryption rounds. Then $Q^{(0)} = (Q^{(0)}_{IT}, Q^{(0)}_1,...,Q^{(0)}_r, Q^{(0)}_{FT})$ is the encryption key, where $Q^{(0)}_j = (Q^{(1,0)}_j,...,Q^{(6,0)}_j)$ is j-th extended round subkey. Let $Q^{(1)} = (Q^{(1)}_{IT}, Q^{(1)}_1,...,Q^{(1)}_r, Q^{(1)}_{FT})$ is the decryption key. To provide possibility to perform encryption and decryption with the same algorithm the following conditions must hold:

$$Q^{(1)}_{IT}= Q^{(0)}_{FT}; \qquad Q^{(1)}_{FT} = Q^{(0)}_{IT}; \qquad \text{and} \qquad "j = 1,...,r$$

$$Q^{(1,1)}_j= Q^{(5,0)}_{r+1-j}; \qquad Q^{(5,1)}_j = Q^{(1,0)}_{r+1-j};$$

$$Q^{(2,1)}_j= Q^{(6,0)}_{r+1-j}; \qquad Q^{(6,1)}_j = Q^{(2,0)}_{r+1-j};$$

$$Q^{(3,1)}_j= Q^{(3,0)}_{r+1-j}; \qquad Q^{(4,1)}_j = Q^{(4,0)}_{r+1-j}.$$

2   Justification of the Function G

Let define mapping $G: \{0,1\}^{32} \times \{0,1\}^{32} \times \{0,1\}^{32} \to \{0,1\}^{32}$ as a vector Boolean function $Y = G(X,A,B)$ in the following form: $(y_1,\ldots,y_{32}) = (f_1(X,A,B),\ldots,f_{32}(X,A,B))$, where f_i are some generating Boolean functions. If $\forall$ i equation $f_i = f_i(x_1,\ldots,x_i,A,B)$ is hold, then such mapping is a sequential model of the controlled substitutional operation. If the functions $f_i = x_i \oplus f_i(x_1,\ldots,x_{i-1},A,B)$ are used as generating functions, then the vector function G performs bijective mapping over X [13]. To simplified general scheme one can consider a single unificated function of fixed number of variables as a prototype of generating functions. In SPECTR-H64 the following unificated function is used:

$$f_i = f(z_1,\ldots,z_8) = f'(z_1,\ldots,z_7) \oplus z_8 = j(z_1,\ldots,z_6) \oplus z_7 \oplus z_8,$$

where $j(z_1,\ldots,z_6) = z_1 z_2 z_3 \oplus z_2 z_4 \oplus z_2 z_5 \oplus z_3 z_6$ is a bent function [14].

Discrete Fourier transformation of the function f [14] consists of 64 elements taking on the value $|U_w| = 32$ for all vectors $w = (z_1,\ldots,z_6,0,0,)$ and of 192 elements having value $U_w = 0$. Taking into account these facts as well as other special properties of the bent functions we have the following characteristic of the function f:

-    non-linearity of the function f is $N(f) = 2^{n-1} - 2^{(n+2)/2-1} = 112$; this value is sufficiently close to maximally possible non-linearity for Boolean functions of 8 variables ($N_{max} = 2^{n-1} - 2^{n/2-1} = 120$);
-    f is correlationally immune function relatively all vectors $(z_1,\ldots,z_8)$, where $(z_7 \oplus z_8) \neq 0$. This means that arbitrary particular function obtained by fixing arbitrary seven or less variables (except simultaneous fixing z_7 and z_8) is a balanced one;
-    f has good self-correlation properties (avalanche effect), since for all non-zero avalanche vectors $(Dz_1,\ldots,Dz_8) \in \{0,1\}^8$, except $(0,\ldots,0,0,1)$, $(0,\ldots,0,1,0)$, and $(0,\ldots,0,1,1)$, propagation criterion of the degree 8 is fulfilled, i.e. the function $Df = f(z_1,\ldots,z_8) \oplus f(z_1 D z_1,\ldots,z_8 D z_8)$ is balanced one;
-    the degree of the algebraic normal form of function f equals 3 ($\deg(f) = 3$).

From general form of the function f one can obtain concrete generating functions of the sequential model of the substitutional operation, using the following substitution of variables:

$$
\begin{array}{cccccccc}
] &] &] &] &] &] &] &] \\
\underset{L}{[} & [\, . & [\, . & [\, . & E & D & [\, . & [\, . \}
\end{array}
$$

$\forall i \in \{1,\ldots,n\}$, where n = 32. Thus, we have the following generating functions:

$$y_i = f_i = x_i \oplus x_{i-1} \oplus x_{i-2} a_i \oplus x_{i-2} x_{i-5} b_i \oplus x_{i-3} x_{i-5} \oplus x_{i-4} b_i,$$

where x_j, a_j, b_j are components of the vectors $X,A,B \in \{0,1\}^{32}$ (i = 1,\ldots,32), respectively, and $(x_{-4},x_{-3},x_{-2},x_{-1},x_0) = (1,1,1,1,1)$ is the initial condition. The function $G \neq G(L \oplus A)$ is bijective over L for arbitrary fixed values of subkeys A,B.

3   Statistic Examination and Discussion

Investigation of statistic properties of SPECTR-H64 has been carried out with standard tests, which have been used in [15] for five AES finalists. The following dependence criteria have been used: 1) the average number of output bits changed when changing 1input bit; 2) the degree of completeness; 3) the degree of the avalanche effect; 4) the degree of strict avalanche criterion.

In the criteria we have used the dependence matrix a_{ij} and the distance matrix b_{ij} are defined as follows:

$$D = {}^{\wedge}; \; \llcorner X . \; \llcorner K \_) ; \quad . \quad ,) ; \quad . \quad {}^{\backprime}$$

for i = 1,...,n and j = 1,...,m and

$$E = {}^{\wedge}; \; \llcorner X . \; \llcorner K \_Z) ; \quad . \quad {}^{-}) ; \quad . \quad = M$$

for i = 1,...,n and j = 0,...,m, where w(W) is the Hamming weight of the components of the vector W, and the vector $X^{(i)}$ denotes the vector obtained by complementing the i-th bit of X. The degree of completeness is defined as:

$$G = - \frac{{}^{\wedge} L \; M\_D = {}^{\backprime}}{QP}$$

The degree of the avalanche effect has been calculated in accordance with the expression:

$$G = - \frac{}{QP} = - \frac{}{X \quad K} = \frac{ME - P}{} \_$$

The degree of strict avalanche criterion has been calculated as:

$$= - \frac{}{} = = - \frac{}{X \quad K} - \_$$

Experiments have been performed in cases "one key and 10000 texts" and "100 keys and 100 texts" (Table 3). In general, results on statistic testing of SPECTR-H64 are analogous to that of AES finalists [15].

Table 3.

Number of round	#K=1, #X=10000;				#K=100, #X=100.			
	(1)	(2)=d_c	(3)=d_a	(4)=d_{sa}	(1)	(2)=d_c	(3)=d_a	(4)=d_{sa}
12	32.044695	1.000000	0.998514	0.991975	32.039656	1.000000	0.998591	0.992178
10	32.040497	1.000000	0.998508	0.992011	32.045756	1.000000	0.998489	0.991799
8	32.049603	1.000000	0.998273	0.991963	32.042373	1.000000	0.998440	0.991953
6	32.047088	1.000000	0.998495	0.991937	32.045458	1.000000	0.998494	0.992088
4	32.013722	1.000000	0.998547	0.991473	32.003152	1.000000	0.998279	0.991663
3	29.213473	1.000000	0.911581	0.872055	29.213597	1.000000	0.912476	0.873493
2	27.931805	1.000000	0.872869	0.513196	27.927716	1.000000	0.872741	0.524154
1	25.220356	0.515459	0.788136	0.140762	25.187814	0.515137	0.787119	0.160758

Because of very simple key scheduling used in SPECTR-H64 it appears to be important to study how a single bit of key influences statistically ciphertext. For this purpose we have used the criteria 1-4 mentioned above with exception that the matrix of dependencies a_{ij} and matrix of distances b_{ij} have been defined relatively key as follows:

$$D = \wedge;\ \llcorner X.\ \llcorner K\_)\ ;\ .\qquad ,,\ )\ ;\ .$$

for i = 1,...,n and j = 1,...,m,

$$E = \wedge;\ \llcorner X.\ \llcorner K\_Z)\ ;\ .\quad {}^-)\ ;\ .\quad = \dot M$$

for i = 1,...,n and j = 0,...,m, where $K^{(i)}$ denotes the vector obtained by complementing the i-th bit of K.

Obtained results (Table 4) show that 7 rounds are quite sufficient to get necessary key's influencing output text (key's propagation property).

Table 4.

Number of round	#K=400, #X=25;				#K=100, #X=100.			
	(1)	(2)=d	(3)=d	(4)=d	(1)	(2)=d	(3)=d	(4)=d
12	32.050219	1.000000	0.998271	0.992005	32.051301	1.000000	0.998197	0.992040
10	32.053772	1.000000	0.998188	0.991940	32.048439	1.000000	0.998298	0.991881
8	32.051826	1.000000	0.998193	0.992168	32.049461	1.000000	0.998261	0.992045
6	31.911817	1.000000	0.995094	0.979446	31.913402	1.000000	0.995040	0.979268
5	31.518421	1.000000	0.983716	0.915390	31.512109	1.000000	0.983649	0.915872
4	30.735910	1.000000	0.960020	0.787367	30.734479	1.000000	0.959858	0.787278
3	28.476102	1.000000	0.889878	0.565557	28.470965	1.000000	0.889718	0.565777
2	26.810414	0.750000	0.837825	0.247683	26.806360	0.750000	0.837699	0.247604
1	21.870804	0.250000	0.683463	0.036559	21.869943	0.250000	0.683436	0.036482

Cryptosystem SPECTR-H64 uses a very fast internal key scheduling which corresponds to the execution of the CP-box permutation above the round subkeys before they are combined with data. The internal key scheduling is not complex, but it changes from one data block to another one strengthening significantly the cryptoscheme against differential and linear cryptanalysis. It introduces no time delay, since it is executed in parallel with some data ciphering operations.

The hardware-realization cost of SPECTR-H64 equals about 100,000 transistors. Experimental cryptochips have been made with 1.2-micron technology, about 1Gbit/s encryption speed being obtained.

The hardware implementation of the proposed cryptoscheme appears to be very fast and inexpensive. One can note that chipmakers can support encryption technique based on CP by adding a CP-box permutation instruction to the CPU. In this case it will be possible to compose very fast (> 400 Mbit/s for Pentium-like processors) software encryption algorithms.

Acknowledgement. This work was carried out as a part of the AFRL funded project #1994P which supported the authors.

References

1. Rivest, R.L.: The RC5 Encryption Algorithm. Fast Software Encryption - FSE'94 Proceedings. Springer-Verlag, LNCS Vol. 1008. (1995) 86–96
2. Kaliski, B.S. and Yin, Y.L.: On Differential and Linear Cryptanalysis of the RC5 Encryption Algorithm. Advances in Cryptology – CRYPTO'95 Proceedings. Springer-Verlag, LNCS Vol.963. (1995) 171–184
3. Biryukov, A., Kushilevitz, E.: Improved Cryptanalysis of RC5. Advances in Cryptology – Eurocrypt'98 Proceedings. Springer-Verlag, LNCS Vol. 1403. (1998) 85–99
4. Rivest, R.L., Robshaw, M.J.B., Sidney, R., Yin, Y.L.: The RC6 Block Cipher. Proceeding of 1st Advanced Encryption Standard Candidate Conference, Venture, California, (Aug 20-22 1998) (http://www.nist.gov/aes)
5. Burwick, C., Coppersmith, D., D'Avingnon, E., Gennaro, R., Halevi, Sh., Jutla, Ch., Matyas Jr., S.M., O'Connor, L., Peyravian, M., Safford, D., Zunic, N.: MARS – a Candidate Cipher for AES. Proceeding of 1st Advanced Encryption Standard Candidate Conference, Venture, California, (Aug 20–22 1998) (http://www.nist.gov/aes)
6. Moldovyan, A.A., Moldovyan, N.A.: A Method of the Cryptographical Transformation of Binary Data Blocks. Russian patent, N 2141729. Bull. N. 32. (Nov 20 1999)
7. Benes, V.E.: Mathematical Theory of Connecting Networks and Telephone Traffic. Academic Press, New York (1965)
8. Waksman, A.A.: Permutation Network, Journal of the ACM. Vol.15, no 1 (1968) 159–163
9. Kawn, M.: The Design of the ICE Encryption Algorithm. Fast Software Encryption - FSE'97 Proceedings. Springer-Verlag, LNCS Vol. 1267. (1997) 69–82
10. Van Rompay, L.R. Knudsen, V.: Rijmen Differential Cryptanalysis of the ICE Encryption Algorithm, Fast Software Encryption – FSE∅8 Proceeding. Springer-Verlag LNCS Vol. 1372. (1998) 270–283
11. Maslovsky, V.M., Moldovyan, A.A., Moldovyan, N.A.: A Method of the Block Encryption of Discrete Data. Russian patent, N 2140710. Bull. N. 30 (Oct 27 1999)
12. Goots, N.D., Izotov, B.V., Moldovyan, A.A., Moldovyan, N.A.: Design of two place operation for fast flexible cryptosystems. Bezopasnosti informatsionnyh tehnologii, Moscow, MIFI, no 4 (2000, in Russian)
13. Goots, N.D., Izotov, B.V., Moldovyan, N.A.: Controlled permutations with symmetric structure in block ciphers, Voprosy zaschity informatsii, Moscow, VIMI, no 4 (2000) 57-65 (in Russian).
14. Nyberg, K.: Constructions of Bent Functions and Difference Sets, Advances in Cryptology – Eurocrypt'90 Proceedings. Springer Verlag (1991) 151–160
15. Preneel, B., Bosselaers, A., Rijmen, V., Van Rompay, B., Granboulan, L., Stern, J., Murphy, S., Dichtl, M., Serf, P., Biham, E., Dunkelman, O., Furman, V., Koeune, F., Piret, G., Quisquater, J-J., Knudsen, L., Raddum, H.: Comments by the NESSIE Project on the AES Finalists. (24 may 2000) (http://www.nist.gov/aes)

CVS at Work: A Report on New Failures upon Some Cryptographic Protocols

Antonio Durante [1], Riccardo Focardi [2], and Roberto Gorrieri [3]

[1] Dipartimento di Scienze dell'Informazione, Universit` a di Roma "La Sapienza",
durante@dsi.uniroma1.it
[2] Dipartimento di Informatica, Universit` a Ca' Foscari di Venezia
focardi@dsi.unive.it
[3] Dipartimento di Scienze dell'Informazione, Universit` a di Bologna
gorrieri@cs.unibo.it

Abstract. CVS is an automatic tool for the veri cation of cryptographic protocols, we have presented in [9], [10], that uses a non-interference based analysis technique which has been successfully applied to many case-studies, essentially most of those belonging to the Clark & Jacob's library [4]. In this paper we report some new failures we have found. More precisely, we have been able to detect attacks upon two unflawed (to the best of our knowledge) protocols: Woo & Lam public key one-way authentication protocol and ISO public key two-pass parallel mutual authentication protocol; and new failures upon two flawed protocols: Encrypted Key Exchange and Station to Station protocols.

1 Introduction

Many cryptographic protocols have been proposed in the literature to achieve several goals, e.g. authentication and secrecy; these protocols are supposed to succeed even in the presence of malicious agents, usually called intruders or enemies.

In recent years, many attacks have been discovered over well known protocols (see e.g., [2], [19]). It is often the case that these attacks are based on unexpected execution sequences, where the enemy intercepts messages and combines them in order to obtain new faked messages. Sometimes, many parallel sessions of the protocol are exploited by the enemy in order to carry out the attack, and this complicates even further the task of detecting such sequences, if they exist. For these reasons, there has been an increasing interest in computer-assisted veri cation tools for the analysis of cryptographic protocols (see, e.g., [23], [20]). Those tools that seem more useful to detect new attacks are essentially model-checkers (see, e.g., the new attacks reported in [23], [9], [6]). On the other hand, theorem provers, such as [21], [24], give better guarantees if a proof is found.

One of the various model checkers is the CVS/CoSeC technology, we have developed in the last three years. It is composed of:

{ the compiler CVS [7], [9], [10] from a high-level protocol speci cation notation (a value passing process algebra with explicit name declarations, called VSP) to a basic process algebra (called SPA), and

V.I. Gorodetski et al. (Eds.): MMM-ACNS 2001, LNCS 2052, pp. 287–299, 2001.

{ the tool CoSeC [13] which takes as input the SPA protocol speci cation and checks on that process a non interference-like property, called   Non Deducibility on Compositions   (NDC) we have proposed for this aim [13], [15], [14].

This paper reports some new failures we have found with the help of CVS and CoSeC. More precisely, we have been able to show attacks upon two unflawed (to the best of our knowledge) protocols: Woo & Lam public key one-way authentication protocol [27] and ISO public key two-pass parallel authentication protocol [18]. Moreover, we have also found new failures upon three flawed protocols: Encrypted Key Exchange [3], Station to Station [5] and Woo & Lam symmetric key one-way authentication protocol [27]. (The last failure is not reported here as it already appeared in [9].) Moreover, most of those belonging to the Clark' & Jacob's library [4] has been successfully analyzed, obtaining the known attacks.

Note that the four protocols mentioned above have been already studied extensively. Hence, on the one hand, our results might appear very surprising. On the other hand, the anomalies we have found are rather subtle and it may be not obvious that they are really harmful. For instance, the   entity authentication attack to the Woo & Lam public key one-way authentication protocol may be considered a real attack depending on the modeling assumptions (packed-switched network vs connection based network [17]). It is thus important, when a particular anomaly is detected, to discuss a possible scenario (i.e., the particular context where the protocol is supposed to run) where the anomaly can show its potential danger (see [8] for a case study fully analyzed in this way). We will discuss this point in the next sections, when presenting the various protocol failures we have found.

2   Non Interference for the Analysis of Security Protocols

CVS/CoSeC has been implemented to support the Non-Interference based analysis of security protocols we have proposed in [13], [10], [14].

Non-Interference [16] has been originally proposed for modeling multilevel security and in particular to solve the problem of   covert channels , i.e., indirect ways of transmitting information from a security level to a lower one. In recent work [12], [13], two of the authors have studied the many non interference-like properties proposed so far, by de ning all of them uniformly in a common process algebraic setting, producing the rst taxonomy of these security properties reported in the literature.

Among the many NI-like properties, we advocated a property called   Non Deducibility on Composition   (NDC for short) that, informally, can be expressed as follows: Given two groups of users   H (high) and   L (low), there is no information flow from   H to   L i there is no way for   H to modify the behaviour of   L.

Analogously, we may think that   L is the set of the honest participants to a protocol and   H is the external, possibly malicious, environment, i.e. the set of possible intruders (or enemies). Following the analogy, no information flow

from high to low means that the intruders have no way to change the low behaviour of the protocol. Of course, this statement may hold depending on what
we mean by low behaviour of the protocol. We should assume that an intruder
may have complete control of the network, and so it is reasonable to assume
that all the activities on the public channels is considered a high behaviour. The
low behaviour is therefore related to extra observable actions included into the
speci cation of the protocol to specify the security properties of interest. For
instance, to model some form of authentication as in [19], it is enough to include
into the speci cation of the protocol special start/commit actions for all the
honest participants; the property holds if the enemy is not able to destroy the
precise correspondences among these start/commit actions.

The interesting point is that many security properties in the literature can
be characterized as specialization of the NDC property, with a suitable insertion
of extra observable low actions: secrecy , some forms of authentication, message
authenticity, some forms of non repudiation and fair-exchange [15], [14]. More
interestingly, we have proved that, once the extra observable events have been
 xed, NDC implies all the other security properties characterized by the extra
observable actions. This means that every attack which is based on certain observable events is for sure detected by our NDC check. Thus, if we observe all
the events corresponding to a set of typical interesting properties, we will in a
sense verify all of them in just one NDC check [10]. Moreover, we can include
a rather generous set of extra observable actions into the protocol speci cation
(containing at least the extra actions of the properties of interest) that may
reveal some suspicious behaviour that we should analyze, as it may reveal a potential anomaly, failure or really dangerous attack. This is indeed the point that
mostly di erentiates the NDC-based approach from the usual ones: we do not x
in advance the speci c security property that we want to check. As a matter of
fact, it is quite typical with this approach to nd \unexpected" protocol failures
that were not considered as important in the initial modeling of the protocol.
In our opinion, the NDC-based approach o ers more opportunities of nding
new attacks with respect to classic approaches where, often, the analyzer sets
the tool (and the security property) in search for an attack that he has already
guessed. This paper is intended to substantiate this claim.

3   The Methodology

Analyzing a security protocol applying NDC consists of the following steps (see
Figure 1).

1. We specify the protocol honest roles in a value-passing version of the Security Process Algebra (SPA [13]), called VSP [7], [9], [10], that includes
   name declarations (e.g., process names), crypto-functions (e.g., symmetric
   and asymmetric encryption, hash functions), and other features that are useful to provide a fully-fedged speci cation of the protocol. In this phase, the
   actual instance of the protocol is to be speci ed de ning the analyzed session

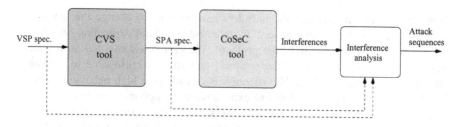

Fig. 1.   The CVS tool and the NDC-based protocol analysis

(e.g., who is playing which role, possibly in how many instances), including some parameters (e.g., initial knowledge) that initialize the structure of the enemy which is then generated automatically by the compiler CVS.

2. The second step is the de nition of the extra observable actions, i.e. those actions that are relevant for specifying and checking the security properties of interest and that can reveal if the enemy has interfered with the correct execution of the protocol. As already stated, a choice of a quite large set of observable events will reveal simultaneously a number of di erent, potentially dangerous, attack sequences. These actions are suitably inserted into the VSP speci cation of the parties.

3. The tool CVS [7] is used to compile a VSP speci cation into SPA code, where the enemy speci cation is provided in full detail. SPA is a slight modi cation of Milner's CCS [22] where actions have been partitioned into two security levels, in order to analyze Non-Interference properties. It basically allows for the speci cation of protocols as the parallel composition of the various parties with the enemy. Each party is de ned by the sequence of messages it can send or receive. The execution of the protocol is then obtained by considering all the possible interleavings of the parties which may also synchronize in order to perform a message exchange.

4. The tool CoSeC [13] takes as input the SPA speci cation representing the protocol that we want to analyze and checks the NDC property on the - nite automaton representing the semantics of that process. If the protocol veri es NDC successfully, then we are sure that the enemy cannot interfere with the correct execution of the protocol in the analyzed session; otherwise we observe some interferences , i.e., execution traces that were not possible without enemy's intervention.

5. The interference traces are then (manually) analyzed. These interferences may reveal some actual attacks upon the protocol.

The next sections are devoted to the presentation of some new attacks we have found with the CVS/CoSeC technology which implements our veri cation anal- ysis technique based on NDC. We will not give the details of how the analysis can be performed. The interested reader can refer to [9], [10], [8] for details on how to use CVS/CoSeC. Indeed, the main aim of this paper is to show that our approach can be useful, by giving some examples of new failures that we have found in some well-known (and deeply analyzed) protocols.

In each analysis we do, we set in the VSP specification as observable actions the start and the correct commit of the protocol.

4 ISO Public Key Two-Pass Parallel Mutual Authentication Protocol

This is one of the protocols proposed in [18] by the ISO/OSI committee. The goal of this protocol is to achieve mutual authentication (in a parallel way). Its formal description is reported below, where the notation $A ! B : msg$ represents the sending of message msg from Alice (A) to Bob (B):

Message 1 $A ! B : A, N_A$		$i\_start \_AB$
Message 1^0 $B ! A : B, N_B$		$i\_start \_BA$
Message 2 $B ! A : N_B, N_A, A, SK_B(N_B, N_A, A)$		r_commit_BA
Message 2^0 $A ! B : N_A, N_B, B, SK_A(N_A, N_B, B)$		r_commit_AB

The protocol consists of four messages, pairwise symmetric. It is based on public key cryptography: PK_U denotes the public key of a generic user U while SK_U is his secret key. Moreover, N_U represents a nonce (i.e., a number used only once) generated by user U. Both A and B prove their identity by signing the respective challenges N_A and N_B with their secret keys.

Each message is decorated by an extra observable action that we have inserted into the specification for analysis purposes. For instance what we observe by executing message 1 is the action $i\_start \_AB$ meaning that A is initiating a session with B. Similarly, in message 2, r_commit_BA signals that B ended the exchange with A as a responder.

For this protocol we have found the particular failure sequence $i\_start \_AA$ r_commit_AA that can be justified by the following attack (we denote the enemy which is impersonating the entity U with $E(U)$ in the position of the sender, and the enemy that intercepts the message directed to U with $E(U)$ in the position of the receiver):

Message 1 $A ! E(A): A, N_A$		$i\_start \_AA$
Message 1^0 $E(A) ! A : A, N_A$		
Message 2^0 $A ! E(A): N_A, N_A, A, SK_A(N_A, N_A, A)$		
Message 2 $E(A) ! A : N_A, N_A, A, SK_A(N_A, N_A, A)$		r_commit_AA

This corresponds to a parallel session attack and can happen when an entity A needs to authenticate itself on a remote computer. The enemy in the middle intercepts all the communications re-sending the messages to the initiator A. At the end of the run the entity A is convinced to communicate with herself. This self-authentication attack might be dangerous when the protocol is used by A, e.g., to transfer files between accounts on different computers both owned by A. This failure is similar to the one reported in [6] for the ISO public key two-pass mutual authentication protocol, where the protocol we analyze here is listed as unflawed.

Note that a simple way for A to avoid this potential attack is simply to verify, in every protocol execution, that the two (signed) nonces in Message 2 are di erent.

5   Woo-Lam Public Key One-Way Authentication Protocol

In [27], Woo and Lam proposed two protocols for achieving one-way authentication. These protocols di er in the cryptosystem they use. Here, we examine the one based on public-key cryptography. The protocol is composed of ve steps:

Message 1 A ! B : A		i _start _A.B
Message 2 B ! A : N_B		r _start _B.A
Message 3 A ! B : $SK_A(N_B)$		
Message 4 B ! S : A		
Message 5 S ! B : $A, SK_S(A, PK_A)$		r _commit_B.A

Message 1 represents the authentication request form A to B. In this message A sends her name as plaintext to B. In Message 2, B replies to A by sending a freshly generated nonce N_B. Once A has received the nonce N_B, she encrypts (signs) it with her secret key SK_A and sends the result to B. B cannot decrypt this message because he does not know the public key of A. Thus, in Message 4, B requests to S a certi cate containing the public key of A. The Server retrieves a certi cate $SK_S(A, PK_A)$ from the key database and sends it to B. B decrypts it using the public key of S, recovers A, PK_A and uses PK_A to decrypt the Message 3. Then, he veri es if N_B is equal to N_B of Message 2.

For this protocol we have found the following failure sequence:

Message 1.a	A ! E	: A	i _start _A.E
Message 1.b E(A) !	B	: A	r _start _B.A
Message 2.b	B ! E(A)	: N_B	
Message 2.a	E ! A	: N_B	
Message 3.a	A ! E	: $SK_A(N_B)$	
Message 3.b E(A) !	B	: $SK_A(N_B)$	
Message 4.b	B ! S	: A	
Message 5.b	S ! B	: $A, SK_S(A, PK_A)$	r _commit_B.A

The failure is very similar to the one we have found for the version based on symmetric key cryptography [9], [10]. The actions performed by the enemy leads the responder B to achieve authentication with A that has never sent any authentication request to him. It is a typical man-in-the-middle attack: indeed, A starts the protocol with E and E is able to exploit this session (denoted by a) with A in order to impersonate A with B in a parallel session (denoted by b).

The fact that this failure is dangerous or not strictly depends on the model assumptions. For a detailed discussion see, e.g., [17]. However, such a failure can be avoided by simply including in the signature of Message 3, the identi er of

the intended recipient, i.e., Message 3 becomes $SK_A(N_B, B)$. With this modi -
cation, E will not be able to forward Message 3 .a (now $SK_A(N_B, E)$) to B as it
contains the identi er of the intended recipient   E. As a matter of fact, B would
immediately recognize that such a message is just the replay of a message sent
to E.

6   The EKE Protocol

The protocol considered is the Encrypted Key Exchange - EKE protocol proposed in [3]. It provides both security and authentication on computer networks,
using both symmetric and public key cryptography. The EKE protocol is proposed for the negotiation of a new session key between two users that share a
common password K_P.

6.1   First Analysis: EKE Based on Di e-Hellman

We consider the version of EKE that is based on the Di e-Hellman protocol.
It is thus useful to brie y summarize how the Di e-Hellman protocol works:
Before starting the protocol, Alice and Bob agree on a large prime   n and on a
number g, such that g is a primitive element mod   n. The prime n is chosen such
that it is infeasible to calculate the discrete logarithm mod   n. These two integers
do not have to be secret; usually they are common among a group of users [25].
The Di e-Hellman protocol is used for establishing a new session key between
Alice and Bob and is composed of the following two messages:

> Message 1 A ! B : $X = g^x \bmod n$
> Message 2 B ! A : $Y = g^y \bmod n$

Alice and Bob respectively choose two random large integers   x and y, and send
one another $X = g^x \bmod n$ and $Y = g^y \bmod n$. Alice and Bob can then compute
the same session key as $Y^x \bmod n = g^{xy} \bmod n = X^y \bmod n$. No one listening
on the channel can compute such a value as it does not know   x and y and can
neither calculate them through logarithm (see, e.g., [25] for more details). At the
end of the protocol, each pair of users shares a unique secret key, and no prior
communication between the two users is required.

   We will denote with $DH_{x,y}$ the Di e-Hellman function, which is used to
compute the key shared between the two principals, i.e., $DH_{x,y}$ will represent
the exponentiation $X^y \bmod n$. The version of EKE based on Di e-Hellman
protocol follows:

> Message 1 A ! B : A, X i _start _A_B
> Message 2 B ·B ! A : $K_P(Y), DH_{x,y}(N_B)$ r _start _B_A
> Message 3 A ! B : $DH_{Y,x}(N_A, N_B)$
> Message 4 B ! A : $DH_{x,y}(N_A)$ i _commit_A_B, r _commit_B_A

In the rst two messages Alice and Bob negotiate the new session key. Indeed,
Alice sends her identi er together with the value $X = g^x \bmod n$, where x is

freshly generated. Then, Bob replies with the value $Y = g^y \bmod n$ encrypted with the shared key K_P, together with the encryption with the Di e-Hellman session key $DH_{x,y}$ of a nonce N_B. In the last two messages Alice and Bob use the obtained session key to authenticate the run through a typical challenge-response mechanism. The encryption of Y with K_P should guarantee that Y has been generated by B since only A and B know such a shared key.

In the analyzed session Alice plays at the same time both the initiator and the responder role. Alice begins two runs with Bob, the rst as an initiator and the second as a responder and she completes both. We describe in detail the actions performed by Alice due to the enemy presence:

Message 1.a	A !	E(B) :	A, X	i _start _AB
Message 1.b	E(B) !	A :	B, X	r _start _AB
Message 2.b	A !	E(B) :	$K_P(Y), DH_{x,y}(N_A)$	
Message 2.a	E(A) !	A :	$K_P(Y), DH_{x,y}(N_A)$	
Message 3.a	A !	E(B) :	$DH_{Y,x}(N_A, N_A)$	
Message 3.b	E(B) !	A :	$DH_{Y,x}(N_A, N_A)$	
Message 4.b	A !	E(B) :	$DH_{x,y}(N_A)$	r _commit_AB
Message 4.a	E(B) !	A :	$DH_{x,y}(N_A)$	i _commit_AB

Alice starts the run a as an initiator with Bob, message 1 is intercepted by the enemy that starts a run b with her pretending to be Bob. The enemy in the middle acts as Bob, intercepting and forwarding the message from Alice to Alice. Alice completes both the runs, thinking in the rst that Bob acts as a responder, and in the second that Bob acts as an initiator. However, Bob has never started any run with Alice and could even not be present on the net.

This kind of weakness is not serious as the negotiated key is not compromised, but it leads to a denial of service attack, as Alice will be convinced to share the (useless) key $DH_{x,y}$ with Bob. However, this execution sequence shows that the protocol should never be used for entity authentication only (no key-exchange). In such a case the attack would be valid as E would be able to completely impersonate B. A way for repairing the protocol is to encrypt with K_P also the identi er of the intended recipient (as similarly done for the previous attack), i.e., message 2 becomes $K_P(Y,A), DH_{x,y}(N_B)$. This allows A to discover that message 2 .a is indeed a replay.

6.2   Second Analysis: EKE Based on RSA

We have analyzed another version of the EKE protocol where the public key cryptography is implemented with the RSA algorithm. Even in this case the two users share a common password K_P and use the challenge response mechanism to perform the authentication goal.

Message 1	A ! B :	$A, K_P(PK_A)$
Message 2	B ! A :	$K_P(PK_A(K))$
Message 3	A ! B :	$K(N_A)$
Message 4	B ! A :	$K(N_A, N_B)$
Message 5	A ! B :	$K(N_B)$

With the rst two messages, the parties negotiate the new session key   K generated by the responder   B. The key exchange is performed using the new public key PK_A that A generates. The last three messages achieve the authentication task. We consider the protocol in a single session, where each agent plays just one role at a time. The failure sequence due to the enemy presence is the following:

Message 1.a	A	!	E	:	$A, K_{Pe}(PK_A)$	i _start _AE
Message 1.b	E	!	B	:	$E, K_{Pe}(PK_E)$	r _start _BE
Message 2.b	B	!	E	:	$K_{Pe}(PK_E(K))$	
Message 2.a	E	!	A	:	$K_{Pe}(PK_A(K))$	
Message 3.a	A	!	E	:	$K(N_A)$	
Message 3.b	E	!	B	:	$K(N_A)$	
Message 4.b	B	!	E	:	$K(N_A, N_B)$	
Message 4.a	E	!	A	:	$K(N_A, N_B)$	
Message 5.a	A	!	E	:	$K(N_B)$	i _commit_AE_K
Message 5.b	E	!	B	:	$K(N_B)$	r _commit_BE_K

The agent E simply acts as a responder running the protocol with   A in the run a, and as an initiator running the protocol with   B in the run b. E renegotiates in the run   a the key K negotiated with   B in the run   b. None of the principals thinks to be connected to an unexpected principal. The anomaly consists in the fact that   A and B share K, at the same time, with   E. Indeed, a key-exchange protocol should guarantee the very basic security property that di erent pairs of principals have di erent shared keys. The e ect of this anomaly is that the attacker   E creates an implicit communication channel between   A and B passing through its location. This kind of failure is common to many cryptographic protocols designed to create a secure channel between two agents A and B without the intervention of a trusted third party.

Of course, the enemy is interested in sharing a common password with both A and B only if this gives some advantage to him. An interesting case could be one where E gets credit for something which is indeed done by   A. For example, suppose that B represents a "teacher server" where students can log-in and have on-line interactive examinations. Suppose also that   E is able to create a clone of the login interface of B on another site (E itself), and to convince some students that such site is the correct one. Now consider a student   A that is logging on E thinking it is the teacher server.   E is now able, through the sequence above, to establish a key K which is shared between A, B and E. This allows E to leave A and E freely interact (with no delays due to man-in-the-middle encryptions and decryptions). A will notice nothing strange since she indeed wanted to talk to the teacher. On the other side,   B will give credit to E of all the (hopefully good) answers of A to the various questions. Moreover   E is able to passively track every communication between the two in order to know what is happening. For a discussion about credit and authentication see, e.g., [1].

7   The Station to Station Protocol

The Station to Station protocol was proposed in [5] with the aim of establishing a secure communication channel between each pair of users of a community using the Di e-Hellman protocol.

The Di e-Hellman key exchange is known to be vulnerable to a man in the middle attack. One way to prevent this problem is to have Alice and Bob signing their message each other, as proposed in [5]. This protocol assumes that each pair of users have a certi cate containing the public key of the other users of the community. The modi ed Di e-Hellman protocol follows:

$$\text{Message 1} \quad A \; ! \; B : X$$
$$\text{Message 2} \quad B \; ! \; A : Y, DH_{X,y} (SK_B (X, Y))$$
$$\text{Message 3} \quad A \; ! \; B : DH_{Y,x} (SK_A (X, Y))$$

Alice and Bob agree on the values   X and Y, and they sign such values in messages 2 and 3 with their private key to avoid the man in the middle attack.

A strange execution sequence for this protocol has been detected by CVS in a multiple session analysis, when Alice and Bob play two times the Initiator and the Responder role, respectively. In detail, the sequence of actions performed by Alice and Bob in the presence of the enemy is the following:

1.a	$A \; !$	$E(B)$:	$X1$	$i\_start \; \_A\_B\_a$
1.b	$A \; !$	$E(B)$:	$X2$	$i\_start \; \_A\_B\_b$
$1.a^0$	$E(A) \; !$	B :	$X2$	$r\_start \; \_B\_A\_a$
$1.b^0$	$E(A) \; !$	B :	$X1$	$r\_start \; \_B\_A\_b$
2.a	$B \; !$	$E(A)$:	$Y1, DH_{X2,y1}(SK_B(X2, Y1))$	
2.b	$B \; !$	$E(A)$:	$Y2, DH_{X1,y2}(SK_B(X1, Y2))$	
$2.a^0$	$E(B) \; !$	A :	$Y2, DH_{X1,y2}(SK_B(X1, Y2))$	
$2.b^0$	$E(B) \; !$	A :	$Y1, DH_{X2,y1}(SK_B(X2, Y1))$	
3.a	$A \; !$	$E(B)$:	$DH_{Y2,x1}(SK_A(X1, Y2))$	$i\_commit\_A\_B\_X1\_Y2\_a$
3.b	$A \; !$	$E(B)$:	$DH_{Y1,x2}(SK_A(X2, Y1))$	$i\_commit\_A\_B\_X2\_Y1\_b$
$3.a^0$	$E(A) \; !$	B :	$DH_{Y1,x2}(SK_A(X2, Y1))$	$r\_commit\_B\_A\_X2\_Y1\_a$
$3.b^0$	$E(A) \; !$	B :	$DH_{Y2,x1}(SK_A(X1, Y2))$	$r\_commit\_B\_A\_X1\_Y2\_b$

Alice, acting as an initiator, begins two runs   a, b with Bob (acting as a respon-der), and she terminates both with him. The enemy acts in the middle inter-cepting and exchanging the order of the messages between the two runs. At the end of the runs Alice and Bob agree on two di erent values of X and Y. In the run a Alice agrees on X 1, Y 2 (message 3.a), and Bob agrees on   X 2, Y 1 (message 3.a') and vice-versa for the run   b. The agents are mutually authenticated but they disagree upon the exchanged values, so upon the new negotiated session key. This kind of weakness does not compromise the secrecy of the negotiated key, but leads to a denial service.

As a matter of fact, in a real implementation of the protocol a pair of users will presumably establish a (non-secure) connection before starting the protocol, e.g., a TCP connection. Suppose that this is done in a preliminary message where

A and B decide to start the protocol. In the case of the execution sequence above, we would have two (TCP) connections between the two pairs of users that are willing to communicate. We call (A_1, B_1) and (A_2, B_2) such two pairs. Basically, the enemy is now able to force the "crossed" pairs (A_1, B_2) and (A_2, B_1) sharing two different secret keys, but A_1 will never be able to communicate with B_2 since they are on two different (TCP) connections. A_1 will indeed try to use the secret key with B_1 with, of course, no success.

This situation is perhaps a typical example in which the refinement of a protocol specifications leads to undesired executions, i.e., denial of service. It may happen in a number of existing protocols, and may be avoided only by including in the messages some information about the low level connections, e.g., IP address and port. An attack which is somehow similar to the denial of service above has been found in Otway-Rees protocol in [11], although in such an attack there is only one session where A and B commit on two different sessions keys. Such an attack is thus effective also at an abstract specification level (without considering low level connections).

8   Conclusion

We think that these case studies support our claim that CVS, and the NDC-based analysis technique for cryptographic protocols it implements, is potentially useful in order to detect new attacks. It is therefore desirable to start investigating more complex protocols, used in real-life applications (such as SET [26]) to see if our technique can uncover possible unknown failures. We are currently working to improve our tools in order to handle such larger protocols. The main issue is the state-space explosion problem for which we are implementing some techniques for reducing the state space, on the one hand, and for managing more efficiently the state memorization, on the other hand.

We are also implementing the automatic filtering of the interference traces which correspond to attacks over some already known security properties like, e.g., those reported in [15], [14]. This will allow us to automatically extract the interferences that correspond to well known (interesting) typologies of attacks. It will then be possible to focus the analysis on the remaining interferences in order to observe if new strange failures are present in the analyzed protocol.

References

1. M. Abadi: Two Facets of Authentication. In: *Proceedings of 11th IEEE Computer Security Foundations Workshop*. IEEE press. (June 1998) 25–32
2. Abadi, M. and Needham, R.: Prudent Engineering Practice for Cryptographic Protocols. *IEEE Transactions on Software Engineering*, 22(1) (January 1996)
3. Bellovin, S.M. and Merritt, M.: Encrypted Key Exchange: Password-Based Protocols Secure Against Dictionary Attacks. In: *Proceedings of 1992 IEEE Computer Society Conference on Research in Security and Privacy*. (1992) 72–84

4. John Clark and Jeremy Jacob: A Survay of Authentication Protocols Literature: version 1.0. Available at http://www.cs.york.ac.uk/ jac/papers/drareview.ps.gz, (November 1997)
5. Di e, W., van Oorschot, P.C., and Wiener, M.J.: Authentication and Authenticated Key Exchanges. Design, Codes and Cryptography , 2:107–125 (1992)
6. Donovan, B., Norris, P., and Lowe, G.: "Analizing a Library of Security Protocols Using Casper and FDR". In   Procee dings of the FLoC Workshop Formal Methods and Security Protocols , Trento (July 1999)
7. Durante, A.: Uno strumento automatico per la veri ca di protocolli crittogra ci. Master's thesis, Universit` a degli Studi di Bologna (sede Cesena). (luglio 1998)
8. Durante, A.: Analysis of cryptographic protocols: the environment and the application. In: Procee dings of 15th IFIP SEC 2000 . (August 2000)
9. Durante, A., Focardi, R., and Gorrieri, R.: CVS: a tools for the analysis of cryptographic protocols. In   Procee dings of 12th IEEE Computer Security Foundation Workshop , (July 1999) 203–212
10. Durante, A., Focardi, R., and Gorrieri, R.: A Compiler for Analysing Cryptographic Protocols Using Non-Interference. To appear in   ACM Transaction on Software Engeneering and Methodology   (TOSEM) (2000)
11. Fabrega, F.J.T., Herzog, J.C., and Guttman, J.D.: Honest Ideals on Strand Spaces. In: Procee dings of 11th IEEE Computer Security Foundation Workshop   . (June 1998) 66–77
12. Focardi, R. and Gorrieri, R.: A Classi cation of Security Properties for Process Algebras. Journal of Computer Security , 3(1):5–33 (1995)
13. Focardi, R. and Gorrieri, R.: The Compositional Security Checker: A Tool for the Veri cation of Information Flow Security Properties.   IEEE Transactions on Software Engineering , 23(9):550–571 (September 1997)
14. Focardi, R., Gorrieri, R., and Martinelli, F.: "Non Interference for the Analysis of Cryptographic Protocols". In   Procee dings of ICALP'00 . LNCS 1853 (July 2000) 744–755
15. Focardi, R. and Martinelli, F.: A Uniform Approach for the De nition of Security Properties. In: Procee dings of World Congress on Formal Methods in the Development of Computing Systems (FM'99), LNCS 1708   . Toulouse (France). Springer-Verlag LNCS. 794–813 (September 1999)
16. Goguen, J.A. and Meseguer, J.: Security Policy and Security Models. In: Proceedings of the 1982 Symposium on Security and Privacy   . IEEE Computer Society Press (April 1982) 11–20
17. Gollman, D.: On the Veri cation of Cryptographic Protocol   s - A Tale of Two Committees. In: Workshop on secure architectures and information flow   . Vol. 32 of ENTCS (2000)
18. International Organization for Standardization. Information technology - Security techniques - Entity authentication mechanism; Part 3: Entity authentication using a public key algorithm. ISO/IEC 9798-3 (1995)
19. Lowe, G.: Breaking and Fixing the Needham-Schroeder Public-key Protocol using FDR. In: Procee dings of Second International Workshop on Tools and Algorithms for the Construction and Analysis of Systems (TACAS'96)   . Passau (Germany). Springer-Verlag, LNCS 1055. 146–166 (March 1996)
20. Lowe, G.: CASPER: A Compiler for the Analysis of the security protocols. In: Procee dings Tenth IEEE Computer Security Foundation Workshop, (CSFW'97) Rockport Massachusetts (USA). IEEE Press 18–30 (June 1997)
21. Meadows, C.: The NRL Protocol Analyzer: An Overview.   Journal of Logic Programming , 26(2):113–131 (1996)

22. Milner, R.: Communication and Concurrency   . Prentice-Hall (1989)
23. Mitchell, J.C., Mitchell, M., and Stern, U.: Automated Analysis of Cryptographic Protocols Using Mur   . In: Procee dings of the 1997 IEEE Symposium on Research in Security and Privacy   . IEEE Computer Society Press (1997) 141–153
24. Lawrence C. Paulson: The Inductive Approach to Verifying Cryptographic Protocols. In: Procee dings Tenth IEEE Computer Security Foundation Workshop, (CSFW'97) , Rockport Massachusetts (USA), IEEE Press. (June 1997)
25. Schneier, B.: Applied Cryptography . John Wiley & Sons, Inc. (1996) second edition
26. SET Secure Electronic Transaction LLC. The SET Standard Speci cation. Available at the URL, http://www.setco.org/set   _speci cations.html (May 1997)
27. Woo. T.Y.C. and Lam, S.S.: Authentication for distributed systems.   Computer , 25:10–10 (March 1992)

On Some Cryptographic Properties of Rijndael

Selçuk Kavut and Melek D. Yücel

Electrical & Electronics Engineering Dept., Middle East Technical University

{kavut, melek-yucel}@metu.edu.tr

Abstract. We examine diffusion properties of Rijndael which has been selected by US National Institute of Standards and Technology (NIST) for the proposed Advanced Encryption Standard (AES). Since the s-box of Rijndael applies a nonlinear transformation operating on each byte of the intermediate cipher result independently, its characteristics have significant effects on the strength of the entire system. The characteristics of Rijndael's s-box are investigated for the criteria of avalanche, strict avalanche, bit independence, nonlinearity and XOR table distribution. We also evaluate the overall performance for different rounds of Rijndael, and compare it to Safer K-64, in terms of Avalanche Weight Distribution (AWD) criterion.

1   Introduction

Rijndael [1] is a symmetric (one-key) block cipher with variable block length and variable key length, designed by Joan Daemen and Vincent Rijmen; and selected as the advanced encryption standard by NIST. The block length and the key length can be independently specified to 128, 192, or 256 bits. The number of iteration rounds given in Table 1 depends on the values corresponding to the input block length and the key length of the cipher.

Table 1. Number of rounds of Rijndael as a function of the block and key lengths

Key length in bits	Input block length in bits		
	128	192	256
128	10	12	14
192	12	12	14
256	14	14	14

All nine combinations of input block length and key length differ from each other only in their key schedules and in the number of rounds used. Therefore, examining the diffusion property of Rijndael with 128-bit input block length and 128-bit key length also gives some information about the diffusion properties of other eight versions of Rijndael.

V.I. Gorodetski et al. (Eds.): MMM-ACNS 2001, LNCS 2052, pp. 300–311, 2001.
Springer-Verlag Berlin Heidelberg 2001

The round transformation of Rijndael is composed of four different transformations which are respectively the byte-sub, shift-row, mix-column transformations and the round key addition. The substitution table (or s-box) is obtained through byte-sub transformation; which is in turn implemented by two transformations:

- First taking multiplicative inverse in $GF(2^8)$. In binary representation '0000 0000' is mapped onto itself.
- Then applying an affine transformation over $GF(2)$ defined by:

$$
\begin{pmatrix} y_0 \\ y_1 \\ y_2 \\ y_3 \\ y_4 \\ y_5 \\ y_6 \\ y_7 \end{pmatrix} =
\begin{pmatrix}
1 & 0 & 0 & 0 & 1 & 1 & 1 & 1 \\
1 & 1 & 0 & 0 & 0 & 1 & 1 & 1 \\
1 & 1 & 1 & 0 & 0 & 0 & 1 & 1 \\
1 & 1 & 1 & 1 & 0 & 0 & 0 & 1 \\
1 & 1 & 1 & 1 & 1 & 0 & 0 & 0 \\
0 & 1 & 1 & 1 & 1 & 1 & 0 & 0 \\
0 & 0 & 1 & 1 & 1 & 1 & 1 & 0 \\
0 & 0 & 0 & 1 & 1 & 1 & 1 & 1
\end{pmatrix}
\begin{pmatrix} x_0 \\ x_1 \\ x_2 \\ x_3 \\ x_4 \\ x_5 \\ x_6 \\ x_7 \end{pmatrix} +
\begin{pmatrix} 1 \\ 1 \\ 0 \\ 0 \\ 0 \\ 1 \\ 1 \\ 0 \end{pmatrix}
\tag{1}
$$

where ($x_7 \ldots x_0$) is the multiplicative inverse of the byte at the input of the s-box.

We present some experimental results, for the criteria of avalanche, strict avalanche, bit independence, nonlinearity and XOR table distribution of the s-box of Rijndael; in addition to its overall Avalanche Weight Distribution (AWD) curves evaluated round by round. Performance of Safer K-64 developed by Massey [2] and investigated by Aras [3] is used for comparison purposes.

2   Definitions

2.1   Completeness and Avalanche Criteria

The property of completeness was defined by Kam and Davida [4]. If a cryptographic transformation is complete, then every input bit depends upon every output bit. If a cryptographic transformation is not complete, it can be possible to find a pair of input and output bits such that flipping the input bit does not cause a change in the output bit for all input vectors.

The idea of avalanche was introduced by Feistel [5]. Avalanche is a desirable cryptographic property, which is necessary to ensure that a small difference between two plaintexts results in a seemingly random difference between the two corresponding ciphertexts. When some input value is concerned, changing one bit in that value, half of the output bits is expected to change; if "avalanche criterion" is to be satisfied.

For a cryptographic transformation which maps n input bits to m output bits; there are 2^n different inputs. If plaintexts P and P_i differ only in bit i ($P_i = P^- e_i$, where e_i is the n-bit unit vector with a 1 in position i) the output difference vector called the "avalanche vector" is computed as:

$$D^{e_i} = f(P)^- f(P_i) = [d_1^{e_i} \ d_2^{e_i} ... d_n^{e_i}], \quad d_j^{e_i} \, {}_{\llcorner} Z_2, \tag{2}$$

whose elements are called the avalanche variables.

One can find the total change of the j 'th avalanche variable over the whole input alphabet of size 2^n, by considering all input pairs P and P_i which differ in the i'th bit:

$$W(d_j^{e_i}) = \qquad d_j^{e_i} . \tag{3}$$

$$\text{for all P} \, {}_{\llcorner} Z_2^n$$

Since there are 2^n terms in the summation given above, (3) should be divided by 2^n to be used as an estimate of probability.

In order to obtain the overall change of all avalanche variables corresponding to the i'th input bit change, we sum normalized form of (3) over all output bits:

$$(1/2)^n \sum_{j=1}^{m} W(d_j^{e_i}) . \tag{4}$$

If (4) is normalized by m it can also be used as a probability estimate.

An n m function is said to satisfy the avalanche criterion if the probability, $k_{AVAL}(i)$, defined below is equal to one half, for all values of i:

$$k_{AVAL}(i) = (1/(m \ 2^n)) \sum_{j=1}^{m} W(d_j^{e_i}) = 1/2 . \tag{5}$$

$k_{AVAL}(i)$ can take values in the range [0,1], and it should be interpreted as the probability of change of the overall output bits when only the i'th bit in the input string is changed. If $k_{AVAL}(i)$ is close to ½ for all i, then the s-box satisfies the Avalanche Criterion within a small error region.

2.2   Strict Avalanche Criterion

Webster and Tavares combined the criteria of completeness and avalanche into Strict Avalanche Criterion (SAC) [6]. A function f: Z_2^n fi Z_2^m satisfies SAC if for all i {1,2,...,n}, j   ,m}, flipping input bit i changes output bit j with the probability of exactly ½. So, each term of the sum given by (5) is required to obey the constraint separately; in other words, an s-box is said to satisfy SAC if for all i, j:

$$(1/2)^n W(d_j^{e_i}) = 1/2 . \tag{6}$$

(6) can also be modified in order to define a SAC parameter, $k_{SAC}(i, j)$ as:

$$k_{SAC}(i, j) = (1/2)^n W(d_j^{e_i}) = 1/2 . \tag{7}$$

$k_{SAC}(i, j)$ can take values in the range $[0,1]$, and it should be interpreted as the probability of change of the j'th output bit when the i'th bit in the input string is changed. If $k_{SAC}(i, j)$ is close to ½ for all (i, j) pairs, then the s-box satisfies SAC within a small error region.

It is easy to demonstrate that an s-box, which satisfies SAC, must satisfy both Completeness and Avalanche Criteria, but the satisfaction of Avalanche Criterion does not necessarily imply the satisfaction of SAC.

2.3   Bit Independence Criterion

The idea of bit independence criterion (BIC) was introduced by Webster and Tavares [6]. For a given set of avalanche vectors generated by complementing a single plaintext bit, all avalanche variables should be pairwise independent. Alternatively, consider two n-bit input vectors which differ only in bit i, with the corresponding avalanche vector D^{ei}. If f meets the bit independence criterion, the j'th and k'th bits of D^{ei} change independently for all i, j, k $(1 \pounds j, k \pounds m$ with j,,k).

To measure the bit independence property, one needs the correlation coefficient between the j'th and k'th components of the avalanche vector D^{ei}, evaluated over all input pairs P and P_i, which differ only in bit i $(P_i = P^- e_i)$. The bit independence parameter corresponding to the effect of the i'th input bit change on the j'th and k'th bits of D^{ei} is the absolute value of this correlation coefficient:

$$BIC^{e_i}(d_j, d_k) = |corr(d_j^{e_i}, d_k^{e_i})| . \tag{8}$$

Then the overall BIC parameter is defined as:

$$BIC(f) = \max_{\substack{1 \pounds i \pounds n \\ 1 \pounds j, k \pounds m \\ j,, k}} BIC^{e_i}(d_j, d_k) . \tag{9}$$

$BIC(f)$ is defined in the range $[0,1]$. It is ideally equal to zero, and in the worst case it is equal to one.

2.4   XOR Table Distribution

XOR table of an s-box gives information about the security of the block cipher against differential cryptanalysis. The essence of a differential attack is that it exploits particular high-valued entries in the XOR tables of s-boxes employed by a block cipher.

The XOR table [7] of an $n \times m$ s-box is a $2^n \; 2^m$ matrix. The rows of the matrix represent the change in the output of the s-box. An entry in the XOR table of an s-box indexed by (d, b) indicates the number of input vectors P which, when changed by d, result in the output difference of $b = f(P)^- \; f(P^-d)$:

$$XOR_f(d, b) = \#\{P| \; f(P)^- \; f(P^-d \;) = b\} \tag{10}$$

where $d \in Z_2^n$ and $b \in Z_2^m$.

Note that an entry in the XOR table can only take an even value, and the sum of all values in a row is always 2^n.

As entries with high values in the XOR table are particularly useful to differential cryptanalysis, a necessary condition for an s-box to be immune to differential cryptanalysis is that, it does not have large values in its XOR table.

2.5   Nonlinearity

The nonlinearity parameter, NLM_f (z), of a cipher f: $Z_2^n \; fi \; Z_2^m$ for a given $z = (a, w, c) \in Z_2^{n+m+1}$ is defined as the number of cases over all cipher inputs $P \in Z_2^n$ such that the affine function $(w \; P^- \; c)$ and the nonzero linear combination $(a . f(P))$ differ from each other [8]:

$$NLM_f (z) = \#\{P | a . f(P),, \; w \; P^- \; c\} \tag{11}$$

where $a \in Z_2^m$, $w \in Z_2^n$ and $c \in Z_2$. The overall nonlinearity measure, NLM_f, of the cipher f is defined as:

$$NLM_f = \min_z NLM_f (z). \tag{12}$$

For a cipher f not to be susceptible to linear cryptanalysis, NLM_f is required to be as close as possible to its maximum value, given by $(2^{n-1} - 2^{-1}2^{n/2})$ for perfectly nonlinear functions. Small values of NLM_f around zero, show that the cipher f is close to affine functions and susceptible to linear cryptanalysis [8].

2.6   Avalanche Weight Distribution (AWD)

Most of the above mentioned criteria are used mainly for testing the s-boxes of block ciphers; and they are not very practical for application to the overall cipher. We propose the Avalanche Weight Distribution (AWD) criterion [3] for fast and rough analysis of overall diffusion properties of block ciphers. This criterion examines whether for quite similar plaintext pairs, P and P_i , which differ only in bit i, the Hamming weight of the corresponding avalanche vector D^{ei}, is sufficiently close to the Binomial distribution around m/2, m being the dimension of output vectors of the cipher.

3   Results

3.1   Avalanche Criterion

Although (5) is not satisfied exactly by the s-box of Rijndael, $k_{AVAL}(i)$ values are found to be very close to one half. Indeed, it is more logical to expect the criterion given by (5) to be satisfied within an error range of $\pm_{\llcorner}$, which we call the "relative error interval"; so that, the modified avalanche criterion within $\pm_{\llcorner}$ for all i is [9, 10]:

$$(1/2)(1-_{\llcorner}) \pounds k_{AVAL}(i) \pounds (1/2)(1+_{\llcorner}).$$ (13)

Corresponding overall relative absolute error $_{\llcorner AVAL}$ of the s-box can be found from (13) as:

$$_{\llcorner AVAL} = \max_{1 \pounds i \pounds n} | 2k_{AVAL}(i) - 1 |.$$ (14)

Relative absolute error for the Avalanche Criterion is obtained as "0.0352", evaluating (5) and (14) successively, for the s-box of Rijndael .

3.2   Strict Avalanche Criterion

The s-box of Rijndael does not satisfy SAC exactly; but it satisfies SAC within a very small relative error interval, as in the case of Avalanche Criterion. So, considering (7) within an error interval as in (13), overall relative absolute error for SAC , $_{\llcorner SAC}$, can be found as [9, 10]:

$$_{\llcorner SAC} = \max_{1 \pounds i,j \pounds n} | 2k_{SAC}(i,j) - 1 |.$$ (15)

SAC is a more specialized form of the Avalanche Criterion, so the relative absolute error for SAC is expected to be larger than that of the Avalanche Criterion.

Computing (6) and (15) successively, ε_{SAC} is obtained as "0.1250" for the s-box of Rijndael.

3.3   Bit Independence Criterion

The situation for BIC is a little bit different from Avalanche Criterion and SAC, as BIC is analysed according to the BIC(f) value of an s-box, which is already defined by (9) as the highest correlation between any two avalanche variables; hence the relative absolute error, ε_{BIC} , for this criterion is:

$$\varepsilon_{BIC} = BIC(f). \qquad (16)$$

ε_{BIC} is obtained, calculating (8) and (9) as "0.1341" for the s-box of Rijndael. In Table 2, we summarize the parameters obtained for the s-box of Rijndael and compare them with those of Safer K-64 [3, 11].

Table 2.   Relative absolute error values of Safer K-64 and Rijndael

Cipher & S-box	ε_{AVAL}	ε_{SAC}	ε_{BIC}
S-box of Rijndael	0.0352	0.1250	0.1341
Exponentiating s-box of Safer K-64	0.5039	1	1
Logarithm taking s-box of Safer K-64	0.0313	0.2500	0.3150

3.4   XOR Table Distribution

The XOR table is a matrix of size 256×256, whose entries are calculated by (10). We divide it into 8 pieces, so that each piece is 32×256, and tabulate the maximum entry for each piece together with the corresponding maximum entry of the exponentiation and the logarithm taking s-boxes of Safer K-64 in Table 3 for comparison. It is apparent that XOR values of Rijndael are much more uniformly distributed than those of Safer K-64; which makes it securer against differential cryptanalysis.

Table 3. Maximum entries of the XOR table for Rijndael and Safer K-64

	Logarithm taking s-box of Safer K-64	Exponentiating s-box of Safer K-64	S-box of Rijndael
1st piece	12	12	4
2nd piece	22	16	4
3rd piece	16	22	4
4th piece	22	12	4
5th piece	12	128	4
6th piece	16	16	4
7th piece	16	22	4
8th piece	128	16	4

3.5   Nonlinearity

We sketch the nonlinearity distribution of the s-box of Rijndael, which we define as the number of z vectors corresponding to a specific value of $NLM_f(z)$. Hence the horizontal axis shows all possible $NLM_f(z)$ values in the range $(0, 2^n)$. The minimum value on the horizontal axis, corresponds to NLM_f defined by equation (11). The nonlinearity distribution of an s-box, which is symmetric around the midpoint $NLM_f(z) = 2^n - 1$, gives us information about the susceptibility of the s-box to linear cryptanalysis. In Figure.1, the nonlinearity distribution curve is sketched for only non-zero z vectors.

Examining Fig. 1, the s-box of Rijndael does not seem to be susceptible to linear cryptanalysis. Because, it is observed that the s-box of Rijndael has got 1275 z vectors with lowest $NLM_f(z) = 112$. For those 1275 z vectors, equation

$$a \ f(P) = w \ P^- \ b \qquad (17)$$

is satisfied for 144 plaintexts. Hence the probability that (17) is correct is equal to $144/256 = 0.5625$, for those 1275 specific vectors. Similarly, one can observe from Fig. 1 that for 4080 z vectors, $NLM_f(z) = 114$, i.e., (17) is satisfied for 142 plaintexts and not satisfied for the remaining 114 plaintexts. Hence, the probability that (17) is correct is equal to $142/256 = 0.5547$, for those specific 4080 vectors. Since the probabilities are very close to ½, the s-box of Rijndael is not susceptible to linear cryptanalysis.

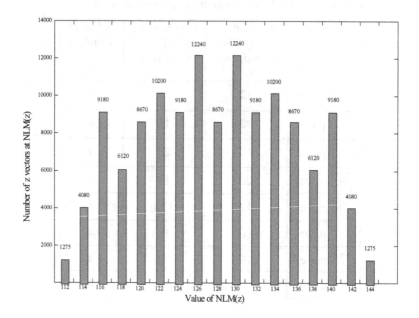

Fig. 1. Nonlinearity distribution curve of the s-box of Rijndael

3.7   Avalanche Weight Distributions (AWD)

We use the following test procedure [3, 11] to examine the diffusion properties of Rijndael with 128 bit input block and key lengths for the criterion of avalanche weight distribution (AWD);

Step 1- A plaintext P is chosen at random and the plaintext P_i is calculated so that the difference between P and P_i is e_i, i.e., $P_i = P^- e_i$; hence, P and P_i differ only in bit i, where $i \in \{1,2,...,128\}$,

Step 2- P and P_i are submitted to r-rounds of Rijndael for encryption under a random key,

Step 3- From the resultant ciphertexts C=f(P) and C_i=f(P_i), the Hamming weight of the avalanche vector w(D^{ei})=k is calculated, where $k \in \{0,1,...,128\}$,

Step 4- The value of the k'th element of the avalanche weight distribution array is incremented by one, i.e., AWD-array[k] = AWD-array[k] + 1,

Step 5- The steps 2-4 are repeated for at least 10000 randomly chosen input vectors P, and the values in the avalanche weight distribution array are sketched versus its index k, as the histogram of possible weights corresponding to e_i.

With the help of the test procedure explained above, 128 figures, each corresponding to an input difference e_i are obtained for each round of Rijndael. It is observed that none of the curves with 1-round of encryption satisfies the AWD criterion, i.e., they do not resemble a Binomial distribution around a mean value of 64. The reason why they are unacceptable is that; after 1-round of encryption all the avalanche vectors have very small Hamming weights, gathered around a mean value of ~16 and none of which exceeding 30. However, after 2-rounds of encryption, all AWD curves start to resemble the expected Binomial distribution around 64. First and second round AWD curves corresponding to an input difference of e_1 are depicted in Fig. 2; as representatives of other one-bit input differences, which all yield very similar curves.

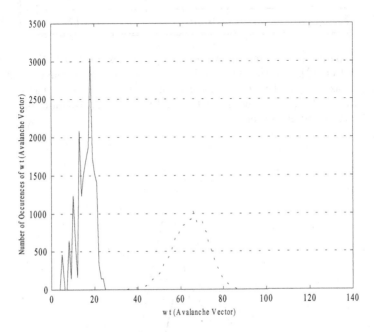

Fig. 2. First round (solid) and second round (dashed) AWD curves of Rijndael for e1

If we now define a normalized measure of closeness between the evaluated AWD curves and the ideal Binomial distribution (subtracting the sum of normalized error magnitudes from unity), the overall first round and second round AWD curves of Rijndael yield a closeness-to-ideal (or resemblance) parameter as given in Table 4.   It is observed from this table that, although the third and higher rounds of Rijndael resemble the ideal Binomial curve (of mean value m/2=64) more than 92%, the resemblance parameter is around 79% at the second iteration and 0% at the first. On the other hand, similarly evaluated curves of Safer K-64 are found to be much closer [3] to the ideal Binomial distribution (of mean value m/2=32) than Rijndael values given in Table 4.

Table 4. Resemblance percentage of the Rijndael AWD curves to ideal Binomial distribution

Round	Input difference vector	
	e_{25}	e_{104}
1	0	0
2	79.44	79
3	92.55	92.84
4	92.52	92.84
10	92.70	93.17

For comparison, a representative of AWD curves of Safer K-64, which is reproduced from [3], is depicted in Fig. 3 for the first and second rounds of the cipher. This figure as compared to Fig. 2, clearly shows that Safer K-64 has much better AWD characteristics, which resembles the ideal Binomial distribution (of mean value m/2=32), even at the end of the first round.

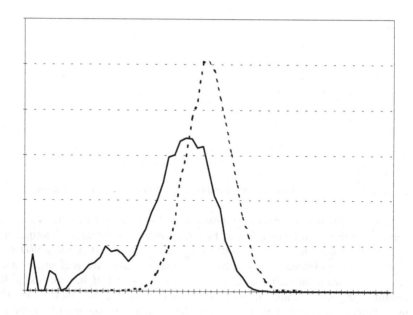

Fig. 3. First round (solid) and second round (dashed) AWD curves of Safer K-64 for e_{57}

4   Conclusions

Through our experimental results for the criteria of avalanche, strict avalanche and bit independence, it is shown that the s-box of Rijndael satisfies these criteria within very small values of relative absolute errors, such as $\varepsilon_{AVAL} = 3.52\%$, $\varepsilon_{SAC} = 12.5\%$ and $\varepsilon_{BIC} = 13.41\%$. Compared to the relative absolute errors of the exponentiating s-box of Safer family given in Table 2, as $\varepsilon_{AVAL} = 50.39\%$, $\varepsilon_{SAC} = 100\%$ and $\varepsilon_{BIC} = 100\%$, it is apparent that Rijndael's s-box parameters are much better than those of Safer. The sensitivity of the exponentiation box of Safer [3,11] to the input vector of e_{128}, which yields ε_{SAC} and ε_{BIC} values of 100% can make the cipher vulnerable to differential cryptanalysis. On the other hand; overall performance of Safer K-64 measured with respect to the AWD criterion is better than that of Rijndael, at least at the end of the first and second rounds, as can be seen by comparing Fig. 2 and Fig. 3.

Finally, we should mention that our observations do not indicate that one of these ciphers is superior to the other; however, they should be interpreted as additional randomness tests [12] to give some ideas for the improvement of the mentioned ciphers; as well as independent confirmation of analogous results.

References

1. Daemen, J. and Rijmen, V.: AES Proposal: Rijndael. NIST Publication (1999)
2. Massey, J.L.: SAFER K-64: A Byte-Oriented Block-Ciphering Algorithm. Fast Software Encryption – Proceedings of Cambridge Security Workshop, Cambridge, U.K., LNCS 809, Springer Verlag (1994) 1–17
3. Aras, E.: Analysis of Security Criteria for Block Ciphers. M.S. Thesis, Middle East Technical University, Ankara, Türkiye (September 1999)
4. Kam, J.B. and Davida, G.I..: Structured design of substution-permutation encryption networks. IEEE Transactions on Computers, Vol. C-28, No.10, (October 1979) 747–753
5. Feistel, H.: Cryptography and computer privacy. Scientific American. Vol. 228, No.5 (May 1973) 15–23
6. Webster, A.F. and Tavares, S.E.: On the Design of S-boxes. Advances in Cryptology: Proceedings of CRYPTO'85, Springer Verlag, New York, (1986) 523–534
7. Adi Shamir and Eli Biham: Differential Cryptanalysis of DES-like Cryptosystems. Journal of Cryptology, Vol. 4, No. 1 (1991) 3–72
8. Meier W. and Staffelbach, O.: Nonlinearity Criteria For Cryptographic Functions. Advances in Cryptology, Proc. EUROCRYPT'89, Springler-Verlag (1989) 549–562
9. Vergili, I.: Statistics on Satisfaction of Security Criteria for Randomly Generated S-boxes. M.S. Thesis, Middle East Technical University, Ankara, Türkiye (June 2000)
10. Vergili, I. and Yücel, M.D.: On Satisfaction of Some Security Criteria for Randomly Chosen S-Boxes. Proc. 20th Biennial Symp. on Communications, Kingston (May 2000) 64–68
11. Aras, E. and Yücel, M.D.: Some Cryptographic Properties of Exponentiation and Logarithm Taking S-Boxes. Proc. 20th Biennial Symp. on Communications, Kingston, Canada (May 2000) 69–73
12. Soto, J. and Basham, L.: Randomnes Testing of the Advanced Encryption Standard Finalist Candidates. NIST Publication (March 2000)

Author Index

Lecture Notes in Computer Science

For information about Vols. 1–1962
please contact your bookseller or Springer-Verlag

Vol. 2002: H. Comon, C. Marché, R. Treinen (Eds.), Constraints in Computational Logics. Proceedings, 1999. XII, 309 pages. 2001.

Vol. 2003: F. Dignum, U. Cortés (Eds.), Agent Mediated Electronic Commerce III. XII, 193 pages. 2001. (Subseries LNAI).

Vol. 2004: A. Gelbukh (Ed.), Computational Linguistics and Intelligent Text Processing. Proceedings, 2001. XII, 528 pages. 2001.

Vol. 2006: R. Dunke, A. Abran (Eds.), New Approaches in Software Measurement. Proceedings, 2000. VIII, 245 pages. 2001.

Vol. 2007: J.F. Roddick, K. Hornsby (Eds.), Temporal, Spatial, and Spatio-Temporal Data Mining. Proceedings, 2000. VII, 165 pages. 2001. (Subseries LNAI).

Vol. 2009: H. Federrath (Ed.), Designing Privacy Enhancing Technologies. Proceedings, 2000. X, 231 pages. 2001.

Vol. 2010: A. Ferreira, H. Reichel (Eds.), STACS 2001. Proceedings, 2001. XV, 576 pages. 2001.

Vol. 2011: M. Mohnen, P. Koopman (Eds.), Implementation of Functional Languages. Proceedings, 2000. VIII, 267 pages. 2001.

Vol. 2012: D.R. Stinson, S. Tavares (Eds.), Selected Areas in Cryptography. Proceedings, 2000. IX, 339 pages. 2001.

Vol. 2013: S. Singh, N. Murshed, W. Kropatsch (Eds.), Advances in Pattern Recognition – ICAPR 2001. Proceedings, 2001. XIV, 476 pages. 2001.

Vol. 2015: D. Won (Ed.), Information Security and Cryptology – ICISC 2000. Proceedings, 2000. X, 261 pages. 2001.

Vol. 2016: S. Murugesan, Y. Deshpande (Eds.), Web Engineering. IX, 357 pages. 2001.

Vol. 2018: M. Pollefeys, L. Van Gool, A. Zisserman, A. Fitzgibbon (Eds.), 3D Structure from Images – SMILE 2000. Proceedings, 2000. X, 243 pages. 2001.

Vol. 2020: D. Naccache (Ed.), Topics in Cryptology – CT-RSA 2001. Proceedings, 2001. XII, 473 pages. 2001

Vol. 2021: J. N. Oliveira, P. Zave (Eds.), FME 2001: Formal Methods for Increasing Software Productivity. Proceedings, 2001. XIII, 629 pages. 2001.

Vol. 2022: A. Romanovsky, C. Dony, J. Lindskov Knudsen, A. Tripathi (Eds.), Advances in Exception Handling Techniques. XII, 289 pages. 2001

Vol. 2024: H. Kuchen, K. Ueda (Eds.), Functional and Logic Programming. Proceedings, 2001. X, 391 pages. 2001.

Vol. 2025: M. Kaufmann, D. Wagner (Eds.), Drawing Graphs. XIV, 312 pages. 2001.

Vol. 2026: F. Müller (Ed.), High-Level Parallel Programming Models and Supportive Environments. Proceedings, 2001. IX, 137 pages. 2001.

Vol. 2027: R. Wilhelm (Ed.), Compiler Construction. Proceedings, 2001. XI, 371 pages. 2001.

Vol. 2028: D. Sands (Ed.), Programming Languages and Systems. Proceedings, 2001. XIII, 433 pages. 2001.

Vol. 2029: H. Hussmann (Ed.), Fundamental Approaches to Software Engineering. Proceedings, 2001. XIII, 349 pages. 2001.

Vol. 2030: F. Honsell, M. Miculan (Eds.), Foundations of Software Science and Computation Structures. Proceedings, 2001. XII, 413 pages. 2001.

Vol. 2031: T. Margaria, W. Yi (Eds.), Tools and Algorithms for the Construction and Analysis of Systems. Proceedings, 2001. XIV, 588 pages. 2001.

Vol. 2032: R. Klette, T. Huang, G. Gimel'farb (Eds.), Multi-Image Analysis. Proceedings, 2000. VIII, 289 pages. 2001.

Vol. 2033: J. Liu, Y. Ye (Eds.), E-Commerce Agents. VI, 347 pages. 2001. (Subseries LNAI).

Vol. 2034: M.D. Di Benedetto, A. Sangiovanni-Vincentelli (Eds.), Hybrid Systems: Computation and Control. Proceedings, 2001. XIV, 516 pages. 2001.

Vol. 2035: D. Cheung, G.J. Williams, Q. Li (Eds.), Advances in Knowledge Discovery and Data Mining – PAKDD 2001. Proceedings, 2001. XVIII, 596 pages. 2001. (Subseries LNAI).

Vol. 2037: E.J.W. Boers et al. (Eds.), Applications of Evolutionary Computing. Proceedings, 2001. XIII, 516 pages. 2001.

Vol. 2038: J. Miller, M. Tomassini, P.L. Lanzi, C. Ryan, A.G.B. Tettamanzi, W.B. Langdon (Eds.), Genetic Programming. Proceedings, 2001. XI, 384 pages. 2001.

Vol. 2039: M. Schumacher, Objective Coordination in Multi-Agent System Engineering. XIV, 149 pages. 2001. (Subseries LNAI).

Vol. 2040: W. Kou, Y. Yesha, C.J. Tan (Eds.), Electronic Commerce Technologies. Proceedings, 2001. X, 187 pages. 2001.

Vol. 2042: K.-K. Lau (Ed.), Logic Based Program Synthesis and Transformation. Proceedings, 2000. VIII, 183 pages. 2001.

Vol. 2043: A. Strohmeier, D. Craeynest (Eds.), Reliable Software Technologies: Ada-Europe 2001. Proceedings, 2001. XV, 405 pages. 2001.

Vol. 2044: S. Abramsky (Ed.), Typed Lambda Calculi and Applications. Proceedings, 2001. XI, 431 pages. 2001.

Vol. 2045: B. Pfitzmann (Ed.), Advances in Cryptology – EUROCRYPT 2001. Proceedings, 2001. XII, 545 pages. 2001.

Vol. 2048: J. Pauli, Learning Based Robot Vision. IX, 288 pages. 2001.

Vol. 2052: V.I. Gorodetski, V.A. Skormin, L.J. Popyack (Eds.), Information Assurance in Computer Networks. Proceedings, 2001. XIII, 313 pages. 2001.

Vol. 2053: O. Danvy, A. Filinski (Eds.), Programs as Data Objects. Proceedings, 2001. VIII, 279 pages. 2001.

Vol. 2054: A. Condon, G. Rozenberg (Eds.), DNA Computing. Proceedings, 2000. X, 271 pages. 2001.

Vol. 2055: M. Margenstern, Y. Rogozhin (Eds.), Machines, Computations, and Universality. Proceedings, 2001. VIII, 321 pages. 2001.

Vol. 2057: M. Dwyer (Ed.), Model Checking Software. Proceedings, 2001. X, 313 pages. 2001.